FIREFLY ENCYCLOPEDIA OF
REPTILES AND AMPHIBIANS

FIREFLY ENCYCLOPEDIA OF
REPTILES AND AMPHIBIANS

General Editor
CHRIS MATTISON

FIREFLY BOOKS

A FIREFLY BOOK

Published by Firefly Books Ltd. 2008

First printing

Publisher Cataloging-in-Publication Data (U.S.)

Firefly encyclopedia of reptiles and amphibians / Chris Mattison, general editor.
2nd ed.
Originally published, 2002.
[240] p. : col. ill., col. maps ; cm.
Includes bibliographical references and index.
Summary: Reference work on reptiles and amphibians, from miniature tree frogs and salamanders to enormous constrictors such as the rock pythons and anacondas.
ISBN-13: 978-1-55407-366-5
ISBN-10: 1-55407-366-9
1. Reptiles — Encyclopedias. 2. Amphibians — Encyclopedias. I. Mattison, Christopher. II. Title.
597.9 22   QL640.7.F57 2008

Library and Archives Canada Cataloguing in Publication

   Firefly encyclopedia of reptiles and amphibians / Chris Mattison, editor. — 2nd ed.
Includes bibliographical references and index.
ISBN-13: 978-1-55407-366-5
ISBN-10: 1-55407-366-9
   1. Reptiles — Encyclopedias. 2. Amphibians — Encyclopedias.
I. Mattison, Christopher
QL640.7.F57 2008   597.9   C2007-907111-2

Published in the United States by
Firefly Books (U.S.) Inc.
P.O.Box 1338, Ellicott Station
Buffalo, New York 14205

Published in Canada by
Firefly Books Ltd.
66 Leek Crescent
Richmond Hill, Ontario L4B 1H1

Developed by The Brown Reference Group plc
(incorporating Andromeda Oxford Limited)
8 Chapel Place
Rivington Street
London EC2A 3DQ
www.brownreference.com

Printed in China

For The Brown Reference Group plc

Editorial Director Lindsey Lowe
Managing Editor Tim Harris
General Editor Chris Mattison
Production Director Alastair Gourlay

2002 Edition

Publishing Director Graham Bateman
Project Manager Peter Lewis
Consultant Editors Tim Halliday and Kraig Adler
Editor Tony Allan
Art Director Chris Munday
Senior Designer Mark Regardsoe
Assistant Designers Frankie Wood, Steve McCurdy
Cartographic Editor Tim Williams
Picture Manager Claire Turner
Picture Researcher Vickie Walters
Production Director Clive Sparling
Editorial and Administrative Assistant Marian Dreier
Proofreader Rita Demetriou
Indexer Ann Barrett

Photos Front Cover: Panther chameleon (*Furcifer pardalis*); Back Cover: Red-eyed tree frog (*Agalychnis calidryas*); Page 1: Yellow juvenile Green tree python (*Morelia viridis*); Pages 2–3: A Flap-necked chameleon (*Chamaeleo dilepis*) – normally a forest dweller – at the edge of the Etosha Pan salt flat, Namibia.

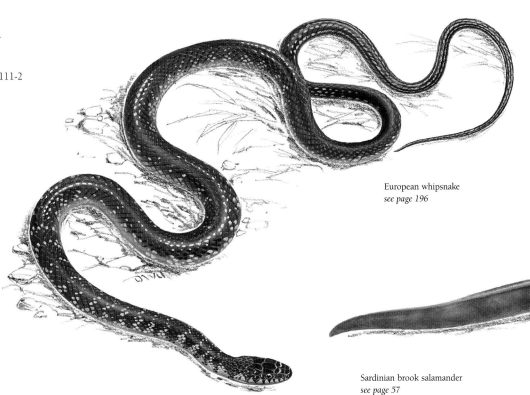

European whipsnake
see page 196

Sardinian brook salamander
see page 57

CONTRIBUTORS

Marginated tortoise
see page 132

AA Anthony Arak, University of Stockholm, Stockholm, Sweden

KA Kraig Adler, Cornell University, Ithaca, NY, USA

RAA Ronald A. Altig, Mississippi State University, Mississippi State, MS, USA

ADB Angus d'A. Bellairs †, St. Mary's Hospital Medical School, University of London, UK

AMB Aaron M. Bauer, Villanova University, Villanova, PA, USA

EDB Edmund D. Brodie, Utah State University, Logan, Utah, USA

GMB Gordon M. Burghardt, University of Tennessee, Knoxville, TN, USA

SDB S. Donald Bradshaw, The University of Western Australia, Perth, Australia

AC Alison Cree, University of Otago, Dunedin, Otago, New Zealand

AJC Alan J. Charig †, Natural History Museum, London, UK

DD Dougal Dixon, Wareham, Dorset, UK

HGD Herndon G. Dowling, New York University, NY, USA

MD Mandy Dyson, The Open University, Milton Keynes, UK

WED William E. Duellman, University of Kansas, Lawrence, KS, USA

SEE Susan E. Evans, University College London, UK

BG Brian Groombridge, World Conservation Monitoring Centre, Cambridge, UK

CG Carl Gans, University of Texas at Austin, TX, USA

HCG H. Carl Gerhardt, University of Missouri, Columbia, MO, USA

HWG Harry W. Greene, Cornell University, Ithaca, NY, USA

LJG Jr Louis J. Guillette Jr., University of Florida, Gainesville, FL, USA

TRH Tim Halliday, The Open University, Milton Keynes, UK

JI John Iverson, Earlham College, Richmond, IN, USA

FJJ Fredric J. Janzen, Iowa State University, Ames, Iowa, USA

AGK Arnold G. Kluge, University of Michigan, Ann Arbor, MI, USA

HBL Harvey B. Lillywhite, University of Florida, Gainesville, Florida, USA

JWL Jeffrey W. Lang, University of North Dakota, Grand Forks, ND, USA

CM Chris Mattison, Sheffield, South Yorkshire, UK

EOM Edward O. Moll, Eastern Illinois University, Charleston, IL, USA

RWM Robert W. Murphy, Royal Ontario Museum, Toronto, Ontario, Canada

SAM Sherman A. Minton †, Indiana University, Indianapolis, IN, USA

GS Gordon Schuett, University of Michigan, Ann Arbor, MI, USA

RDS Raymond D. Semlitsch, University of Missouri, Columbia, MO, USA

LT Linda Trueb, University of Kansas, Lawrence, KS, USA

PPvD Peter Paul van Dijk, TRAFFIC Southeast Asia, Petaling Jaya, Selangor, Malaysia

DAW David A. Warrell, Centre for Tropical Medicine, University of Oxford, UK

MHW Marvalee H. Wake, University of California at Berkeley, CA, USA

GRZ George R. Zug, Smithsonian Institution, Washington, DC, USA

KRZ Kelly Zamudio, Cornell University, Ithaca, NY, USA

Artwork Panels

David M. Dennis

Denys Ovenden

Common frog
see page 87

CONTENTS

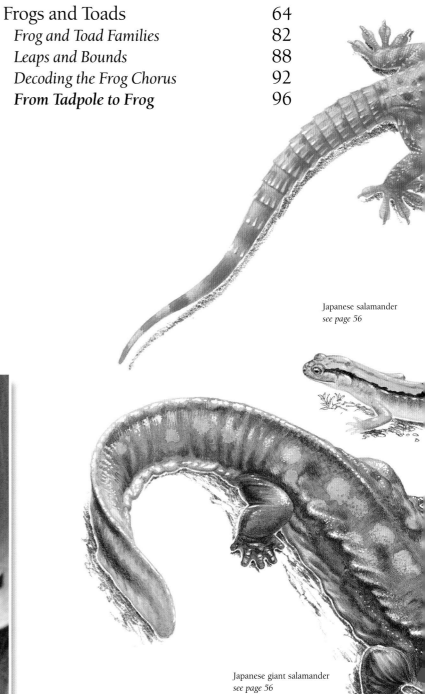

Japanese salamander
see page 56

Japanese giant salamander
see page 56

Masked puddle frog
see page 86

Gharial
see page 213

Turkish gecko
see page 170

REPTILES 98

Sea krait
see page 207

PREFACE

*a*mphibians and reptiles form two separate classes of animals, but they are traditionally studied together, a situation that arose partly because the distinctions were not well understood in the early days of zoology. In practice, it is still convenient to study them together because they often live in the same places and searching for them often involves the same techniques. The study of reptiles and amphibians is known as herpetology, from the Greek word herpeto, meaning "to creep."

In the early days, herpetology had a smaller following than other biological disciplines such as ornithology or botany. Part of the reason lay in distribution patterns, with reptiles and amphibians tending to be most numerous in the parts of the world where scientists were least numerous, notably South America, Africa, South East Asia, and Australasia. In Europe and temperate North America, the traditional seats of learning, they remain unseen for much of the year. In addition, they are usually inedible, if not downright poisonous, and therefore had little economic importance. And, it must be said, some are not beautiful in the conventional sense of the word; at least, not in the eyes of the uninformed. Little wonder, then, that the naturalists of the eighteenth and nineteenth centuries did not think they played a very important part in the grand scheme of things. Linnaeus called them "foul and loathsome creatures," while Charles Darwin described the Marine iguana as "a hideous-looking creature, of a dirty black color, stupid, and sluggish in its movements."

Although some species are still the objects of fear, superstition, and prejudice, for the most part attitudes changed gradually over the twentieth century. Furthermore, because they are relatively easy to house and observe in captivity, reptiles and amphibians often make good subjects for studies whose implications spill over into other branches of the biological sciences.

Through the media of books, photography, and film, there is a growing appreciation nowadays of the ways in which reptiles and amphibians have adapted for life in a variety of different environments, notably rain forests, where amphibian and reptile diversity reaches its greatest heights, and in deserts, where reptiles (especially lizards) are often the most obvious form of vertebrate life. Inherent variation in lifestyle has much to do with their success. To name just a few of these variables, reptiles and amphibians may be oviparous or viviparous; several are parthenogenetic, and many species can shut down their physiological systems for long periods of time to avoid prolonged heat, cold, or drought. Reptiles can uncouple the activities of mating and fertilization, allowing females to store sperm for weeks, months, or even years, while many amphibians have developed breeding systems that enable them to be independent of standing water.

This book discusses all these subjects, and more. It follows the traditional threads that represent the way in which species, genera, and families are arranged into hierarchical groups depending on their similarities and differences—the science of taxonomy. There is an extensive introduction to each of the two classes, describing their origins and giving an overview of their biology. Within each class, separate accounts of the three orders of amphibians and the four orders of reptiles deal in more detail with their specific biology and provide interesting facts about their members, and there is a short description of each family. The accounts of the five largest groups are summarized with annotated lists of the families, giving statistics for the numbers of species and genera, their most important characteristics, some examples, and a distribution map.

Superimposed on this outline plan are other threads, which cut across the taxonomic divisions and sub-divisions and deal with the disciplines of anatomy, physiology, ecology, and behavior, subjects that are common to all species, including other animals, thus emphasizing their similarities as well as their differences. Throughout the book there are also articles on topics of special interest. These highlight aspects of the study of reptiles and amphibians that have made herpetology an increasingly important discipline within zoology. They can be read independently of the rest of the text. In addition, captions for many of the diagrams and photographs are extensive, providing interesting supplementary material.

Recent developments in the techniques by which animals can be studied have given our understanding of reptiles and amphibians new impetus in recent years. Advances in DNA technology and other types of biochemical analysis, for example, are helping to solve many of the puzzles surrounding their relationships. All this takes time but, as each unit (species, genus, family, or order) is investigated, new schemes, such as that recently established for a radical system of frog classification, will occur across the board. The classification of the Colubridae, a family that currently contains an unlikely 60 percent of all snakes, readily springs to mind as an example of just one other area that needs urgent attention. While all this is taking place in the laboratory, field workers are benefiting from the technological revolution in electronics, where miniaturization and telemetry give them the tools to track animals that spend a large part of their time out of sight. Activities and social interactions that were previously unknown, or mere guesswork, are now being monitored, and the resulting information is fascinating; amphibians and reptiles are not the mindless creatures that many people thought they were but often lead complex and effective lives to ensure their survival.

Unfortunately, animals whose survival strategies have evolved over thousands of generations and in response to gradual changes are ill-equipped to deal with the sudden and dramatic changes that are now occurring. In the previous editions of this book, the authors and editors frequently stressed the need for conservation and the dangers facing the future of reptiles and amphibians. Since the 2002 edition the situation has deteriorated further. A steady stream of reptiles and amphibians are going extinct in most parts of the world. More than 20 species of reptiles have gone in living memory, due to habitat destruction and competition and predation from introduced species, especially domestic animals. Another 200 species are either Endangered or Critically Endangered, and many species that were once common are now scarce and their populations fragmented. Amphibians are the worst affected, for reasons that are discussed elsewhere, and the statistics speak for themselves. The IUCN estimates that 737 species of amphibians are Endangered, another 441 are Critically Endangered, and 34 have recently become extinct. These 1,212 species represent a staggering 18 percent of the total. To put it another way, nearly *one species in five* is likely to become extinct in the next couple of decades. And as there are huge areas where surveys have not, or cannot, be carried out, the real figure is probably even higher. Since amphibians are sensitive indicators of the state of the environment as a whole, the implications of this do not auger well.

The forerunner of this book, the *Encyclopedia of Reptiles and Amphibians*, was published in 1986 with the aim of encouraging an interest in reptiles and amphibians by presenting accurate information about them. Its success was followed in 2002 by an updated version, the *'New' Encyclopedia of Reptiles and Amphibians*, which incorporated much new material based on advances in the study of reptiles and amphibians as well as new graphics and photography. The present edition is the result of a further revision, especially in the area of taxonomy, where great changes have recently taken place. It says much for the excellence of the earlier editions that very little of the original text has had to be touched. The concept of the book, and the information at its core, has stood the test of time remarkably well. Great credit for this must go to the editors, Tim Halliday and Craig Adler, and to the individual authors listed on the previous pages.

CHRIS MATTISON
SHEFFIELD

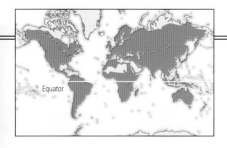

AMPHIBIANS

tHE AMPHIBIA ARE AN ASTONISHINGLY DIVERSE *class of vertebrates that have existed for more than 230 million years. Probably since at least sometime in the Permian period (295–248 million years ago) their lineage has been evolving independently. It is erroneous to think of modern amphibians as transitional forms between fish and reptiles, although they do possess some anatomical features that are intermediate. There is sometimes a tendency to regard living amphibians as evolutionary losers, partly because of their small size and generally inconspicuous nature. Instead, one should think of them as descendants of an ancient lineage of tetrapods that was extraordinarily successful in exploiting an extremely wide range of habitats. Modern amphibians exhibit an enormous array of life histories, and they often constitute a dominant element in many natural communities. Without habitat preservation and other steps to conserve them, however, amphibian species will continue to disappear at very high rates.*

The living amphibians – frogs, salamanders, and caecilians – display a stunning variety: some animals with tails and others without, some looking like snakes or lizards, others hopping on long hind legs or burrowing without legs at all, and with colors ranging from drab browns to iridescent blues, greens, and reds.

Of 40,000 known species of vertebrates (animals with backbones), about 6,180 are amphibians. Next to the mammals, they are the smallest class of living vertebrates, descendants of a once-dominant group of land animals, the first land vertebrates, some the length of a moderate-sized crocodile, that flourished several hundred million years ago.

Amphibians are an important group for study because they are early descendants of the first vertebrates to conquer land, a group that later gave rise to reptiles (and they, in turn, to mammals and birds). Living amphibians are divided into three orders: Caudata (the salamanders, including newts and sirens, 556 species); Anura (the frogs, including toads, 5,453 species); and Gymnophiona (the eel-shaped caecilians, 173 species) – at the time of writing, a total of 6,182 species. In fact, the number of newly-described species has greatly increased in recent years, due to a number of factors: surveys in previously unexplored regions; the use of non-morphological (for example, molecular and behavioral) characters to distinguish species; and the rush to describe species

before they vanish due to environmental changes.

The word "amphibian," from the Greek *amphibios*, means "a being with a double life"; specifically, one that lives alternately on land and in water. Such a double life is the rule for Amphibia, but there are exceptions; some species are permanently aquatic and others completely terrestrial. All are ectotherms, using environmental temperature to regulate body temperature.

No structure uniquely defines all amphibians as feathers do birds, so one must resort to a combination of characteristics. Further complicating any definition is the fact that living forms have diverged significantly from the primitive fossil ones, and there is no information at all on certain key features in the fossil forms. Indeed, the class Amphibia as here defined does not include the earliest land vertebrates, but in order to understand the origins of amphibians we must consider the origins of the first four-legged vertebrates (tetrapods).

The Transition to Land
EVOLUTION AND FOSSIL HISTORY
The oldest known tetrapods are from Upper Devonian deposits (374–354 million years ago). All were recovered from freshwater sites, except for *Tulerpeton*, which was found in marine deposits in Russia. The best known of these early tetrapods are *Ichthyostega* and *Acanthostega*, both found in East Greenland and dating from about 365 million years ago, but others are slightly older

△ **Above** *The fire salamander (Salamandra salamandra) is a successful modern amphibian species, being widespread across Europe. In terms of body shape, salamanders most closely resemble the early fossil tetrapods.*

and come from a geographically wide array of localities: *Elginerpeton* of Scotland, *Hynerpeton* from the northeastern United States, *Metaxygnathus* of southeastern Australia, and *Obruchevichthys* of eastern Europe. Greenland in particular may seem an unlikely place for any early tetrapod to have lived, but its location and climate were very different during the Devonian period (417–354 million years ago), when it straddled the Earth's equator and lay within a moist and warm tropical region extending from present-day Australia through Asia to northeastern North America. Until the early Jurassic (about 190 million years ago), all of the Earth's land mass was united into a single supercontinent called Pangaea. Thus, it is not surprising to find evidence that the earliest tetrapods rapidly spread to now distant lands including Europe, Australia, and eastern North America, and, by the early Triassic (about 230 million years ago), even to Antarctica.

The ancestors of these earliest tetrapods were members of a group of bony fish (class Osteichthyes) in the order Sarcopterygii (the lobe-finned fishes). Unlike most other bony fish, which had fins supported by cartilaginous rays (order Actinopterygii, the ray-finned fishes that comprise most of the living fish), the fins of sarcopterygian fish had bony elements comparable to those of the limbs of land vertebrates.

Furthermore, the sarcopterygian fish that gave rise to tetrapods had lungs (although they are only distantly related to that group of fish called lungfishes, the Dipnoi) and some had internal nostril openings (nares), so air could be taken into the lungs when the mouth was closed or when only the external nares were above water. Internal nares are characteristic of land vertebrates. In most fish the external nares serve only a sensory function; they lead to blind pockets not connected to the mouth cavity.

Ichthyostega was similar to an extinct fish, *Eusthenopteron* (family Osteolepididae), found in Upper Devonian deposits in Quebec, Canada. Both had lungs and internal nares and shared two traits found only in some other osteolepidid fish and in early tetrapods: a brain case divided transversely into anterior and posterior portions and, secondly, an infolding of the enameled surface of the teeth that creates, in cross section, a complex labyrinthine pattern. The extinct fish most closely related to tetrapods, but less well known, are two genera once regarded as osteolepidids: *Elpistostege* of Quebec and *Panderichthys* of eastern Europe (family Panderichthyidae).

Ichthyostega, although unquestionably a tetrapod, retained a number of fishlike characteristics, among them the opercular bones – remnants of bones that in fish connect the gill covering to the cheek – and a tail fin supported by bony rays. But the limb and girdle structure of ichthyostegids had already fully reached the early tetrapod condition; thus, the earliest land vertebrates – the ichthyostegalians, which include early relatives of *Ichthyostega* – remain undiscovered and must be sought in even older deposits. Panderichthyid fishes, the nearest relatives of tetrapods, flourished about 380–375 million years ago and the earliest tetrapods appeared only about 5–10 million years later. The genera *Elginerpeton* and *Obruchevichthys* may be particularly close to the fish-to-tetrapod transition and are considered stem-tetrapods: they are early tetrapods but are not amphibians.

How did the transition to land come about? The classic explanation was that the Devonian was a period of severe droughts. Fish with sufficiently strong fins could avoid stranding and death by crawling to available pools. According to this idea, land vertebrates could have evolved as a by-product of selection originally for increased agility in finding water, not land! New evidence casts doubt upon the scenario of periodic droughts, however, and it seems likely that the Devonian was a time of relatively continuous moist environments, at least in tropical regions.

It is possible that some of the features that are associated with the first tetrapods actually evolved in the aquatic environment. For example, the development of a functional neck, and the separation of the skull from the pectoral girdle to accomplish this step, may have evolved in prototetrapods, permitting sudden sideways movement of the head when stalking and capturing prey in water. Perhaps this change was a preadaptation later facilitating the capture of prey on land.

One or more of the following factors are believed to have led to the evolution of land vertebrates. During Devonian times aquatic environments, with their enormous diversity of fish and other organisms, contained many more competitors and predators than did the land, and land also may have been a safer place to deposit eggs and for juveniles to survive. The water of the warm Devonian swamps in which tetrapods arose was probably poor in oxygen, especially in the shallows, but the fish ancestors of land vertebrates must have had lungs, as all of their living descendants do. Possibly these fish congregated in shallow waters and ventured occasionally onto land. It might have been the more agile juveniles that did so, in order to exploit insect and other invertebrate food. Although this transition doubtless occurred over a period of millions of years, there is no known fossil record of these stages, but the consensus view is that the fish-to-tetrapod transition occurred only once and that all tetrapods are, therefore, a monophyletic group.

In becoming terrestrial, these tetrapods overcame numerous challenges, although some changes could have occurred even in a shallow-water habitat. On land, gravity became a key factor molding the development of the skeleton. Without the buoyancy of water, the body was suspended from the vertebral column, which in turn

◐ **Below** The Amazonian egg-eating treefrog (Osteocephalus oophagus) was first described scientifically only in 1995. The female of this species (seen here with an identification band around her waist) lays trophic eggs when clasped by the smaller male. Their tadpoles hover around and eat the eggs as soon as they are extruded. Without the nutrition provided by these trophic eggs, the tadpoles would die.

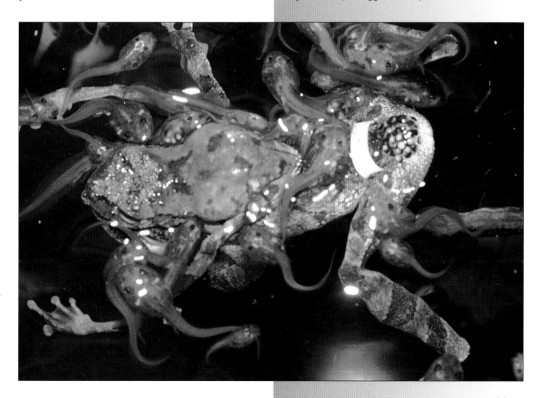

had to be supported by the limbs and limb girdles. When the animal rested on the ground, a well-developed rib cage, as was present in *Ichthyostega*, prevented injury to internal organs. The elongated neural arches and articulating surfaces of the vertebrae distributed the gravitational forces more evenly along the vertebral column.

Practically nothing is known about the skin of the earliest tetrapods. It has often been assumed that they had soft, naked skin like that of modern amphibians, but fossil evidence suggests instead that scutes covered their undersides. Some types had osteoderms on the upper surfaces of the body. Many species were aquatic and possessed gills, whereas others were adapted to land.

ANATOMY OF EARLY TETRAPODS

In the Upper Devonian, the early tetrapods (from which living amphibians are thought to be descended) evolved from sarcopterygian fishes, and so initiated the conquest of the land by vertebrates. The evidence for this development derives from marked similarities between bone structure in, respectively, the fins and limbs of the two groups:

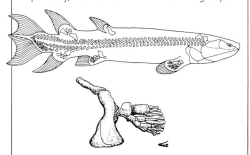

⬤ *Above* *The fleshy, lobed fins of sarcoptery-gian fishes such as* Eusthenopteron *(top) pointed outward from the body and served as a prop. The shoulder girdle and fin (detail below) contain bony elements that correspond broadly to those found in the limbs of early tetrapods (bottom).*

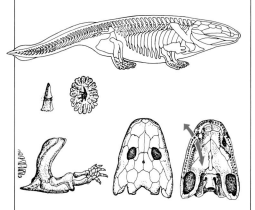

⬤ *Above* Ichthyostega *(top) had a robust ribcage to protect its internal organs against the stresses of terrestrial life. Its labyrinthodont teeth (center; whole and in cross-section at the base) are characteristic of early tetrapods. Strong limbs (below left) enabled it to hold itself off the ground; its skull (below right) reveals the presence of internal nares, another key adaptation to terrestriality.*

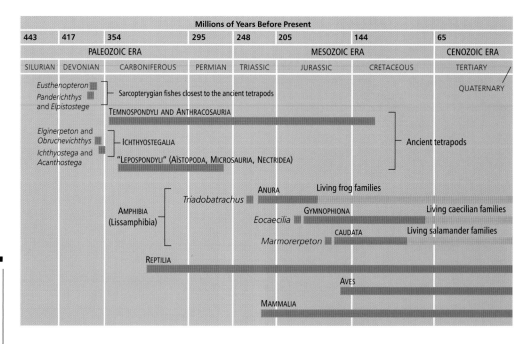

Millions of Years Before Present							
443	417	354	295	248	205	144	65
PALEOZOIC ERA				MESOZOIC ERA			CENOZOIC ERA
SILURIAN	DEVONIAN	CARBONIFEROUS	PERMIAN	TRIASSIC	JURASSIC	CRETACEOUS	TERTIARY

QUATERNARY

Eusthenopteron
Panderichthys and *Elpistostege* — Sarcopterygian fishes closest to the ancient tetrapods

TEMNOSPONDYLI AND ANTHRACOSAURIA

Elginerpeton and *Obruchevichthys* — ICHTHYOSTEGALIA
Ichthyostega and *Acanthostega* — "LEPOSPONDYLI" (AÏSTOPODA, MICROSAURIA, NECTRIDEA)

Ancient tetrapods

ANURA   Living frog families
Triadobatrachus

AMPHIBIA (Lissamphibia)   GYMNOPHIONA   Living caecilian families
Eocaecilia

CAUDATA   Living salamander families
Marmorerpeton

REPTILIA

AVES

MAMMALIA

Although many were heavy-bodied and lizardlike in build, there were some truly bizarre types including legless, eel-shaped forms (Aïstopoda) and others with extremely wide heads drawn back into peculiar horns (Nectridea).

After the appearance of these first tetrapods in the late Devonian, a period of rapid evolution occurred resulting in an enormous diversity of types. By the end of the Triassic, however, about 160 million years later, nearly all of them had become extinct. Some were truly enormous in size. The largest, *Mastodonsaurus* of the Triassic of Russia, had a skull 125cm (49in) long and a total length estimated at 4m (13ft); the largest living amphibians, by comparison, are the Asiatic giant salamanders, which reach a total length of 160cm (63.5in).

These ancient tetrapods were found on all land masses and were the dominant land animals of their day. Mammals and birds did not evolve until after most of these ancient vertebrates had become extinct, but the first reptiles evolved from them midway through the Carboniferous period (about 320 million years ago). It is important to stress that reptiles did not evolve from the modern amphibians, which arose sometime after the Carboniferous.

▮ Missing Links
MODERN AMPHIBIANS

In contrast to our knowledge of the origin of reptiles, the ancestry of modern amphibians is a puzzle, largely because there are no fossils linking the ancient tetrapods to any of the three orders of living amphibians. This is one of the largest gaps in the history of terrestrial vertebrates. The earliest known fossil amphibian (*Triadobatrachus*, of the Middle Triassic of Madagascar; 230 million years ago) is already froglike in some of its features, although it still retains a short tail and has twice as

many vertebrae as do modern frogs. An early frog-like amphibian from Poland (*Czatkobatrachus*, only five million years younger than *Triadobatrachus*) provides evidence that frogs were already widely distributed during the Triassic. The earliest salamanders (*Marmorerpeton*, of the Middle Jurassic of Britain, dating from 165 million years ago) and caecilians (*Eocaecilia*, a legged form from the Lower Jurassic of Arizona, USA, from 190 million years ago) are already as specialized, each in its own way, as modern forms. It is likely, therefore, that the first amphibians arose much earlier, probably sometime in the Permian but possibly even earlier.

Thus, the incomplete fossil record provides little help. Indeed, it prompts the question why these animals were not readily fossilized, since even very small and fragile labyrinthodont larvae with external gills have been recovered. The reason may be ecological: the ancestors of living amphibians probably occupied very shallow waters or rushing mountain streams where the large species of ancient tetrapods could not pursue them and, coincidentally, places where fossilization is relatively rare. We know, for example, that *Marmorerpeton* was a fully aquatic salamander.

Without critical fossil material, evolutionary relationships must be inferred from comparisons of the living species. For a long time, given the enormous differences between frogs and salamanders, it was believed that each had descended from different orders of Paleozoic tetrapods. It was later proposed that despite their different appearances, frogs, salamanders, and caecilians have many basic features in common, particularly (1) the types of glands in the skin, and also the fact that the skin is used as a respiratory organ; (2) the structure of the inner ear and the retina of the eye; and (3) an unusual pedicellate tooth structure in which each tooth has a base (or pedicel) fixed in

◑ **Left** Diagram showing the geological occurrence of extinct and living amphibians, the ancient tetrapods, and other vertebrate groups. Key genera are shown in italics. The fossil record is tantalizingly incomplete; as yet, no fossils of transitional forms have come to light unequivocally linking the ancient tetrapods to the living amphibians. The many similarities among the three orders of modern amphibians lead scientists to believe that they are monophyletic, i.e. that they share a common ancestor.

the jaw to which a replaceable crown is attached (see Amphibian Body Plan).

The possibility that all of these and other common features evolved independently is so unlikely, it has been argued, that it is more reasonable to assume a monophyletic origin. Therefore, most biologists place the three living groups in one subclass, the Lissamphibia.

Adaptations to the Environment
FORM AND FUNCTION

Many of the features first evolved by the early tetrapods relate to the crucial transition from water to land and were inherited by their descendants, the amphibians. As such, amphibians possess true tongues (to moisten and move food), eyelids (which, together with adjacent glands, wet the cornea), an outer layer of dead cells in the epidermis that can be sloughed off, the first true ears (and a voice-producing structure, the larynx), and the first Jacobson's organ, a chemosensory structure adjacent to the nasal cavities that reaches its developmental zenith in lizards and snakes (see Lizards). Presumably these characteristics also

existed in the now-extinct tetrapod ancestors of amphibians.

There are also striking changes in the nervous system related to life in a more complex terrestrial environment. The spinal cord is enlarged in the regions adjacent to the limbs, correlated with the more intricate movement of limbs compared to the fins of their fish ancestors. Invasion of the outer layer of the cerebral hemispheres by nerve cells exists in amphibians, but not nearly to the same degree as the tremendous enlargement of the mammalian cerebrum.

The skin of living amphibians, which is moistened by the secretions of numerous mucus glands, is not a passive outer layer but plays a vital and active role in water balance, respiration, and protection. Some frogs possess antibiotic substances (magainins) in their skin. It is highly permeable to water, especially in terrestrial species. Aquatic forms have reduced permeability to offset the inflow of water by osmosis.

Although most amphibians are restricted to moist habitats, there are specializations that permit many species to live in otherwise inhospitable environments. For example, desert toads create an osmotic gradient across their skin by retaining urea in their urine, thus permitting water uptake from extremely dry soils. Most terrestrial frogs possess a patch of skin, rich in blood capillaries, in the pelvic region that allows uptake of water even from a thin surface film. Other frogs and a few salamanders form a cocoon of shed skin to reduce water loss, and some tree frogs reduce evaporative water loss by wiping fatlike skin

secretions over the body surface.

On the other hand, loss of body water through the skin is used in some species as a method of temperature regulation through evaporative cooling. In most species the moist skin and surfaces in the mouth cavity also serve a respiratory function, since dissolved gases pass across them; the numerous members of one family of salamanders (Plethodontidae) have lost lungs altogether and depend entirely upon this mode of gas exchange.

The Biogeographical Background
DISTRIBUTION PATTERNS

Amphibians today are found on every continent except Antarctica. They live from sea level – and sometimes even below it, in caves and lightless underground streams – to high mountain peaks. No amphibians are adapted to life in seawater, and they are generally not found on oceanic islands.

Beyond these general statements, the patterns of distribution (or biogeography) of species, genera, and families of amphibians vary greatly from one taxon to another due to events that occurred in the geologically distant past (historical biogeography) and to environmental conditions in the recent past and at present (ecological biogeography). Historical biogeography allows us to understand otherwise inexplicable distributions. For example, the existence of caecilians on the isolated Seychelles Islands in the Indian Ocean, some 1,000km (620mi) northeast of Madagascar, is due to the fact that these islands are fragments left behind by the Indian tectonic plate when it broke away from Africa during the Mesozoic era.

◑ **Right** Giant tetrapods of another age. The Triassic (248–205 million years ago) saw the emergence of crocodile-sized land vertebrates such as **1** Mastodonsaurus, which measured 4m (13ft) from snout to tip of tail, **2** Diadectes, 3m (10ft), and **3** Eryops, 1.5m (5ft).

A more recent example is an explanation for the low diversity in South America (2 genera and 25 species) of plethodontid salamanders, a family that had its center of dispersal in North America and exhibits a large diversity in Middle America (13 genera and about 200 species). Due to erosion and continental drift, the connection between Middle America and South America was interrupted during the Eocene – producing the Panamanian Portal – and was not reconnected for another 50 million years, during the Pliocene, due to volcanic activity and associated uplifting. The gap was closed only about 3 million years ago, thus restoring the land route for salamander dispersal to South America.

Other distributions of amphibians can be explained by their physiological capabilities and the physical and biotic environments in which they live today. For example, there are no marine amphibians because they cannot remain in osmotic balance with seawater; they would continuously lose water, dehydrate, and die. A few salamanders and frogs can withstand high salinity but only one species, the Crab-eating frog of Southeast Asia, is adapted to live regularly in brackish, mangrove-filled estuaries. It does so by maintaining high levels of urea in its bloodstream, thus remaining in osmotic equilibrium with its environment.

Many species of frogs and salamanders live at high latitudes or altitudes where freezing temperatures could present a threat to their existence. They must hibernate below the frost line or return to spring breeding ponds after temperatures rise above freezing; but in fact some species are freeze-tolerant and thus are distributed in regions otherwise unavailable to them. A dozen species of amphibians (both frogs and salamanders) are known to be able to survive freezing temperatures by releasing glucose or glycerol into their blood stream to lower the freezing point of water in the cells. Not surprisingly, the Wood frog and Spring peeper, which are two of the earliest species to enter the breeding ponds in North America, are among the species that use glucose as an antifreeze.

Fire is another environmental factor that can limit the distributions of amphibian species since fires occur regularly in many habitats. Brush fires are frequent in coastal California but the California newt is able to survive them by rapidly secreting mucus over its body and walking directly through

the flame front. Newts have been observed to walk quickly through the flames, when the slime on their bodies foams up to a consistency like an egg meringue; after they pass through the flames, the white crust is easily wiped off their wet bodies. In West Africa where savannah fires are prevalent, recent research shows that reed frogs respond to the crackling sound of approaching fires by fleeing to protective cover, thus enabling them to survive in an otherwise inhospitable habitat.

Carnivores and Cannibals
DIET AND FEEDING

Amphibians are carnivores. They eat animal prey whole, without chewing it into pieces. The major exception are tadpoles, the aquatic larval stage of frogs, which feed on algae and protists by scraping them off underwater surfaces or by filtering them from water (see Swimming, Eating, Growing Machines). As a rule, small amphibians eat insects and other invertebrates but larger ones occasionally eat vertebrates, including members of their own species. Cannibalism is, in fact, widespread in amphibians, especially among larval stages. Cannibal morphs are known to develop regularly in some species of salamanders and frogs.

Most species are generalists. Individual Red-backed salamanders of North American forests may feed on hundreds of different species of invertebrates, limited only by the size of their mouths, but they prefer soft-bodied prey and avoid distasteful species. In contrast, some species are specialists, like the Mexican burrowing toad, which has a tiny mouth and feeds only on termites.

Many amphibians are adapted to habitats in which food is available only seasonally. In temperate zones, they may be active for only a few months each year and in some desert regions only for a few weeks. Thus, they must feed rapidly and store a lot of food (mostly as fat) which they then use to survive lean times and/or to yolk their eggs. Amphibians are highly efficient at extracting full

● *Right* Internal organs such as the liver and pancreas are visible through the translucent skin on the abdomen of this Fleischmann's glass frog (Hyalinobatrachium fleischmanni). Note also the greenish egg mass in the ovisac awaiting deposition.

AMPHIBIAN BODY PLAN

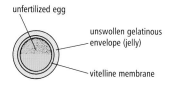
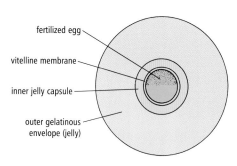

● *Above* Embryos (fertilized eggs) of amphibians, like those of fish, have gelatinous envelopes but lack the protective membrane (amnion) found in all higher vertebrates. Amphibian eggs also lack shells and therefore must be laid in fresh water or in moist places to avoid drying out, although a few species give birth to fully-formed young. Larvae possess external gills, and in frog tadpoles these become enclosed inside a chamber by a flap of skin (the operculum). The larva undergoes an abrupt metamorphosis (see pp.10–11) to the adult stage, from which it often differs markedly in structure.

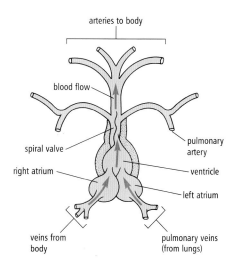

● *Above* The heart has three chambers, two atria, and a ventricle, which may be partly divided. Amphibians have paired lungs, but in four families of salamanders these are sometimes reduced or completely absent; in caecilians, the left lung is greatly reduced.

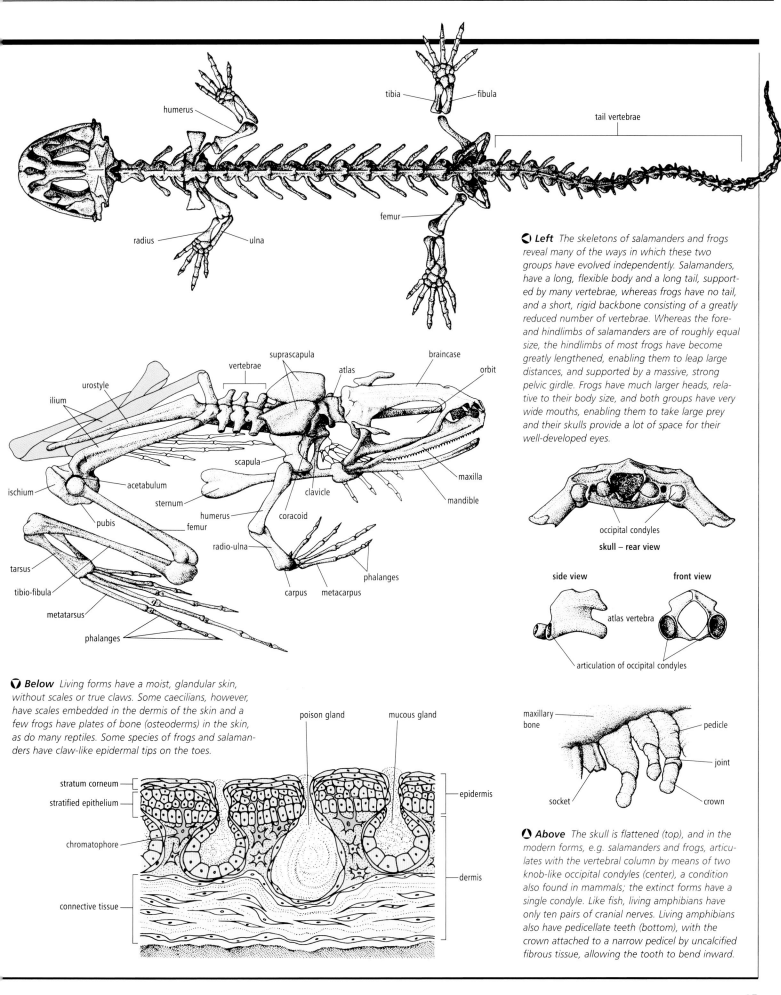

humerus

tibia   fibula

tail vertebrae

radius   ulna

femur

Left *The skeletons of salamanders and frogs reveal many of the ways in which these two groups have evolved independently. Salamanders, have a long, flexible body and a long tail, supported by many vertebrae, whereas frogs have no tail, and a short, rigid backbone consisting of a greatly reduced number of vertebrae. Whereas the fore- and hindlimbs of salamanders are of roughly equal size, the hindlimbs of most frogs have become greatly lengthened, enabling them to leap large distances, and supported by a massive, strong pelvic girdle. Frogs have much larger heads, relative to their body size, and both groups have very wide mouths, enabling them to take large prey and their skulls provide a lot of space for their well-developed eyes.*

suprascapula   braincase

vertebrae   atlas   orbit

urostyle

ilium

scapula

ischium   maxilla

acetabulum   clavicle

sternum   mandible

pubis   coracoid

humerus

femur

radio-ulna

tarsus

tibio-fibula   phalanges

metatarsus   carpus   metacarpus

phalanges

occipital condyles

skull – rear view

side view   front view

atlas vertebra

articulation of occipital condyles

Below *Living forms have a moist, glandular skin, without scales or true claws. Some caecilians, however, have scales embedded in the dermis of the skin and a few frogs have plates of bone (osteoderms) in the skin, as do many reptiles. Some species of frogs and salamanders have claw-like epidermal tips on the toes.*

poison gland   mucous gland

maxillary bone   pedicle

joint

socket   crown

stratum corneum

stratified epithelium   epidermis

chromatophore

dermis

connective tissue

Above *The skull is flattened (top), and in the modern forms, e.g. salamanders and frogs, articulates with the vertebral column by means of two knob-like occipital condyles (center), a condition also found in mammals; the extinct forms have a single condyle. Like fish, living amphibians have only ten pairs of cranial nerves. Living amphibians also have pedicellate teeth (bottom), with the crown attached to a narrow pedicel by uncalcified fibrous tissue, allowing the tooth to bend inward.*

nutrient value from their food and many species have extremely low food requirements, at least under cool conditions. Some plethodontid salamanders, for example, can survive seemingly unaffected after being kept in cold conditions for one year or more, even without food.

Feeding modes differ fundamentally in aquatic and terrestrial forms. Like fish, aquatic species and aquatic life stages feed by suddenly enlarging the oral cavity, thereby creating a negative pressure and drawing food that is suspended in the water into the mouth. This suction method is found in aquatic salamanders (both larvae and adults), tadpoles, and even adult pipid frogs, which are aquatic. Hellbenders and their Asiatic relatives, the giant salamanders, have the unique ability to suck prey into only one side of their mouths because the halves of their mandibles move independently.

Terrestrial amphibians, including caecilians and some frogs and salamanders, use a bite-and-grasp method that also involves teeth and a simple, non-projectile tongue. Most frogs and terrestrial salamanders, however, have projectile tongues. In both, the tongue is projected using a system of cartilaginous rods (the hyobranchial skeleton) and associated muscles in the floor of the mouth, but the mechanism in frogs and salamanders is quite different. In frogs, the anterior end of the tongue is attached at the front of the lower jaw; thus, when the posterior end of the tongue is fully extended it must be flipped upside down, and so it is the dorsal surface that actually sticks to the prey. In the Marine or Cane toad, the tongue strikes the prey in about 37 milliseconds (msec) and the entire prey-capture cycle takes only about 143 msec. In salamanders having the most advanced tongue-projection systems, the hyobranchial skeleton is much longer and the mushroom-shaped tongue sits on its anterior tip, not on the floor of the mouth. When the hyoid muscles contract, the hyobranchial skeleton is propelled forward extremely rapidly, sticking the tongue to the prey. In some neotropical salamanders (*Bolitoglossa*), the full prey-capture cycle takes only 4–6 msec (see Miniature Salamanders box in Salamanders and Newts).

◐ **Above** *The carnivorous nature of amphibians is amply demonstrated by Knudsen's frog (Leptodactylus knudseni) from the Amazon rain forest. This large leptodactylid species is fully capable of taking substantial vertebrate prey such as bats. In common with many of the large anurans that share its habitat, Knudsen's frog employs the active foraging mode of predation, and hunts at night.*

◑ **Right** *Only four species of salamanders normally produce living, fully metamorphosed young. In this Turkish salamandrid (Lyciasalamandra luschani), the newborn young is quite large compared to the mother.*

Amphibians forage for food in two basic ways. Some sit and wait for prey to come to them; these species are generally diurnal and cryptically colored and may even use lures, such as species of South American horned frogs that twitch the tips of their toes to attract other frogs within range. Others are active foragers that seek out prey; these species are often nocturnal but, if diurnal, are aposematically colored and toxic, like poison frogs or the eft stage of the Red-spotted newt.

Undulations and Bounds
LOCOMOTION

There has been a striking adaptive radiation among amphibians in terms of locomotion and reproduction. Salamanders and caecilians swim like fish, with side-to-side sinusoidal movements. Frogs, on the other hand, swim (and jump) in a totally different way. The vertebral column has become progressively shortened, the hindmost vertebrae have fused into a single element (the urostyle), and the bones of the hind legs have become elongated. Thus, frogs have relatively inflexible bodies and swim by means of simultaneous thrusts of the legs (see Leaps and Bounds).

Terrestrial salamanders move by means of lateral undulations, advancing diagonally opposite feet each time the body bends; some species use the tail as a fifth leg. Caecilians are legless and, except for a few completely aquatic species, live in burrows. Since the burrow walls greatly restrict lateral undulations, caecilians move by an alternating fold-and-extension progression in which only the vertebral column bends, producing momentary points of contact with the substrate which allows extension of other parts of the body, superficially resembling the locomotion of an earthworm.

Sending Sperm in Packets
REPRODUCTION

Amphibians exhibit the greatest diversity of reproductive modes of any vertebrate group. Fertilization can be external or internal. In the most primitive families (the giant and Asiatic salamanders) it is external, the sperm being shed into the water near the eggs. However, most salamanders transfer sperm in small packets called spermatophores that are picked up by the female with her cloacal lips during courtship. The sperm can then be used at once or, in most species, stored in specialized glands (spermathecae) in the cloaca for use during the following season. In some North American mole salamanders (genus *Ambystoma*) the sperm merely activate the developmental process and, in a form of parthenogenesis, do not contribute genetically (see Unisexuality: The Redundant Male?).

With few exceptions, frogs fertilize externally. Usually sperm is deposited as the eggs are laid, with the male clasping the female with his forelegs (amplexus). Some poison frogs have no amplexus at all, the males fertilizing the eggs after deposition. In some narrow-mouthed toads the bodies are temporarily glued together, and in a few other species the male's cloaca is held next to the female's while sperm is transferred, so that fertilization is internal, but there is no intromittent organ. In the North American tailed frogs, however, the tail is, in fact, an extension of the male's cloaca that is inserted into the female's cloaca to transfer sperm. All caecilians fertilize internally, the male everting his cloaca and using it as an intromittent organ.

Most amphibian species are oviparous, laying eggs in fresh water or on land. Others are viviparous, with the mother retaining the eggs in her body and the embryos being nourished either by food stored in their own yolk sac or by materials obtained directly from the mother. Clutch size in frogs varies from species to species and ranges from a single egg to about 25,000; in salamanders, the number of eggs is generally no more than a few dozen, but some newts lay up to 400.

Fertilized eggs may be laid singly, in clusters, or in long strands, but are invariably enclosed within gelatinous envelopes. If laid in water (or near enough that hatched larvae can crawl or be swept into it by floods), the larvae have gills and lead an aquatic existence, eventually metamorphosing into miniature adults (see A Key Amphibian Event).

Amphibians use a variety of sites for laying eggs, including still or running water, mud basins constructed by the male, cavities beneath logs or stones, debris or burrows, leaves overhanging water, or the water-filled axils of plants. However, each species generally has one preferred site. Those that lay eggs on land typically have no free-living larval stage and undergo direct development; this is true in many tropical frogs and in virtually all terrestrial species of salamanders. In many frogs and salamanders the eggs are defended by one of the parents, and in several species of frogs the eggs or tadpoles are carried (see Conscientious Parents).

A few frogs are viviparous. In the Puerto Rican live-bearing frog, the embryo's own yolk provides the nutrients necessary for development; the embryo's tail is thin and rich in blood supply and may function in gas exchange. Several species of African toads (in the genera *Nimbaphrynoides* and *Mertensophryne*) are viviparous; in one, the West African live-bearing toad, the fetuses ingest a mucoprotein (uterine milk) secreted by the oviduct when their yolk supply is exhausted.

Four European and Southwest Asian salamanders are normally viviparous. In a Turkish salamander (*Lyciasalamandra luschani*) and two species of high-elevation Alpine salamanders (*Salamandra atra*, *S. lanzai*), one or two fully-formed offspring are typically produced. In *S. atra*, only 1–2 young survive from as many as 30 fertilized ova. The survivors cannibalize their own siblings and they metamorphose before birth after a gestation period of 2–4 years. In the European Fire salamander, a member of the same genus, the young are deposited as larvae into the water, but in montane regions they may be born as fully-metamorphosed juveniles. In addition, the Olm of southeastern Europe may also be viviparous when water temperatures are warm, in which case two gilled larvae are produced. Although many caecilians lay eggs which are guarded by the mother, almost one-half of the species are viviparous. After utilizing their own yolk, the large fetuses feed on "uterine milk" and also scrape material from the oviduct walls with their specialized teeth. Gas exchange occurs between the greatly-enlarged gills and the wall of the oviduct, but the gills are resorbed before birth. These are all of the truly live-bearing amphibians, so far as is known.

Subtle Signals
SOCIAL BEHAVIOR AND COMMUNICATION

Social behavior between individuals requires some form of communication: chemical, acoustic, visual, or tactile. Amphibians exhibit a wide array of social behaviors, but some are very subtle and have not been recognized until recently. Among caecilians, some species produce squeaks and clicking sounds of unknown function, but most communication is chemical and tactile and involves a protrusible organ (the tentacle) that is unique among vertebrates. These paired structures lie midway between the eyes and nostrils and are extended using hydrostatic pressure. Although little is known about the tentacle's function, it may be used for prey location either underground or underwater, depending upon species, when the nostrils are closed.

Salamanders are nearsighted and must use visual cues close up, as, for example, when male Red-backed salamanders signal aggression by elevating the body. Many salamanders utter popping or clicking sounds; Ensatinas hiss and Lesser sirens emit shrill distress calls. During courtship, males of many species use tactile cues (rubbing, nudging, butting) to increase a female's receptivity.

The predominant social cues for salamanders, however, are chemical. Amphibians in general have two separate chemosensory areas in the nose that send signals to different parts of the olfactory lobes of the brain: the olfactory epithelium, which detects volatile (airborne) odors, generally small molecules; and the vomeronasal (or Jacobson's) organ, which detects non-volatile odors, mainly large molecules. Using odors, salamanders can discriminate between species and sexes, determine the reproductive status of other individuals, and stimulate sexual activity in others. Some plethodontid salamanders tap their snouts on the

substrate, allowing chemical signals left by others to move by capillary action up the nasolabial groove to the vomeronasal organ. In some salamandrids, like the Red-spotted newt, males exude chemicals from glands on their heads and rub them on the female's snout; European newts release chemicals in the water and, using their tails, waft them at the female. In both cases these social signals make the female receptive to male courtship.

Frogs primarily use acoustic social signals. Most species produce a species-specific call that is used in mate location, but in several species a repertoire of different calls is produced: courtship calls to attract females, territorial calls used by males defending territories, release calls when males are accidentally amplexed by other males, and distress calls given when grasped by a predator. Some calls

Some frogs also use visual and tactile cues. In the Brazilian torrent frog, males vocalize but also foot-flag, extending a hind leg above the body and spreading the toes. The female may foot-flag in response, and the male also foot-flags other males that intrude into his territory. In other species, hand signals are used, or the toes can be subtly undulated. Tactile cues are used by some poison frogs during courtship, when the female strokes the male with her forefeet to signal her receptivity and to stimulate the male to release sperm. Tactile cues are also used by tadpoles, to signal the female to deposit nutritive eggs on which they feed or to follow her to water (see Conscientious Parents).

In Peril Worldwide
CONSERVATION AND ENVIRONMENT

Amphibians are in peril worldwide. Although extinction of species is a natural event over geological time scales, amphibians are now disappearing at an alarming rate. At least 100 species are thought to have gone extinct in the last 30 years alone in the neotropics, Australia, and New Zealand. Recent studies point to many causative factors, mostly environmental, including pesticides and other pollutants, introduction of predators and competitors, commercial exploitation, increased exposure to UV light, and infections by parasitic chytrid fungi. Most important, however, is the loss of suitable habitat, especially vital breeding sites.

The consequences of these losses can be profound. Amphibians represent a major part of terrestrial ecosystems and are an essential trophic link between their tiny invertebrate prey and the larger vertebrates that, in turn, eat them. Similarly, in aquatic habitats tadpoles not only are major consumers of unicellular algae and protists but, by metamorphosing to terrestrial adults, contribute to nutrient flow from freshwater to terrestrial habitats.

Reversal of these trends will require basic research to provide critical knowledge of a species' biology, education of the local people to protect amphibians, legislation both for local protection and to govern international trade, and reestablishment of populations within the natural range. But without preservation of suitable habitats, none of these actions ultimately will be successful.   KA

Above In a dramatic instance of cooperative social behavior, male Gray tree frogs (Chiromantis xerampelina), high in a tree on the South African savanna, beat a secretion produced by the females into a foam nest. The outside of the foam nest dries to a crust but the inside remains moist, allowing the eggs to develop.

Left After hatching within their protective foam nest, Gray tree frog tadpoles wriggle free and drop into a seasonal pond that forms below the nesting tree during the rainy season. Here, they will complete their development.

Right A female African pig-nosed frog (Hemisus marmoratus) digs a mud slide to guide her newly hatched tadpoles from the underground breeding chamber to water. The tadpoles use tactile cues to follow their mother in this vitally important activity.

are extremely loud – comparable to the sound pressure levels of a train locomotive passing within a few meters – and others are so soft that they cannot be heard from more than 1m (3.3ft) away.

Female frogs can select individual males based on various call parameters. A Central American frog (Engystomops pustulosus) makes a two-part call, consisting of a whine and a chuck; males can produce a whine alone or a whine plus 1–6 chucks. Females prefer more complex calls, but since these last longer, the males making them may be located by, and fall prey to, frog-eating bats. Females of some species prefer calls with lower fundamental frequencies (pitch), as these signal larger and more desirable males. Sometimes smaller, satellite males do not call at all, but sit near calling males and intercept females and mate with them.

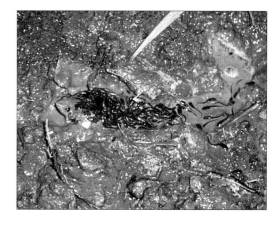

CLASSIFICATION AND TAXONOMY

Evolutionary relationships of amphibians and reptiles

SINCE TIME IMMEMORIAL, HUMANS HAVE CREATED categories of kinds of organisms by grouping the most similar ones together and giving each of them a unique name. Aristotle, in the 4th century BC, recognized the major groups of vertebrates – fish, reptiles, birds, and mammals – although he failed to note the significance of metamorphosis for classification and lumped the amphibians with the reptiles. At the end of the 17th century, an English clergyman and naturalist, John Ray, suggested that each kind of organism, or species, had a biological basis that was based on common parentage. It was the Swedish botanist Carl von Linné (originally Linnaeus) who systematized Ray's suggestion and wrote books giving Latin names for every species of organism then known. He developed a system in which each species was given a two-part name – the first word called the genus, which is capitalized, and the second, the species epithet, which is not – and in Latin, since vernacular names for each organism usually varied from one country to another. Thus, the formal Latin naming of all animals was begun officially by Linné, and we date some of our best-known species to his 1758 book: *Bufo bufo* (European common toad), *Ichthyophis glutinosus* (Ceylon caecilian), and *Salamandra salamandra* (Fire salamander) among amphibians; and *Chelonia mydas* (Green seaturtle), *Caiman crocodilus* (Common caiman), *Hemidactylus turcicus* (Mediterranean house gecko), and *Python molurus* (Indian python) among the reptiles.

Linné viewed species as fixed and unchangeable entities created by God and, for convenience, devised a higher-order, purely artificial classification to organize the known species. By 1820, the amphibians were eventually recognized as distinctly different from reptiles. The traditional classification of vertebrates thus became as follows:

Phylum Chordata (animals having an embryonic notochord, a flexible support along the back)

  Subphylum Vertebrata (all backboned animals)

    Class Agnatha (jawless fish such as lampreys and hagfishes)

    Class Chondrichthyes (cartilaginous fish such as sharks and rays)

    Class Osteichthyes (all bony fish, including ray-finned fishes and the lobe-finned fishes such as coelacanths and lungfishes)

    Class Amphibia (amphibians)

    Class Reptilia (reptiles)

    Class Aves (birds)

    Class Mammalia (mammals)

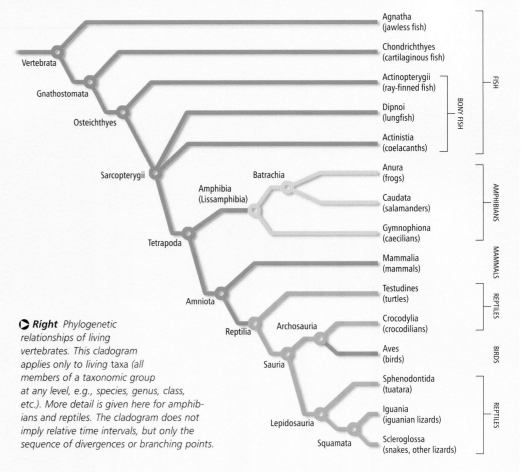

◐ **Right** *Phylogenetic relationships of living vertebrates. This cladogram applies only to living taxa (all members of a taxonomic group at any level, e.g., species, genus, class, etc.). More detail is given here for amphibians and reptiles. The cladogram does not imply relative time intervals, but only the sequence of divergences or branching points.*

● **Above** *Together with four other genera, African clawed toads of the genus Xenopus – shown here the platanna (Xenopus laevis) – are classified as the family Pipidae, partly on the grounds that all lack tongues.*

◗ **Left** *The French biologist Georges Cuvier adopted and extended Linné's system of classification. This early 19th-century plate depicts Cuvier's groups of animal phyla, running from cellular organisms (top left) to human beings (bottom right). By this stage, amphibians (at the end of the fifth row down) were clearly separated from reptiles (rows 6–8).*

Note that while this classification is hierarchical, it does not convey evolutionary relationships. Of course, the idea that the species included in this classification might have some historical relationship to one another was unknown to Linné and other naturalists of those times. Only with the advent of Charles Darwin and his concept of evolution by natural selection, which implied that all living species were related to one another and that common ancestors lived (and died out) long ago, could evolutionary relationships be considered. But how can these relationships be indicated in a classification?

It was not until the 1960s that the ideas of a German biologist, Willi Hennig, began to revolutionize the way we analyze evolutionary relationships among organisms. His method, which is called "phylogenetic systematics" (or "cladistics"), depends upon the recognition of a monophyletic

evolutionary origin. Specifically, a monophyletic lineage (or clade) is composed of an original ancestor species and all of its descendant species. Only such monophyletic lineages are recognized in phylogenetic systematics and given a formal name in classifications. By this method, clades are recognized by identifying so-called derived characters – those that differ from the ancestral condition – and, by determining the order in which the clades branched during evolution, a phylogenetic classification can be established. For example, all pipid frogs (including the Surinam toad and four genera of African clawed frogs) lack tongues. All other frogs – indeed, all other amphibians – have tongues, thus it is most reasonable to believe that the tongue was first lost in the original ancestor of the pipid frogs and therefore all descendants are tongueless. The presence of a tongue, therefore, is the primitive condition and its absence is the derived condition. Its absence in this group of frogs is a key feature convincing biologists that these five genera of frogs represent a monophyletic group that should be given a formal systematic name, the family Pipidae.

Unlike the classical hierarchical classification, a phylogenetic classification does encode evolutionary relationships. Using this modern method of classification, the relationships of the vertebrate classes are surprisingly different, as indicated in the diagram on the opposite page.

What does this phylogenetic diagram (or clado-

gram) tell us? First, it says that all land vertebrates (tetrapods) are, cladistically speaking, bony fish (Osteichthyes). Second, it implies – correctly it is believed – that all living amphibians had a common ancestor and represent a monophyletic lineage. (Unfortunately, we cannot include all fossil species long classified as amphibians in this category, because some of the most important derived characteristics of the soft anatomy of living amphibians cannot be determined from the fossil record and lissamphibians share derived traits known to be lacking in the fossil forms.) Third, the cladogram shows that crocodilians and birds are each others' closest relatives. Fourth, this cladogram tells us that living reptiles are monophyletic only if birds are considered to be reptiles, since they are descendants of a group of reptiles called archosaurs. If birds are regarded as a separate class, reptiles are not monophyletic, because with birds left out not all of the descendants of ancestral reptiles would be included.

Reptiles and birds, however, are still most often classified as co-equal classes of vertebrates. Tradition dies hard! Nevertheless, biologists regularly use the terms reptiles and birds in their traditional ways, while recognizing that their phylogenetic relationships are more complex than this usage implies. In this book, we have chosen to use the more scientifically correct classifications of amphibians and reptiles as based upon phylogenetic systematics. KA/HWG

A KEY AMPHIBIAN EVENT

The role of metamorphosis in frogs, salamanders, and caecilians

METAMORPHOSIS – THE ABRUPT TRANSFORMATION from larva to adult – is one of the defining characteristics of all amphibians, the only four-limbed animals in which the phenomenon occurs. These morphological changes, and the modifications in physiology and behavior that accompany them, are far more remarkable in frogs than in salamanders or caecilians.

The lifestyles of larval frogs and toads (the only larvae properly called "tadpoles") and those of larval salamanders and caecilians are profoundly different. Tadpoles are mostly herbivores adapted either to a suspension-feeding way of life by filtering particles of food from the water or by tearing and scraping plant material. Giant cannibal tadpoles are known in a few species. Tadpoles of some species can extract food items (such as blue-green bacteria) as small as one-tenth of a micrometer (0.000004in) in diameter – an efficiency comparable to the best mechanical sieves. A few can filter water at a rate of more than eight times their body volume per minute. Salamander larvae, however, are active carnivores that hunt tiny zooplankton and, later, large, individual prey items including other salamander larvae. Larval salamanders look like miniature adults with external gills. Thus, besides the loss of gills and lateral line sensory organs, and some internal changes in skeleton, teeth, and musculature, metamorphosis in their case is a relatively subtle process, involving the resorption of the tail fins, differentiation of the

eyelids, and changes in the thickness of the skin and its permeability to water. Caecilian larvae hatch at an even more advanced stage and resemble adults except in the size and presence of gills.

In frogs and toads, metamorphic events are more dramatic because tadpoles are so different from the carnivorous adult stage; for example, they have large, propulsive tails that are completely resorbed at metamorphosis. The larval "teeth" (in fact, not true teeth but keratinous denticles) are shed, and the mouth enlarges greatly. The hind limbs, which later become the primary means of locomotion, are tiny and nonfunctional in tadpoles until shortly before metamorphosis. The forelimbs cannot be seen externally except at hatching, because they exist inside a chamber that develops by the overgrowth of a flap of skin (the operculum) that houses the gills. Internally, the differences between larva and adult frog are equally extreme, especially in the digestive system. At metamorphosis, the tadpole's long, coiled intestine – a vital physiological feature that is necessitated by its largely vegetarian diet – becomes greatly shortened, in some species to as little as 15 percent of its original length. Metamorphosis is the time of greatest vulnerability to predators, since metamorphs can neither swim as effectively as tadpoles nor hop as effectively as frogs. Studies conducted with garter snakes have shown that their stomachs contain disproportionate numbers of metamorphs.

◗ **Left** *The prime example of a paedomorphic amphibian species is the axolotl (Ambystoma mexicanum). Certain larval features, such as the prominent external gills, are retained. Such forms are often termed neotenic (from the Greek neos, meaning "youthful.")*

Metamorphosis is under hormonal control, as is development generally. Hormones produced in the pituitary (prolactin) and thyroid (thyroxine) glands are involved. Increased amounts of thyroxine, and changing sensitivity of the tissues to thyroxine, trigger metamorphosis and can be caused by environmental factors such as crowding, low oxygen, or other stress factors.

The time to metamorphosis can vary enormously. In some species of frogs and salamanders the larvae overwinter and may not transform until the next summer or even later, whereas tadpoles of some desert-dwelling spadefoot toads complete the process in as little as eight days, an adaptation to the temporary nature of desert pools. However, by no means all salamander larvae transform into a typical adult form; some

⬦ Above *The dramatic physiological changes that take place when a tadpole metamorphoses into a frog are evident in this Eastern dwarf treefrog froglet* (Litoria fallax). *The duration of the larval stage varies considerably among anurans, from a few days to a year or more.*

retain larval characteristics, even though they become reproductive adults. Retention of larval or juvenile traits in adults is due to changes in growth rates (or heterochrony) of a type called paedomorphosis. This phenomenon greatly complicates our understanding of salamander evolution based on morphology.

Paedomorphic characteristics in salamanders include a functional lateral-line system, an absence of eyelids, and the retention of external gills. One or more of these traits has been found in some species or populations of all families of living salamanders. In some families (the giant salamanders, the mudpuppies and olm, the amphiumas, and the sirens) all species are paedomorphic. In lungless salamanders and in the olm, paedomorphosis is associated with adaptation to life in caves. For all

of these families it is usually a fixed genetic trait, and the application of thyroxine by researchers does not induce metamorphosis.

In the other families of salamanders paedomorphosis is found only in some individuals or populations within a species, and the application of thyroxine does cause metamorphosis. For example, axolotls living in Lake Xochimilco in central Mexico become sexually mature in an otherwise larval state, although transformed adults have been found. Somehow the Xochimilco environment favors paedomorphosis, perhaps because its waters contain an insufficient quantity of iodine, which is necessary to produce the hormone thyroxine. Or the effect may be due to the lake's cold temperatures in which, laboratory studies would suggest, thyroxine has little effect.

In the American Red-spotted newt (*Notophthalmus viridescens*), some coastal populations bypass the normal terrestrial (eft) phase, retain gills, and become reproductively mature. Paedomorphosis is often found in high-elevation populations but is unknown in lowland populations of the same species. Several species of American mole

salamanders, the European Alpine newt and related species, and a Japanese salamander, *Hynobius lichenatus*, all show the same pattern.

Ecologists suggest that aquatic habitats surrounded by hostile terrestrial ones ought to favor paedomorphosis. This circumstance is often true for species living in caves, in desert ponds, in streams running through arid areas, and in high-elevation ponds. Some species, however, do not fit this pattern. Sirens and amphiumas, for example, have lost the genetic capacity to metamorphose, but have other adaptations such as the sirens' ability to estivate in mud cavities or the amphiumas' to move overland whenever their aquatic habitat dries.

Unlike salamander and caecilian larvae, frog tadpoles have apparently sacrificed reproduction in favor of feeding and rapid growth rates. They are literally "feeding machines," with head skeletons and enormous guts adapted to a herbivorous lifestyle (see Swimming, Eating, Growing Machines). Space for the complete differentiation of reproductive organs becomes available only at metamorphosis.   KA

KALEIDOSCOPIC ADAPTATION

The many uses of amphibian colors

AMPHIBIANS DISPLAY A KALEIDOSCOPE OF COLORS. The colors as they appear to the human eye arise from a combination of the differential absorption and reflectance of light (chemical color) as well as diffraction and other interference phenomena of light (physical color). They are produced by pigment granules in the epidermal (upper) layer of the skin, by specialized pigment-containing cells, collectively called chromatophores, in the dermal (lower) layer, and occasionally by pigment in even deeper tissues. The green color of most frogs is produced by the combination of separate blue and yellow pigments, but in certain neotropical species it results from the deposition of a green excretory product, the bile pigment biliverdin, in soft tissues and bones.

Many amphibians can change color by concentrating or dispersing melanin (colored black or dark brown) or other pigments in the chromatophores, but since these changes are largely under hormonal control they occur rather slowly, on a time scale of seconds to minutes. By changing color in this way amphibians, like many reptiles, can regulate body temperature, since dark-colored bodies absorb radiant energy more rapidly than light-colored ones do. Thus, frogs are pale or nearly white in the hot sun. A dramatic example occurs when a frog's body is partly shaded and exhibits a two-tone color divided at the edge of the shadow. Melanin pigment is also an extremely effective filter for ultra-violet light, those wavelengths of sunlight potentially most damaging to body tissues and to genetic material and which are implicated in declines of some species in nature.

Color and pattern in amphibians are often used for concealment (crypsis), either to allow them to avoid detection by would-be prey or, especially, to avoid becoming prey themselves. Although many amphibians possess skin toxins that are noxious or even fatal to predators, most species are cryptically colored. Many employ camouflage that allows them to match their background to varying degrees, or disruptive coloration that breaks up their body outline.

Certain forest-floor dwelling frogs (for example, the Malaysian horned toad, the Neotropical leaf toad, and *Hemiphractus* of South America) look like dead leaves, even down to the detail of having fleshy projections that resemble the edge of a leaf and a midrib-like stripe down the midline. In some species, different individuals have different color patterns (polymorphism), making it difficult for a predator to have a reliable search image. Detailed studies show that the reflectance of a frog's back and that of its normal background correspond very precisely, even to minute variations at particular

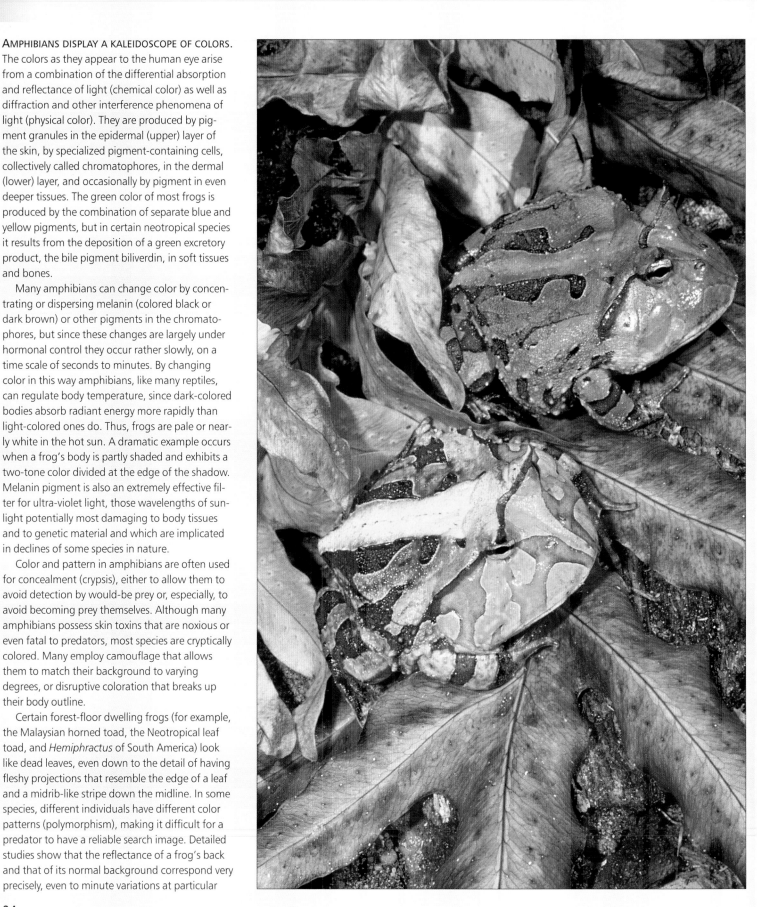

◐ **Left** *Amazonian horned frogs (here, two different color morphs of Ceratophrys cornuta) are renowned for their rapacious predatory behavior. In their case, effective camouflage is not for self-preservation, but is used as part of their "sit-and-wait" hunting strategy, as they lie in ambush in the leaf litter.*

◑ **Right** *In its Great Smoky Mountains habitat on the Tennessee–North Carolina border, the Imitator salamander (Desmognathus imitator; right) mimics the red cheek and leg coloration of the unpalatable Jordan's salamander (Plethodon jordani; left). This is an example of Batesian mimicry.*

◗ **Below** *An immature example of the Golden poison frog (Phyllobates terribilis). The bright, iridescent hues of this South American rainforest dweller hint at its extreme poisonousness; it is the most toxic frog in the world.*

wavelengths. Certain species of neotropical frogs that sit on leaves match their background both in the visible spectrum as well as in the near infrared, which may conceal them from detection by tree-dwelling predators like monkeys and even by pit vipers searching with infrared-sensitive pit receptors. Many species are effectively countershaded, with a dark upperside but with light-colored bellies that render them less visible in water when viewed from below against the light-colored sky.

In contrast, some species are boldly colored and easily detected. Many of these, such as the poison frogs of the neotropics and the fire salamanders

of Europe, have toxic skin secretions, and their bright coloration is believed to have evolved as a warning to potential predators not to molest them (aposematic coloration). Some palatable species win protection by mimicking distasteful ones (Batesian mimicry); in eastern North America the Eastern red-backed salamander avoids predation by color-matching red efts, the land stage of the Red-spotted newt, which have skin toxins that are lethal to predators. At the same time, the efts have a different relationship (Müllerian mimicry) to the toxic Red salamander. Their similar warning colorations take advantage of the fact that predators

generalize bad experiences with one species to other similarly-colored and -patterned species.

A few species combine crypsis and warning coloration, being cryptically colored when viewed from above but having a brightly colored belly that is exposed only when the animal is severely threatened by a predator. The fire-bellied toads of Europe and Asia, and Pacific and western American newts both have skin toxins and exhibit a so-called "unken" reflex when stressed (see Repellent Defenders).

Other examples of intimidation of predators involving color and pattern include the flash colors of frogs. These bright colors are restricted to body surfaces such as the flanks and posterior side of the thigh that are hidden when the animal is at rest, but which appear and disappear as it jumps again and again, possibly confusing predators that might attempt to capture the colored objects. The South American False-eyed frog has two large, eyelike spots on its rump; when threatened, it aims the rump toward the predator. Some frogs have patterns that cannot be seen at all in the human visual spectrum. For example, White's tree frog from Australia has a large spot on its snout that can be seen only in the infrared range; its function is unknown.

Colors and patterns are also used for recognition between and within species. Different species sharing the same range tend to have distinctively different color patterns. In some species the sexes differ in color, sometimes markedly so (as in the Yosemite toad and the Golden toad); this sexual dimorphism might function in sex recognition in day-active species. In the Stream frog, a poison frog of Trinidad and Venezuela, calling males turn pitch black and fight only with other black males, and losers turn brown within a few minutes.   KA

CONSCIENTIOUS PARENTS
Parental care in amphibians

ANY BEHAVIOR BY ONE OR BOTH PARENTS THAT increases the survival of their fertilized eggs (technically, embryos) or offspring at a potential cost to them is referred to as parental care. The costs to the parents can include increased predation or reduced intake of food or, by devoting a prolonged period of time to one clutch or group of offspring, simply the inability to produce additional young. In contrast, the benefits to the parents of increasing the survival of offspring by providing them care must actually outweigh any cost to the parents, otherwise parental care could not evolve.

Among amphibians, parental care has evolved numerous times, but its distribution among species is not random. About half of all caecilian species lay eggs, and in probably all of them the female remains with the eggs until they hatch. In salamanders, egg attendance by females or males is reported in all families, but is as yet known from only about one-quarter of the species. By far the greatest diversity of forms of parental care in amphibians is found among frogs, in about two-thirds of the families but less than 10 percent of the species, based on available information.

Parental care in amphibians is associated with increased terrestriality. Clutches of fertilized eggs

Right One of the most unusual forms of amphibian parental care is found in the mouth-brooding frogs of South America. The male takes fertilized eggs into his vocal sac, where they complete their development. Here, a Darwin's frog (Rhinoderma darwinii) has just disgorged two froglets.

Below Between September and December, depending upon latitude, the Marbled salamander (Ambystoma opacum), which inhabits the eastern United States, lays around 40–230 eggs in a nest cavity. While waiting for autumn or winter rains to fall and stimulate hatching, the female guards the eggs by wrapping herself around them.

and soft-bodied larvae in streams and ponds often fall prey to numerous predators, but on land the chances of survival are typically greater, especially if one or both parents provide care. Exceptions to this trend occur, however, in some fully aquatic species, such as the giant salamanders and Surinam toads, which exhibit parental behavior.

Amphibians display the widest array of reproductive modes of any of the vertebrate classes, and possess almost every conceivable type of parental care. In large measure, this diversity reflects the different trade-offs that different species make between numbers of offspring produced and the amount of care invested.

Since females produce fewer gametes (germ cells) than males, and at a greater energy expense, they normally invest more time in caring for their offspring, but there are many intriguing exceptions among the Amphibia. In the most basal living salamander families (the giant and Asiatic salamanders) and in many frogs, nearly all of which have external fertilization, the eggs are laid within the territory of the male and are defended by him. This is the most common form of parental care, reducing predation and desiccation of eggs. Sometimes the male will simultaneously guard the clutches of several females. In the most derived group, the lungless salamanders, and the egg-laying species of caecilians, all of which fertilize internally, and in some frogs, the females guard the eggs on land or in the water.

Animals are rarely cared for by their fathers, but among vertebrates paternal care is most common in fish and amphibians. This may be due to the fact that fertilization is external; males can be certain which eggs they fertilized, and their reproductive success can therefore be increased by caring for their offspring. Or it may be simply that males must be present for external fertilization and thus have the opportunity for care; in internally-fertilizing species mating and egg laying may be separated by a month or more, so male parental care is a less likely option.

Several modes of parental care have been observed among Amphibia. Perhaps the mode that is most transitory in duration, yet one that nonetheless exposes the parent to some risk, occurs when the female takes great care to deposit the fertilized eggs. In several species of European newts, the female carefully wraps each fertilized egg in the leaves of aquatic vegetation, which protects the developing embryos from predation and the negative effects of UV radiation. In many other species, the eggs are attended for extended periods by either or both parents. In fully aquatic salamanders, it is either the female (in mudpuppies) or

male parent (hellbenders) that attend the egg clutches, primarily for physiological purposes – to increase gas exchange around the clutch, by rapid movements of the gills and by rocking motions of the body, respectively. Sometimes, mudpuppies even lay their eggs in the nest cavities of hellbenders, a parasitic strategy for nest defense that requires further study. Some terrestrial breeding frogs that attend nests, such as the coqui of Puerto Rico, provide hydration of the embryos through direct transfer of water across the male's skin. Other nest-attending species provide care by protecting the eggs from pathogens such as fungi, perhaps by secreting fungicidal substances. In some other species of salamanders and frogs, nest attendance is accompanied by nest-guarding behavior in which the attending parent attacks intruders, be they members of the same species or other would-be predators. In field experiments with a New Guinean microhylid frog, the nests were attacked and consumed by arthropods if the guarding male was removed.

Brooding Frogs

All other modes of parental care in amphibians are restricted to frogs. The first of these – brooding behavior – is found in a number of aquatic and terrestrial species. In the aquatic Surinam toads of South America, males fertilize the eggs during a somersaulting bout in which the eggs are placed on the female's back. Each egg develops in a separate pocket, and the tail of the tadpole, rich in capillaries, serves the same function as a placenta. The young then emerge either as tadpoles or fully metamorphosed young, depending on the species.

In about 70 species of Neotropical frogs the female carries the eggs on her back. In three genera – *Hemiphractus* (family Hemiphractidae) and *Cryptobatrachus* and *Stefania* (family Cryptobatrachidae) – the eggs adhere to the back. In one other, *Flectonotus* (family Amphignathodontidae), the eggs are carried in open pouches. The larvae of all 17 species in these four genera develop into froglets. In the marsupial frogs (*Gastrotheca*), also members of the Amphignathodontidae and containing more than 50 species, there is a pouch on the back that opens just above the cloaca. After fertilization by the male, the eggs are placed in the pouch, thus holding the developing embryos constantly in a moist environment, and the young emerge as tadpoles or froglets, depending on species.

Perhaps the most bizarre case of egg-brooding in frogs is found in the Australian Gastric-brooding frogs. The female ingests as many as 20 fertilized eggs, which undergo development in her stomach and are then "vomited up" as tadpoles and

Continued overleaf ▷

froglets. During this time the parent does not feed; in fact, the digestive system is inhibited by the release of a hormone-like substance, a prostaglandin, secreted in the oral mucus of the larvae. It shuts down hydrochloric acid secretion and the peristaltic waves of the gut.

Eggs or tadpoles of many frogs are transported by one of the parents to protect them from temperature extremes, desiccation, predation, or parasitism. In the midwife toads of Europe, the male carries strings of eggs, sometimes from more than one female, entwined around his hind legs, occasionally taking them back to water to keep them moist. When the larvae are ready to hatch, he carries them to water. In the Australian Pouched frog the tadpoles wriggle into the bilateral brood pouches on the male's flanks and later emerge as tiny froglets. In the South American mouth-brooding frogs of the genus *Rhinoderma*, the male of one species carries as many as 20 larvae in his

> Right *The slide-digging behavior of the South African bullfrog (*Pyxicephalus adspersus*) is well documented. In the hot summer, when this species becomes active after seasonal rains, the male bullfrog digs a channel between bodies of water to allow his tadpoles to escape shallow pools before they dry out. Channels can exceed 15m (48ft) in length.*

vocal pouch, which must elongate to the full length of his body as the tadpoles grow in size. Experiments show that the male actually provides nutrients to the developing embryos. The males may carry the eggs from several different females at the same time. In the other species, the male simply carries the tadpoles in his vocal pouch from the terrestrial nest to the water.

In some Neotropical poison frogs, the Seychelles frogs (*Nesomantis* and *Sooglossus* spp.), and Hamilton's frog of New Zealand (*Leiopelma hamiltoni*) the eggs are laid on land and, after hatching, the tadpoles wriggle onto the parent's back and are then carried to water, a journey that can last several days. In some species it is the male that performs the task and in some the female; in still others it may be both parents.

Eggs that Provide Nutrition

A particularly unusual form of parental care is the provisioning of trophic eggs (unfertilized or fertilized) by the mother to her tadpoles. This behavior has evolved at least six times among frogs, but in each case only where tadpoles develop in water-holding wells (bromeliad axils, tree holes, or bamboo segments) that have little or no food.

In poison frogs of the genus *Oophaga* the female deposits tadpoles in water-filled axils of understory herbs or in epiphytic bromeliads high off the ground, but never more than one tadpole per axil. She then returns at intervals of 1–8 days,

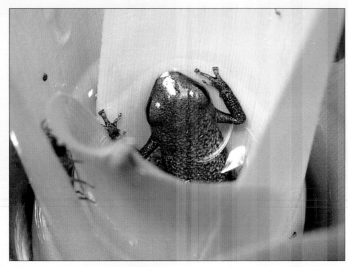

>◑ Right and above *In the rainforest of Costa Rica, the Strawberry poison frog (*Dendrobates pumilio*) has evolved an ingenious method of ensuring the safety of developing offspring. Clutches of 4–6 eggs are laid in the leaf litter on the forest floor. After the eggs hatch, the female carries the tadpoles up to high water-filled tree crevices or the axils of bromeliads. In these aquatic microhabitats, fed periodically with trophic eggs, the tadpoles grow into froglets.*

visually inspects the axil, backs into the water, and lays a small clutch of unfertilized eggs on which her tadpole feeds and without which it would die. The tadpole signals its presence by placing its head near the female's vent and then vibrating its body and tail. Tadpoles feed by biting through the jelly coat and sucking out the yolk. In a monogamous Brazilian species of this genus, both the male and female play key roles. Tadpoles are transported singly from the egg deposition site by the male and placed into water-filled treeholes. The pair courts on average every five days;

courtship apparently stimulates the female to ovulate eggs. The male then guides her to the tree holes containing their solitary tadpoles, whereupon she deposits one or two trophic eggs for the tadpole to consume.

Very similar behavior is reported in a totally unrelated mantellid frog, the Madagascan climbing mantella, which breeds in water-filled wells (tree holes or broken bamboo segments). During courtship, the male leads the female to a well, where they mate and a single egg is laid. Males defend the wells, while the females return to the water and lay a single trophic egg. The convergence between these Neotropical and Madagascan frogs, in terms of behavior and tadpole oral morphology, is remarkable.

Frogs may also guard or attend young by staying with the tadpoles for long periods after they hatch. In some, the parent sits in or near the tadpole school and will attack animals that disturb them. In one Panamanian leptodactylid frog, a school of tadpoles moved with the female parent along a water-filled ditch, apparently in response to repeated pumping movements of the female's body that produced surface waves that traveled toward the tadpoles. In the African pig-nosed frog (*Hemisus marmoratus*) the female, while still in amplexus with the male, digs an underground breeding chamber in the bottom of a future pond; the male leaves after fertilizing the eggs. Tadpoles hatch one week later and, after flooding, the female emerges at the surface and frees the tadpoles into open water. Sometimes the females will dig a slide on the surface, with the tadpoles swimming closely behind, thus guiding the tadpoles to water. KA

DECLINING AMPHIBIAN POPULATIONS

The causes and implications of a growing worldwide threat

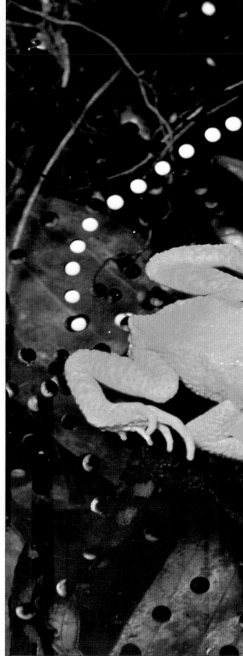

SINCE 1989, WHEN HERPETOLOGISTS FIRST became aware of the problem, there has been increasing concern about dramatic population declines of amphibians worldwide. The declines have led to the apparent extinction of more than 120 species, and scientists recently announced that at least one-third of the world's species are threatened with extinction. This is symptomatic of the global deterioration of the environment, and amphibians are by no means uniquely affected; there is similar concern about reptiles, birds, and all other forms of life. What has particularly raised concern in relation to amphibians is that declines and extinctions have occurred in nature reserves, national parks, and other supposedly protected areas set aside to preserve biodiversity.

A classic example is the loss of several frog and salamander species unique to the Monteverde Cloud Forest Reserve in Costa Rica, including the Golden toad (*Ollotis periglenes*), which has become a potent symbol of the declining amphibian phenomenon. Other examples include the harlequin toads (genus *Atelopus*), which range from Panama to the Andes of South America; between one-half and two-thirds of the 113 or so species have disappeared recently.

In Australia, both gastric-brooding frogs (genus *Rheobatrachus*) are probably extinct just a few years after their discovery, at least one other species is extinct, and four more have not been

seen for 15 years. Ten species are listed as critically endangered and 18 as endangered.

Common features of amphibian population declines in protected areas in such widely separated parts of the world as eastern Australia, the Pacific northwest of the USA, and in Central and South America are, first, that they have been very rapid, with species vanishing over two or three years; and, secondly, that they have affected some sympatric amphibian species but not others in the same habitats. These facts have stimulated research to hunt for one or more factors in the environment that could affect amphibians on a global scale, and to which some species might be more susceptible than others.

One such factor is the increase in the amount of ultraviolet B (UV-B) radiation that now reaches the Earth's surface as a result of the thinning of the ozone layer by atmospheric pollutants. Research, both in the field and in the laboratory, has shown that the eggs, embryos, and larvae of amphibians are generally highly sensitive to elevated UV-B, which breaks up their DNA and thus causes them to develop abnormally and die. Moreover, some species have been found to be unaffected by increased UV-B, raising hopes that the global factor affecting only some amphibians had been found.

This optimism was short-lived, however. There are many amphibians that have declined, especially in the tropics, in localities where levels of UV-B

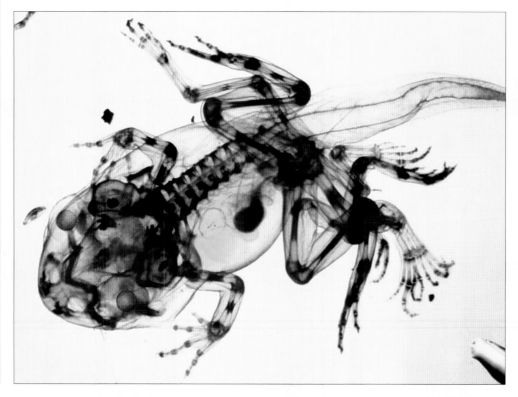

◐ **Above** *Two pairs of Golden toads spawn in Costa Rica's Monteverde Cloud Forest Reserve – a spectacular sight that may never be seen again. Officially listed as Critically Endangered, the Golden toad may in fact be extinct; although 1,500 individuals were counted in the reserve as late as 1987, there have been no reported sightings since 1989.*

◐ **Left** *Subjected to a process known as clearing and staining, this photograph shows in X-ray fashion the extra limbs of a deformed Pacific treefrog (Hyla regilla). The culprits responsible for such abnormalities are tiny, amoebalike parasites called trematodes, which burrow into the skin of tadpoles, interfering with the natural development of the limbs. Deformities are also caused by unnatural factors in the environment, such as pollution and UV radiation.*

be the result of predatory attacks, and there are parasites that burrow into the limb buds of frog tadpoles, causing two or more legs to develop where there should only be one. Non-natural causes of deformities (usually missing limbs or parts of limbs) include various manmade chemicals, increased ultraviolet radiation, and inbreeding in small, isolated populations. Deformities are sometimes very common in individual populations and so may have a negative impact on amphibian numbers at a local level; they may also represent a response to sublethal levels of environmental factors than can kill amphibians.

In many parts of the world, industrial activity creates acid rain, which may fall hundreds of kilometers away from the immediate source of the pollution; for example, the burning of fossil fuels in Britain is a major cause of acid rain in Scandinavia. Acidification of water has a negative effect on the egg and embryo stages of amphibians, and can cause amphibian population declines over wide areas.

Many amphibians are highly dependent on ephemeral ponds or streams for breeding, and their mating activity is closely linked to climatic changes that herald the advent of suitable conditions. Amphibians in Britain are now breeding several weeks earlier in the year than they were 20 years ago, a trend that is commonly seen as a symptom of global warming. Climate change can impact amphibians in many different ways and has been implicated in several instances of population decline. Notably, the dramatic loss of several frog species at Monteverde, Costa Rica, has been linked to a succession of El Niño events that have resulted in a marked reduction in the amount of land that becomes enveloped by low cloud each year. It has been suggested that the drier conditions that have resulted from the reduced cloud cover have forced amphibians to concentrate in fewer underground hiding places, increasing the spread of parasites and diseases.

The Threat from Disease

The most dramatic impact on amphibians in the last ten years has been disease. In the 1990s, mass mortalities among Common frogs (*Rana temporaria*) occurred over a wide area of southern Britain; these were caused by virus infections. Of much greater concern has been an apparently global outbreak of the disease known as chytridiomycosis. Caused by a single-celled fungus called a chytrid, which invades the skin of amphibians, this infection appears to have been responsible

radiation have not increased, and in species whose eggs and embryos are not exposed to sunlight. While this rules out UV-B as the cause of all amphibian declines, it nonetheless remains a significant threat to some species, particularly those that breed in shallow water at high altitude, where levels of UV light are highest. Recent research also indicates that, while elevated UV-B does not always cause mass mortality, it does have a harmful effect on developing amphibians, limiting their growth and causing physical deformities, and thus reducing the reproductive output of populations.

In many parts of the world, amphibians are threatened by one or more manmade chemical compounds, released into the environment as herbicides, pesticides, and fertilizers or else as the by-product of industrial processes. The list of compounds known to harm amphibians is very long

indeed. Of particular concern are nitrates, used as agricultural fertilizers, which accumulate in ponds and streams, and also a variety of compounds known as endocrine disruptors that interfere with amphibians' natural hormones. These substances have two major harmful effects. Firstly, they can cause amphibians to develop abnormally, with deformed mouthparts or, in extreme cases, with missing or additional limbs. Secondly, even at tiny concentrations, they can have a feminizing effect on males, reducing their reproductive success.

Deformities among amphibians have excited a great deal of public interest in the USA, but their relevance to population declines as a whole is unclear. They tend to be concentrated in particular areas; Minnesota, in particular, is a "deformed frog hotspot." Deformities are caused by several factors, some of which are entirely natural. They can

Continued overleaf ▷

○ **Above** *Now listed as Vulnerable, the Golden-striped salamander* (Chioglossa lusitanica) *of Spain and Portugal is one of a number of salamander species considered under threat. Globally, the main danger comes from habitat loss, although climate change and increased UV-B radiation also play a part.*

◖ **Right** *The threat to Giant Titicaca frogs* (Telmatobius culeus) *comes mainly from fishing; they are sold to the restaurant trade to feed a demand for frog's legs and, latterly, for a supposedly aphrodisiac frog juice. The loose skin of this largely aquatic species is an adaptation to maximize oxygen uptake.*

for the catastrophic collapse of amphibian faunas in central America, eastern Australia, and parts of the western USA. First described in captive animals, chytridiomycosis has been found in nearly every continent of the world. In particular, a previously unknown form, *Batrachochytrium dendrobatidis*, was linked to die-offs in Australia and throughout the Americas. The fungus feeds on the frog's skin, preventing cutaneous respiration and causing death in one to two weeks.

A strain of chytrid fungus was first discovered in an African clawed frog (*Xenopus laevis*), a species that was widely used in laboratories around the world. The fungus was found in Australia in 1978 and was probably present in America even earlier. There is evidence that the spread of the disease is therefore a result of the movement of amphibians between continents and can perhaps be carried by humans on clothes and footwear. The widespread infection of frogs in nature reserves and national parks, many of which are widely visited by scientists and tourists, seems to support this.

Much of the research carried out to investigate possible causes of amphibian declines inevitably involves considering a single factor in isolation

although, in reality, amphibians are threatened by many different factors at the same time. Some research has looked at interactions between two or more factors and has shown that there can be very significant synergistic effects between them. For example, in the western USA, climate change, increased UV-B radiation, and disease have acted together to cause amphibian declines. Climate change has reduced water levels in breeding ponds, with the result that amphibian eggs are less protected by deep water from UV light. This fact in turn makes the eggs more susceptible to the pathogenic fungus *Saprolegnia*, which invades and kills amphibian eggs.

The eggs and larvae of most amphibians have poor defenses against predators such as fish, and many amphibian populations have been devastated by the artificial introduction of fishes to ponds, lakes, and streams. For example, mosquito fish have been released into many parts of the world in an attempt to control malaria-carrying mosquitoes, while trout are commonly introduced to provide sport. Both kinds of fish find amphibian larvae easy and attractive prey. The loss of several amphibian species from mountain lakes in Califor-

nia is largely due to predation by introduced trout. Fish are not the only introduced enemies of amphibians; even other amphibians, moved to places where they do not belong, can threaten native species. The North American bullfrog (*Lithobates catesbeiana*) is a case in point. It has been introduced to many parts of the world to sustain a trade in frog-legs. Its larvae grow to enormous size and often outcompete the larvae of native species.

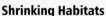

◁ **Left** *Once one of the most common frogs along North America's Pacific coast, the Red-legged frog (Rana aurora) has been badly affected by the introduction of non-native species, including trout and larger bullfrogs, to the wetlands where it lives. Nitrate pollution is also thought to have contributed to its decline.*

▷ **Right** *The Tinkling frog (Taudactylus rheophilus), confined to a small mountainous region of north-eastern Queensland in Australia, apparently disappeared in 1991 and was not seen again until 1998, when a few individuals were discovered in part of their former range. The species had declined dramatically over two years leading up to 1991 and may have fallen victim to an outbreak of chytridiomycosis, a fungus that invades the skin of adult frogs and reproduces repeatedly, apparently feeding on keratin, or a viral infection.*

Shrinking Habitats

The pressure generated by the world's ever-expanding human population creates an insatiable demand for land that results in the destruction of the natural habitat of plants and animals. This process is offset, to a very small degree, by the creation of nature reserves, but these can become prisons rather than havens for animals such as amphibians. Many amphibians live in small, local populations, the longterm survival of which depends on the occasional immigration of animals from other such populations elsewhere. Increasingly, amphibians are being forced to live in fragmented landscapes in which roads, built-up land, and agriculture separate one population from another. There is growing evidence that this isolation leads to inbreeding and a consequent loss of genetic diversity, manifested by decreased survival rates and by an increased incidence in anatomical deformities.

As animals become rare, their value in the international pet trade increases, and collecting can in turn become a further threat to their survival. This danger poses a risk to several of the world's most colorful frogs, including the poison and harlequin frogs of Central and South America and the mantellas of Madagascar.

Although amphibian population declines have attracted a great deal of scientific and media interest, there is no reason to think that they are unusual or unique. All the factors that adversely impact upon amphibians pose a threat to other forms of wildlife. In particular, the kinds of freshwater habitats upon which many amphibians depend – ponds, marshes, and wetlands – are under severe threat all over the world, with serious consequences for countless fish, insects, and other animals that frequent them.

What may be special about amphibians is that they are providing an early warning of an ecological disaster that is just beginning. Amphibians possess a number of features that make them especially sensitive to a wide variety of environmental insults. As eggs, larvae, and adults, they lack any kind of of protective body surface that can shield them from radiation or chemical pollution. Their young stages often lack protection against predators and can only develop safely in ephemeral water bodies that are threatened by climate change and habitat destruction. Compared with many animals, amphibians have very poor powers of dispersal, with the result that habitat fragmentation prevents the exchange of genetic diversity on which the longterm survival of individual populations depends.   TRH

SAVING THE WORLD'S AMPHIBIANS

Confronting global and local threats to conservation

THE WORLD'S AMPHIBIANS FACE A VARIETY OF threats to their continued existence (see Amphibian Population Decline). The geographical scale at which these perils apply ranges from global phenomena like climate change to very local factors, such as toads being killed by traffic as they cross a road on their way to a breeding pond. When it comes to asking what can be done to protect amphibians, and by whom, the answer depends on the scale at which a conservation initiative is directed. If it is indeed the case that amphibians are declining because of climate change, elevated ultraviolet radiation, or acid rain, then the solution lies in the hands of politicians and of global organizations that must seek the appropriate remedies through international treaties and agreements. There is little that individuals or local conservation groups can do to counter such threats, other than adding their voice to the pressure on political leaders to move environmental issues up the agenda.

At the local level there is, however, a great deal that small groups of people can do to protect and encourage amphibians. In many parts of Britain, mainland Europe, and North America, groups go out at night in spring to protect migrating amphibians as they cross busy roads. In some places, such groups have succeeded in persuading local authorities to close crucial stretches of road

for an appropriate period of time. Another strategy that addresses the same threat is the construction of tunnels under roads, which, if appropriately designed and positioned, enable amphibians to reach their breeding sites in safety.

Habitat loss can be offset, to a small extent, by habitat creation. Research carried out in Britain and the USA has shown that new ponds, created on agricultural land, are quickly colonized by newts, frogs, and toads. Even tiny ponds in gardens will support good populations of amphibians, provided that they are not also stocked with fish, and it is estimated that a larger proportion of Britain's Common frog population now lives in garden ponds than in natural habitats. Amphibians can be a bonus in gardens; the Common toad has been called "the gardener's friend" because of its appetite for slugs and insect pests.

Conservationists must remember, however, that most amphibians spend only a small part of their lives in water and that the creation of suitable terrestrial habitat is just as important as making new ponds. Since the ecology of terrestrial amphibians is very poorly known, creating suitable habitat for them is often a matter of guesswork.

In many developed countries, endangered amphibian species are afforded varying levels of legal protection. In Britain, for example, it is illegal

to collect or kill a Crested newt or a Natterjack toad. More importantly, their breeding sites are also often protected, and developers who wish to destroy a pond have to pay for mitigation measures, such as the creation of a pond elsewhere to which the threatened population can be moved.

Some amphibians have been successfully conserved by programs involving captive breeding and release of animals back into the wild. This procedure has great potential for many amphibians – always provided that it is combined with measures to protect their natural habitat – thanks to their high fecundity. In captivity, the heavy mortality by predation that is typical in nature can be prevented, with the result that very large numbers of animals can be produced. The Mallorcan midwife toad (*Alytes muletensis*) has been conserved in this way, and in Australia a similar project is seeking to protect the Endangered Corroboree toadlet (*Pseudophryne corroboree*).

The role of disease in amphibian declines requires its own set of conservation measures. Individual amphibians infected with the fungal disease chytridiomycosis can be cured with a preparation that is used to relieve athlete's foot in humans! This treatment is unlikely, however, to be of any help in protecting natural populations. There is a real possibility that herpetologists, the

◗ **Right** On a busy highway outside Brno, in the Czech Republic, environmental agencies have built an elaborate underpass to help conserve local toad populations. A female toad carrying a male approaches a wide-gauge grid, through which they will fall and continue their journey safely under the road.

◐ *Above* In Europe, warning signs alert motorists to the seasonal presence of frogs and toads on the road. Traffic may cause significant mortality among amphibians; certain frog and toad species are especially vulnerable, since they use the same route year after year to return to their breeding site.

◐ *Left* Amphibians are drawn en masse every spring to breeding sites in lakes, ponds, streams, and even puddles. The most favored conditions for such mass migrations are warm, rainy nights. Here, common toads are seen emerging from a specially-constructed road tunnel in the United Kingdom.

very people who seek to conserve amphibians, may have helped to spread diseases by carrying spores on their rubber boots or collecting gear. Many organizations, including the Declining Amphibian Populations Task Force, have issued guidelines to try to prevent the local spread of amphibian diseases. At the international level, there are moves to control and reduce the movement of amphibians around the world, to try to reduce the chance that diseases will be spread from one country or one continent to another.

Nature reserves are an obvious way of conserving amphibians, though they cannot protect them from many of the threats they face. One important issue is how protected areas should be designed to provide optimal conditions for amphibians. Populations based on a single breeding site are likely to face eventual extinction despite protection, because they become inbred. Many amphibians seem to require a network of breeding sites, connected by habitat that they can cross reasonably easily, in order for the population to maintain a high level of genetic variation.

While a great deal can be, and is being, done to conserve amphibians at local, national, and international levels, much of it is carried out more in the hope than the expectation of success. Truly effective conservation requires a deep understanding of ecology, and, sadly, there are many aspects of the ecology of amphibians about which we remain profoundly ignorant. For most amphibians we do not even know the answer to the simple question: Where do they go when they are not breeding?   TRH/RDS

SWIMMING, EATING, GROWING MACHINES

The tadpole's struggle to survive in a hostile world

WE COMMONLY INTERPRET HOW THINGS ARE from a mammal-oriented viewpoint, and from this perspective the changing of a tadpole into a frog is a truly unconventional event. Although about 20 percent of the 5,453 species of frogs in the world lack a tadpole stage, the rest feature a tadpole as part of their developmental cycle for anything from a few days to several years. This non-reproducing creature is little more than a swimming, eating, growing machine, whose main ecological goal is to grow as quickly as possible so as to send the largest possible metamorph into the reproductive part of the frog's lifecycle. Many factors of tadpole ecology influence the success of the enterprise, and because of the many biological and environmental hazards involved, perhaps only about 1 percent of tadpoles ever reach metamorphosis; considerably fewer than that end up as reproductive adults.

The early development of all the frogs that have a tadpole stage is broadly similar; it is also relatively well-known, because easy access to large eggs developing outside the mother has made frog reproduction a handy subject for students of vertebrate embryology. Tadpoles occur in every imaginable freshwater habitat, from a few milliliters of water in the axil (angle between leaf and stem) of a bromeliad through puddles and ponds to large lakes and torrential streams. In addition, anything

from one to a dozen species may be found at any given site. A small number of species in South America, Africa, and India actually spend most or all of their time out of free water, although always in moist conditions.

Tadpoles occur in many different microhabitats, and have enough locomotor and feeding adaptations to provide exceptions to any general statements about them. Even so, some aspects of their biology can be summarized.

In terms of morphology, the typical tadpole has a series of ornate feeding structures around the mouth that occur in no other vertebrates, and the structure and operations of the jaws are unique. The mouth is typically surrounded by an oral disc with papillae along the margins in various patterns, and hundreds of tiny, labial teeth that are keratinized (composed of the material that human fingernails are made of) rather than ossified like the teeth of adult frogs; these are arranged in transverse rows on the upper and lower labium to serve as raspers. Keratinized sheaths on the jaw cartilages serve as surfaces for biting, gouging, and cutting. The oral disc of those tadpoles that attach to rocks in fast water is huge; those of pond-dwellers are smaller. The mouthparts of midwater suspension feeders lack all keratinized and most soft tissues. Water pumped in through the mouth passes over the gills and over intricate

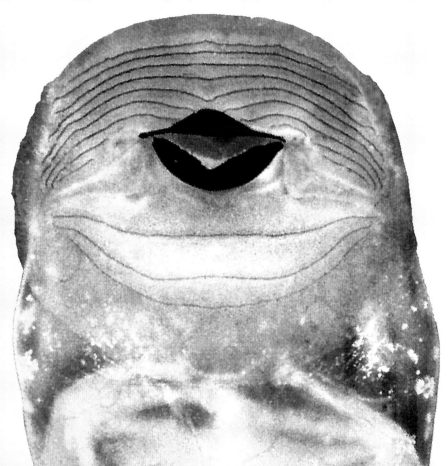

⟩ Above Dead or dying animals are a rich source of nutrients for tadpoles. There are some fully carnivorous tadpoles, but even mainly herbivorous tadpoles will also occasionally scavenge on carcasses. Here, hylid tadpoles of the Meadow tree frog (Isthmohyla pseudopuma) are cannibalizing a conspecific.

⟨ Left Greatly magnified, the mouthparts of a Cascade frog (Boophis sp.) tadpole show the large oral disc surrounding the mouth. These free-swimming tadpoles develop in mountain streams in Madagascar, using the disc to attach themselves to rocks while feeding in fast-flowing water.

⟩ Right A tadpole of the African clawed toad (Xenopus laevis) showing its undulating tail tip. Larvae of this highly successful species are midwater suspension feeders that filter phytoplankton. However, they need to rise to the surface regularly to take in air; they break the surface of the water extremely quickly, in 80 milliseconds or less, yet only open the mouth when it is above the water's surface and close it again before the mouth is fully submerged.

waters are able to move and feed while attached to rocks by their large oral discs. Many tadpoles are opportunistic scavengers on dead animals, but a few are specialized for nipping pieces out of living tadpoles or even ingesting entire individuals. Even if these carnivorous tadpoles are sometimes cannibalistic, they prefer to feed on nonsiblings.

Because tadpoles are nonreproductive, they lack the colorations commonly related to reproduction in other groups. Most tadpoles exhibit rather somber hues that act as camouflage, and countershading featuring dark colors on top and light below, which makes an individual less easy to detect against the normally downwelling light, is a common feature. A few are conspicuously colored, apparently either to enhance social cohesion or else to advertise noxious or toxic materials in the skin (aposematic coloration). Both the body and tail muscle may be striped or banded, and although the fins are usually clear, some species have prominent spots or contrasting colors in the fins. A recent finding that body shape and coloration are rather plastic in the presence of certain predators is exciting.

Because tadpoles themselves do not reproduce, most of their behavior serves in some way to enhance survival, like various sorts of escape modes and social interactions. Tadpoles commonly aggregate, gathering in groups in response to environmental stimuli such as areas of warmer temperatures, but a few species form stationary or mobile schools – social groupings that exhibit complex interactions. In some cases, a parent may lead these schools to good feeding zones or to less hazardous areas. Tadpoles of some *Bufo* and *Rana* species can distinguish kin from nonkin, seemingly by chemical cues, and prefer to associate with the former.   RAA

food-trapping structures that in some species can collect particles the size of bacteria, and then exits via a spiracle that most commonly is single and on the left side. The long gut, typically arranged in a double-spiraled coil, is the most prominent internal structure.

Body shapes vary with habitat. Bottom-dwelling tadpoles, like those of toad (*Bufo*) and true frog (*Rana*) species, are somewhat depressed, while those attached to rocks via the oral disc in fast water, such as *Ascaphus* in North America and *Heleophryne* in South Africa, are even more so. Tadpoles that live in bromeliads, leaf axils, and tree holes (phytotelmata) are commonly quite attentuate. In all such cases the eyes are on top of the head. Tadpoles that spend most of their time in midwater, including many hylids and some hyperoliids, have compressed bodies and eyes that are located on the sides of their heads. In the latter group, the tail fins are tall and extend well forward onto the body, while in bottom-dwelling and especially fast-water forms, the fins are low and may end at the tail–body junction or further toward the tip of the tail.

When feeding, a typical pond tadpole uses its mouthparts to graze small particles from the prolific fauna that grows on all submerged surfaces. Some forms have the oral disc oriented upwards and harvest particles caught in the surface film by tilting head-up, whereas others are midwater suspension feeders. Some midwater forms float quietly in a horizontal position, but others such as *Xenopus laevis* maintain a head-down posture with the help of a constantly undulating tail tip. Species that inhabit rapidly flowing

Caecilians

OFTEN MISTAKEN FOR LARGE EARTHWORMS, *caecilians are long-bodied, limbless amphibians with little or no tail. Scientists are only beginning to understand the lives of these secretive tropical burrowers, which are difficult to observe and have a limited fossil history.*

Most caecilian fossil material is of vertebrae from the Holocene back to the Paleocene (65 million years ago). Recently, however, material from the Cretaceous of North Africa and the Jurassic of North America (Arizona) has been discovered. The Arizona specimens provide major clues about caecilian evolution. They have small limbs and a tail, and their bodies are only moderately elongated, although the skull structure indicates that they are unmistakably caecilians.

Wormlike Burrowers
FORM AND FUNCTION

The bones, teeth, fat bodies, and other structures of caecilians, as well as molecular data, show that they are related to salamanders and frogs, and are thus members of the class Amphibia. They apparently underwent a major change early in their evolutionary history. With the elongation of their bodies they lost their limbs (no living caecilian species have limb or girdle rudiments) and their tails (only phylogenetically basal caecilians have tails, of 4–20 vertebrae). Their skulls became massively bony and their eyes were reduced as they assumed an underground lifestyle.

Today, caecilians appear to live in low densities in many localities, but are abundant in others, especially in southern China, Ghana, and some areas of Central America. Yet they have radiated throughout the tropics. They are difficult to observe, emerging infrequently from their burrows, although the aquatic species of South America are occasionally taken in fishermen's seine nets. The resemblance to large earthworms springs from the segmental rings around their bodies. Aquatic caecilians have been mistaken for synbranchid eels.

Caecilians are very diverse in size. The smallest (*Idiocranium russeli*, from West Africa) is mature at 7cm (2.8in); the longest is *Caecilia thompsoni* of Colombia at 1.6m (5.2ft). Other forms are shorter but stouter; *C. nigricans* is 80cm (31in) long with a body diameter of 4cm (1.6in).

Caecilians are burrowing animals, whether in soil or in the substrate of a body of water. They use their heads as trowels for digging, or to poke in mud for food. Their skulls are heavily boned for this use and the skin is very adherent, its underlying layers fused to the bones so that the skin does not shear away during digging. Locomotion is by body undulation, a wave of muscular activity from the head backward. The curves formed by the body resist the soil or water and forward movement occurs. The segmental rings around the body have not been demonstrated to play a role in locomotion.

Above The segmental body rings or "annuli" that are common to all caecilians are especially prominent on the appropriately named Siphonops annulatus of South America.

Left A notable feature of caecilians is the retractable tentacle sensor, located on each side of the head between the eye and the nostril. This organ, which is unique in the vertebrate world, transmits chemical messages from the environment to the nasal cavity, and aids the caecilian in searching out prey. Shown here is the tentacle of Ichthyophis glutinosus, a species from Sri Lanka.

CAECILIANS

Order: Gymnophiona (Apoda)

173 species in 33 genera and 3 families:
Caeciliidae (28 genera, 120 species);
Ichthyophiidae (3 genera, 44 species);
Rhinatrematidae (2 genera, 9 species).
The taxonomic relationships of caecilians are
currently under study, and it is likely that more
families, genera, and species will be revealed.

DISTRIBUTION SE Asia from India and Sri Lanka,
S China, Malay archipelago to S Philippines; E and W
Africa; Seychelles; C and much of S America.

Equator

HABITAT Moist, loose soil and ground litter of tropical
forests and plantations, often near streams; typhlonectids in lowland rivers and streams.

SIZE Variable: smallest species 7–11.6cm (2.8–4.6in)
in length when mature, largest 1.6m (5.2ft); many
30–70cm (12–28in); most stout-bodied, some slender.

COLOR Many species uniform blue-gray, some bright
yellow, brown, or black; some with lighter lateral or segmental stripes, others with blotched patterns; heads
often lighter, to pink or light blue.

REPRODUCTION Internal fertilization; some species lay
eggs, producing free-living larvae, others have direct
development through metamorphosis before hatching;
several species in 3 families are viviparous – the mother
nourishes 2–25 young in her oviducts for 9–11 months.

LONGEVITY 12–14 years in *Dermophis mexicanus*.

CONSERVATION STATUS Sagalla Hill caecilian
(*Boulengerula niedeni*) is listed as Critically Endangered;
Mahe caecilian (*Grandisonia brevis*) is Endangered; and
Cooper's black caecilian is Vulnerable. The status of
many species of caecilians is unknown because they
are poorly studied.

The skin is smooth, its outer layers somewhat toughened with keratin. The inner layer contains many mucus glands and variable numbers of poison glands, whose secretions can be quite toxic to predators, including humans. In many species there are also patches of scales in the segmental rings. These resemble fish scales in structure and embryonic origin, and are unlike those of reptiles. There is a trend toward loss of the scales – phylogenetically basal (ancestral, or "primitive") species have them along the entire length of their bodies, others only in rear segmental rings, while the most derived species lack scales altogether.

Caecilians have well-developed immune systems. Recent work on the neuroanatomy of the animals indicates that their brains and peripheral nerves are diverse in structural patterns. Free-living larvae have mechanoreceptor and electro-receptor organs in their skins. Most caecilians have only one well-developed lung; some aquatic typhlonectids have both lungs developed, but one species is lungless. All caecilians have considerable gaseous exchange via their skins and the linings of their mouths.

Living off the Soil

DIET AND FEEDING

Most caecilians are generalized opportunistic feeders preying, for example, on earthworms, termites, crickets, and grasshoppers. Some appear to be specialists, on termites in the case of *Boulengerula* or earthworms (*Dermophis*, when worms are abundant), or beetle pupae (*Typhlonectes natans*). Small lizards are occasionally eaten by *Dermophis*, and *Siphonops* will eat baby mice in the laboratory. Caecilians themselves are preyed upon by snakes and birds.

Feeding involves a modified "sit-and-wait" strategy. Typically, caecilians slowly approach their prey, then quickly seize it with a strong grab of the jaws. The lower jaw is underslung, so as not to interfere with the burrowing function of the head. All caecilians have two rows of teeth on the upper jaw, and one or two rows on the lower. These inwardly curved teeth hold the prey fast as successive bites and expansions of the mouth and throat propel it into the gut. The teeth are modified for cutting and holding, with sharp edges and sometimes two cusps. Some muscles have changed function, from contracting the oral cavity to closing the lower jaw, to effect a strong bite.

Larvae and Live Young

BREEDING AND DEVELOPMENT

All caecilians have internal fertilization. The male extrudes the rear part of his cloaca and inserts it into the vent of the female, thus directly transferring sperm to the female's reproductive system, where fertilization takes place as in reptiles, birds, or mammals. Virtually nothing is known of mate recognition and courtship, although aquatic species have been observed in an undulating "dance" before mating. Basal species of caecilians lay eggs in burrows near streams; gilled larvae hatch and wriggle into the water, becoming terrestrial after metamorphosis. Some species are direct

developers; the eggs are laid underground and the young develop within them through metamorphosis, so that a miniature adult hatches. In apparently all the egg-laying species, the female provides parental care, lying coiled around her eggs.

Viviparity (live birth) is the most striking feature of caecilian reproduction. Perhaps half of all species retain the developing young in the female's oviducts through metamorphosis, nourishing the young with "uterine milk," rich secretions from glands in the oviducts, after the yolk supply is exhausted. Very few frogs and salamanders have evolved such a mechanism. The developing fetuses have many tiny teeth, of a different

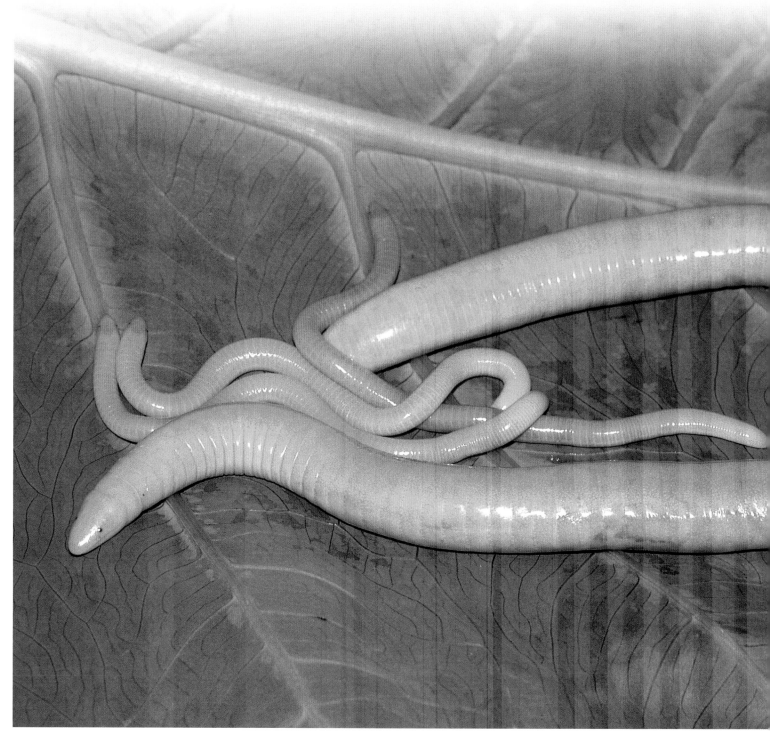

Right *A wormlike body and covered eyes are typical of caecilians (here,* Caecilia tentaculata). *The eyes are covered by skin, or skin and bone, and adhere to these covering layers. The lens of the eye is fixed, and some species have no eyeball muscles. The lens and retina are reduced in some species, but almost all have an optic nerve, suggesting that the function of the eyes is to sense light.*

Below *Most caecilians have inconspicuous coloration, but* Schistometopum thomense *is a striking exception. Its bright orange-yellow skin may serve to warn predators that it secretes noxious chemicals that make it unpalatable.*

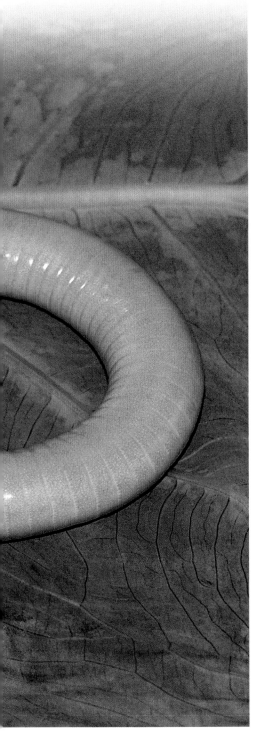

shape in each species, associated with feeding on these glandular secretions of the oviduct. They acquire adult teeth just at birth. The fetal gills are thought to function in gaseous, and perhaps also nutrient, exchange with the mother, through transport of materials across the capillaries to the circulating blood of both fetus and mother. Gestation is long, from 7 to 11 months depending on the species, and the young are nourished by the maternal secretions for most of that time. A clutch numbers 2–25 young, and the mass of each fetus increases enormously during gestation, so the energy demand on the mother is very great.

Three Main Groupings
CAECILIAN FAMILIES

The three families of caecilians are distinguished by combinations of characters including those of reproductive biology, external body form, bones, teeth, and muscles. Molecular data are increasing, and are helping to resolve our understanding of the relationships and identity of caecilian taxa. These amphibians are so little known that there are no common names for the families and genera, nor for most of the species.

The Rhinatrematidae and the Ichthyophiidae include genera with several rings in each body segment (rather than just one or two), caudal vertebrae (a "tail") extending beyond the anus, larger numbers of skull bones, and an egg-laying mode of reproduction with free-living larvae. Members of the genus *Uraeotyphlus* differ slightly in having two rings in most body segments, fewer skull bones, and differences in musculature.

Members of the family Caeciliidae have only one or two rings per body segment, no true tail, and fewer skull bones. Some bones appear never to develop; others fuse to form fewer elements. They may be either egg-layers (with larvae or direct-developers) or live-bearers. This family is divided into two subfamilies. The live-bearing Typhlonectinae have one ring per body segment and a slight to moderate dorsal "fin." They are aquatic and semi-aquatic, the only caecilians that are not terrestrial burrowers. The Scolecomorphinae also have only one ring per segment. Several species are live bearers and have a reduced number of skull bones, including the absence of the stapes, or middle ear bone.

Subject to Stress
CONSERVATION AND ENVIRONMENT

Caecilians are becoming rarer at many of the localities at which they have been observed. Land use changes (particularly the destruction of forests for agriculture and ranching) are causing the restriction and even the extinction of many populations and some entire species. Infections by a chytrid fungus, as has been found in a number of frog and some salamander species, are also bringing about the death of caecilians in several locations. Such infections apparently are harbingers of general environmental degradation, and consequently of stressed and vulnerable animals. It is most unfortunate that caecilians are dying, and possibly even facing extinction, just as scientists are beginning to appreciate their biology and their place in ecosystems. MHW

Salamanders and Newts

SALAMANDERS AND NEWTS ARE TAILED *amphibians that typically lead secretive lives. Generally living in cool, shady places and active only at night, few are more than 15cm (6in) long. Unlike frogs and toads, they do not advertise their presence by making loud sounds.*

Certain species can, however, be extremely abundant. In some mountain forests of eastern North America, it is estimated that the total mass of woodland salamanders exceeds that of all the birds and mammals put together. The newts of Europe and North America migrate from over a wide area to breed in ponds each spring, and very large breeding populations can occur. In recent years, research has revealed a wealth of fascinating information about their habits. New species are still being discovered and described for the first time, particularly in the tropical forests of Central America.

Smooth-skinned Carnivores

FORM AND FUNCTION

Salamanders and newts comprise the order of amphibians known as the Caudata (or Urodela). They typically have elongated bodies, long tails, and two pairs of legs of roughly similar size, although some forms have lost the hind limbs. They thus resemble more closely the earliest fossil amphibians in terms of overall body shape than any other present-day amphibian group.

The term "salamander" derives from the Latin *salamandra*, in turn derived from the Greek for "fire-lizard." Salamanders were associated with fire because they crawled out of logs thrown onto a fire, and it was thought they could crawl through fire; asbestos was once called "salamander's wool." "Salamander" is applied generally to any tailed amphibian, but more especially to those with terrestrial habits. The term "newt," derived from the Anglo-Saxon *efete* or *evete* which became *ewt* in Medieval English, refers to those genera that return to water each spring for a protracted breeding season. These include the European genus *Triturus*, the North American genera *Taricha* and *Notophthalmus*, and the Japanese genus *Cynops*. The term "eft," from the same Anglo-Saxon derivation, refers to the young, terrestrial stage of *Notophthalmus* species.

Like other amphibians, salamanders and newts have a smooth, flexible skin that lacks scales and is usually moist. The skin acts as a respiratory surface, at which oxygen enters the body and carbon dioxide is released. For this reason, salamanders and newts are restricted to damp or wet habitats. The outermost layer of skin is often shed. In some species it comes off in bits and pieces, in others it peels off whole. The shed skin is usually eaten, but sometimes the complete skin of an aquatic newt can be found hanging from water weed. Many salamanders and newts are protected against predators by secretions from numerous poison glands in the skin (see Repellent Defenders).

All salamanders and newts are carnivorous, feeding upon small, living invertebrates such as insects, slugs, snails, and worms. They possess a tongue that is used to moisten and move food in the mouth; in some species it can be flicked some distance forward to capture small prey. The tail is

❯ **Right** *A female Great crested newt (Triturus cristatus) feeding on her recently sloughed skin. Some populations of European newts spend almost the entire year in water once they have become adults. The numerous small pits on the head are lateral-line organs, which allow the newt to detect tiny disturbances in the water, such as those caused by its prey.*

❯ **Below** *Displaying the quintessential characteristics of salamanders and newts, the Spotted salamander (Ambystoma maculatum) has a tubular body with shiny, pliable skin, a long tail, and prominent costal grooves. This species is relatively abundant, though rarely seen, since it spends the daylight hours underground. Like many other salamanders, its skin exudes a toxin when it is attacked.*

FACTFILE

SALAMANDERS AND NEWTS

Order: Caudata (Urodela)

Over 556 species in 64 genera and 9 families

DISTRIBUTION N America, C America, northern S America, Europe, Mediterranean, Africa, Asia, including Japan and Taiwan.

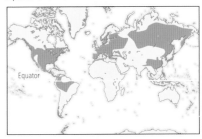

HABITAT Aquatic, terrestrial, or amphibious.

SIZE Length from snout to tip of tail from 2cm–1.4m (0.8in–4.6ft), but most are 5–15cm (2–6in).

COLOR Highly variable, including green, brown, black, red, orange, yellow.

REPRODUCTION Fertilization mostly internal; most lay eggs, but in some species eggs develop inside female.

LONGEVITY In captivity, sometimes 20–25 years, occasionally more than 50; very little known about longevity in the wild, but some species do not breed until they are several years old.

CONSERVATION STATUS 55 species, including the Desert slender, Lake Lerma, and Sardinian brook salamanders, are Critically Endangered; 110 are Endangered, and 91 Vulnerable.

well developed for swimming in many aquatic species, being laterally compressed, and may bear a dorsal fin. The two pairs of limbs both have digits, often fewer on the front than on the hind limbs. In some species, such as the amphiumas of North America, the limbs are extremely small and perform no locomotory function. In others, such as the Palmate newt, the toes of the hindlimbs have webbing between them that facilitates rapid swimming.

The eggs of tailed amphibians do not have a shell, but are coated with layers of protective jelly. Each layer forms a distinct capsule and the eggs of some salamanders are contained within as many as eight capsules. Usually the egg hatches into a wholly carnivorous larva, which grows until it metamorphoses into the adult form. Metamorphosis involves a number of complex changes that equip the salamander or newt for its adult life. In newts, metamorphosis coincides with a dramatic change of habitat from water to land.

The moist skin of newts and salamanders is, for most species, only one route by which oxygen and carbon dioxide enter and leave the body. Typically, larvae have feathery external gills and adults have lungs. Some species, including both aquatic and terrestrial forms, also use the inner surfaces of their mouths, rhythmically sucking in and expelling water or air through the mouth or nostrils. This "buccal pumping" is very apparent as rapid vibrations of the soft skin under a newt's or a salamander's chin. It serves not only as a respiratory mechanism but also enables the animal continually to sample its external environment for odors. A sophisticated sense of smell is very important for tailed amphibians, particularly in communication.

The largest family, the lungless salamanders, have entirely lost the lungs of their ancestors and breathe only through their skin and mouth lining. Some live in fast-flowing streams, where oxygen is abundant; others are totally terrestrial. Salamanders and newts living in static water, where oxygen levels can become very low, have to breathe through their lungs, coming to the water surface at frequent intervals to take in a supply of fresh air, or through external gills, which they retain from the larval stage into adult life.

The aquatic newts have three ways of obtaining oxygen when they are living in water during the breeding season: through their skin, their mouth lining, and their lungs. For most of the time, oxygen obtained from the water via the skin and mouth is sufficient to sustain their activity. However, if they become more active, for example during sexual behavior, they have to make frequent ascents to the water surface to gulp in air. Sometimes a newt will gulp in so much air that it is unable to sink back to the bottom of the pond. It can restore its negative buoyancy by expelling air bubbles, or "guffing," in the manner of a diver. Breathing ascents are potentially dangerous because newts are particularly conspicuous as they rise to the surface, and it is then that predatory birds such as herons are most likely to catch them. Newts also have to breathe more often on very warm days, when the high temperature causes them to be more active but when oxygen levels in ponds become very low.

The exact number of salamander and newt species in the world is not yet known, because new species are being described all the time. A recent list contained 556 species, whereas, 20 years ago, only 358 species of tailed amphibians were known. This dramatic increase may be ascribed to two factors. First, exploration of undocumented regions has uncovered several new species. This is especially true of Central and South America, where several new species of lungless salamanders, many of which are very small, have been discovered (see Miniature Salamanders). Second, as a result of the application of new techniques for analyzing mitochondrial and nuclear DNA, a number of wide-ranging salamanders have been found to include several genetically distinct forms, each with a clearly-defined and limited range.

On Land and In the Water
DISTRIBUTION PATTERNS

Salamanders and newts are largely confined to temperate climates in the northern hemisphere, although one group, the lungless salamanders, has invaded the tropical zone in Central and South America. They live in a variety of habitats and include fully aquatic and fully terrestrial forms, as well as species that divide their time between water and land. Aquatic forms are found in rivers, lakes, mountain streams, ponds, swamps, and underground caves. Terrestrial species commonly live under rocks and logs, but some burrow deep into the soil and some may climb to considerable heights in trees. Because salamanders and newts have a skin that is permeable to water, they cannot tolerate hot, dry conditions and, for many species, the summer is a time when they retreat into damp refuges, emerging only on cool nights. In cold weather they become inactive, and species that live in temperate climates bury themselves in the ground or hide beneath large rocks and logs and become torpid.

Different Lives, Different Life Cycles
REPRODUCTION AND DEVELOPMENT

Because there is such a great deal of diversity, there is no "typical" life cycle for members of this order. However, three general types can be distinguished: wholly terrestrial, wholly aquatic, and amphibious. A typical terrestrial species is the Red-backed salamander, found in woodland areas

○ **Below** Spotted salamander eggs with developing larvae. Often, a green alga grows within the egg capsule, to the benefit of both species; eggs containing algae have larger embryos that hatch earlier and survive better. Many spotted salamander breeding ponds in the US northeast are polluted by acid rain.

of eastern North America. Mating takes place on land, and the female retires to lay her eggs within a partly rotten log. A small number of large eggs (20–30) is produced and the embryos develop rapidly within them. The entire larval stage of development is passed within the egg, which hatches to produce a miniature version of the adult salamander. In some terrestrial salamanders, for example the American Dusky salamander, one of the parents guards the developing eggs. In a few species, such as members of the European genus *Salamandra*, the female retains her fertilized eggs within her body while they develop, later giving birth to live young, which may be either larvae or juveniles. In one particular Spanish population of the Fire salamander, it has been found that the faster-growing young eat their smaller siblings while still inside their mother.

In wholly aquatic species, adults mate in water, and some practice external fertilization. Clutches tend to be large (up to 500 eggs), and the eggs hatch into externally-gilled larvae. The adults of several wholly aquatic species, such as the olm of the Balkan coast and the mudpuppy of eastern North America, retain certain larval structures, such as external gills, into adulthood, a condition known as paedomorphosis. The retention of gills

○ **Above** *In common with other species that spend much of their time in water, the Great crested newt's tail is laterally flattened, enabling it to use it as a rudder and a means of propulsion. The prominent crest identifies this individual as a male.*

○ **Below** *A comparison of the life cycles of salamanders and newts. Many salamanders are completely terrestrial, while newts return to water each year to breed. The axolotl of central Mexico, a member of the mole salamander family, is totally aquatic.*

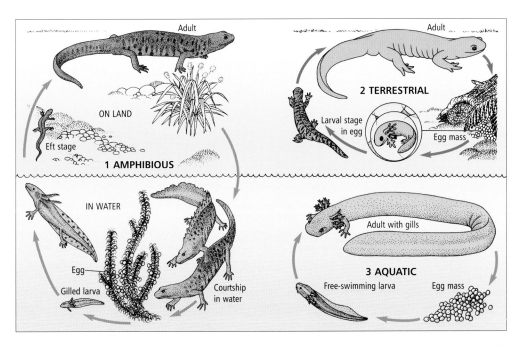

Adult

ON LAND

Eft stage

1 AMPHIBIOUS

IN WATER

Egg

Gilled larva

Courtship in water

Adult

2 TERRESTRIAL

Larval stage in egg

Egg mass

Adult with gills

3 AQUATIC

Free-swimming larva

Egg mass

throughout life is most common in those species that, like the sirens, live in stagnant, oxygen-deficient water. Paedomorphosis may be a permanent feature of all members of a species, as in the sirens, the olm, and the mudpuppy, or it may be a temporary condition occurring in some individuals lacking a critical nutrient. The best-known example of this last option is the Mexican axolotl, which retains its larval gills if it is deprived of the iodine it needs to produce the hormone thyroxine.

In the amphibious life cycle, the adult spends most of its life on land, but migrates to water in order to breed. Courtship may occur in dense aggregations of individuals, all coming to breed in the same pond at the same time, as in the Spotted salamander of North America. By contrast, the breeding season in some European newts (genera *Triturus, Lissotriton,* and *Mesotriton*) lasts for several months, with individual animals varying considerably in terms of how long they remain in the water. The female lays eggs that develop into aquatic larvae. Clutches are often large (100–400 eggs), but eggs tend to be small. The larvae metamorphose into juveniles, or efts, which leave the water to live on land until they have grown to reproductive age, a process that takes 1–7 years. Newts of the family Salamandridae, such as the European Smooth newt and the North American Red-spotted newt, typify this type of life cycle.

Giant Salamanders

FAMILY CRYPTOBRANCHIDAE

Much the largest of all the salamanders, the three giant salamander species live in rivers and large streams. They are generally nocturnal in their habits, spending the day beneath rocks. Despite their large size and ugly appearance, they are quite harmless to humans. In Japan and China, they are caught with a rod and line and are considered a delicacy. Their diet consists of virtually any animal that lives in rivers and streams and includes fish, smaller salamanders, worms, insects, crayfish, and snails. Prey is usually caught from the cover of a rock or overhang by means of a rapid sideways snap of the mouth. They feed mostly at night and rely on smell and touch to locate their prey. Their vision is poor; the eyes are small and, being positioned far back on the sides of the head, cannot both focus on the same object at the same time.

Giant salamanders never leave the water and, although they lose their gills quite early in life, they never fully lose all larval characteristics. They are probably very long-lived; one specimen survived in captivity for 52 years. The Japanese giant salamander can reach a length of over 1.5m (5ft) and the hellbender has a maximum length of about 70cm (28in). Their large size and lack of gills probably confine them to flowing water, where oxygen is in good supply. A conspicuous fold of skin along their flanks increases the surface area through which oxygen can be taken in. They can breathe through their lungs, although not

very efficiently, and animals kept in an aquarium will make frequent ascents to gulp in air.

In the hellbender, breeding occurs in late summer. The male excavates a large space beneath a rock and defends it against other males. He also excludes females who have laid their eggs, but allows any egg-bearing female to enter. The female lays her eggs in the male's nest in two long strings; the eggs are held together by a sticky thread that adheres to rocks and hardens soon after it comes into contact with water. A large female can lay as many as 450 eggs, and several females may lay their eggs in the nest of a single male. The male fertilizes the eggs, covering them with a cloud of milk-like seminal fluid, and guards them through-

⚫ *Above* One of the largest living salamanders, the Chinese giant salamander (Andrias davidianus) is named for the 19th-century French missionary and naturalist Armand David. It lives in clear, fast-flowing mountain streams, where the oxygen supply is plentiful, and has heavily wrinkled skin that helps respiration in the absence of gills.

⚫ *Below* The California giant salamander (Dicamptodon ensatus) inhabits the boreal coniferous forest on the Pacific northwest coast of America, from British Columbia down to northern California. Along with the three other species in its family, it once thrived in the virgin temperate rainforest conditions of the region, but its habitat is increasingly coming under threat as the area is opened up to exploitation by commercial logging.

out their development (10–12 weeks). The larvae then leave the male's nest and lead a wholly independent existence, feeding mainly on small aquatic animals. The skin of larvae and adults produces a noxious slime that deters potential predators.

All three species are threatened in many areas by the silting up of their habitat by forestry. There is concern in the USA about a decline in numbers for the hellbender throughout its range, and the Japanese giant salamander is classed as Vulnerable.

Asiatic Salamanders
FAMILY HYNOBIIDAE

Restricted to central and eastern Asia, the Asiatic salamanders are poorly known. From both their form and their reproductive biology, they are clearly some of the most primitive tailed amphibians. Terrestrial for much of their lives, they tend to breed in mountain streams where oxygen is abundant, and have very small lungs or no lungs at all; large lungs could make it difficult to avoid becoming very buoyant and being swept away by the current. Some species have sharp, curved claws, whose function is unknown.

During mating, the female produces paired sacs, each containing 35–70 eggs. The male grasps these as they emerge from her cloaca and, pressing them to his cloaca, sheds sperm onto them. In at least some species, the male guards the eggs until they hatch. The Semirechensk salamander has a very unusual form of mating. The male deposits sperm in a simple spermatophore and the female places her egg sacs on top of it.

There is concern about the conservation status of several Asiatic salamanders. Many are threatened by habitat destruction and deterioration, and several species are collected in large numbers to supply the worldwide pet trade.

Olm, Mudpuppies, and Waterdogs
FAMILY PROTEIDAE

The two genera in the family that includes the olm, mudpuppies, and waterdogs lead entirely aquatic lives and possess a number of larval characteristics, notably feathery external gills, although they also have lungs. The olm lives in water-filled underground caves in limestone mountains along the Balkan coast of the Adriatic and spends much of its time buried in mud or sand. Its skin lacks pigment, giving its body a white, pasty appearance, and its gills are bright red. It has short, feeble limbs and, with only vestigial eyes, is totally blind. Olms grow up to 30cm (12in) long and are becoming increasingly rare, partly because of pollution, but also because they are collected for the aquarist trade.

All that is known of their reproduction is based on aquarium observations. The eggs are fertilized inside the female's body, and after fertilization she may do one of two things with them. Sometimes (probably as a response to high temperatures) she lays 12–70 eggs beneath a stone, where both male and female guard them during their development; or just one or two eggs may develop within her body, the rest breaking down to provide nutrients for these. In this case the mother eventually gives birth to well-developed larvae.

The mudpuppy and the waterdogs of eastern North America are much stouter animals, with pigmented skin, that live in ponds, lakes, and streams. The mudpuppy is in the northern part of the range; waterdogs are found farther south. Both names reflect the erroneous belief that these animals produce a noise like barking.

They have red or purple external gills that resemble miniature ostrich plumes; these vary in size depending on the oxygen content of the water. In stagnant water where oxygen is scarce, the gills are very large, but in well-oxygenated running streams they are quite small. Mudpuppies and waterdogs feed on a variety of animals, including insects, crayfish, and fish, and they take several years to reach maturity. They mate in the fall, the eggs being fertilized internally but not laid until the following spring. A female lays 20–190 eggs, sticking them individually to logs or rocks, and the male guards them for the 5–9 weeks it takes them to develop.

Sirens
FAMILY SIRENIDAE

There are only four species of sirens, found in the southern and central United States and northeastern Mexico. They live in shallow water in ditches, streams, and lakes, and are active predators, feeding on such animals as crayfish, worms, and

MINIATURE SALAMANDERS

Among the plethodontid salamanders are animals that are some of the smallest terrestrial vertebrates; during their evolution, they have undergone a process of miniaturization. Most notable are some species in the genus *Bolitoglossa* and all species of *Thorius*, new species of which are still being discovered. *Bolitoglossa* includes at least 93 species, living over a large area from northern Mexico to Bolivia and Brazil. Twenty-four species of *Thorius* have been named, all from the forested mountains of southern Mexico. These two genera demonstrate the fact that the diversity of amphibians in Central and South America is enormous, but is still poorly documented.

One *Thorius* species measures only 25–30mm (1–1.2in) in total length. Miniaturization involves simplification of the skeleton, especially of the skull; several bones present in larger, related salamanders simply fail to develop. The sense organs, however, do not show a corresponding decrease in size, so that miniature species have very large eyes relative to their size. Several species of *Bolitoglossa* are adapted for arboreal life. They have prehensile tails that they can coil around twigs and suckerlike feet that enable them to walk over the smooth surfaces of leaves and to climb vertical surfaces.

Both genera feed by means of a projectile tongue, similar in function but not in detailed structure to that of chameleons; it can be shot out at great speed to strike an insect over a distance representing 40 to 80 percent of the salamander's length. Very small animals need good antipredator defenses; *Bolitoglossa* species arch their backs and wave their tails when attacked, and both genera can drop (autotomize) their tail if it is grasped by a predator.

These salamanders have remarkable life histories. In *Bolitoglossa subpalmata*, the male reaches sexual maturity at 6 years old, but the female does not breed until she is 12! They lay small clutches of very large eggs that take several months to develop. Some of these miniature salamanders have no obvious breeding season but can be found breeding at any time of year, evidently because they live in wet, tropical forest habitats where the environment is more or less constant throughout the year.   TRH/JH

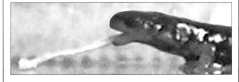

♦ **Above** In a video sequence taken by Stephen Deban of the University of California at Berkeley, a Bolitoglossa subpalmata *captures a prey item.*

♦ **Above** *Within milliseconds, the salamander's sticky, projectile tongue has drawn the prey in to be ingested.*

snails, and spending much of their time buried in mud or sand. The front of the mouth lacks teeth; instead, sirens have a horny beak and they feed by suction, drawing water and prey into the mouth. They resemble overgrown larvae with their long eel-like bodies and external gills; they lack hind-limbs but have small, weak forelimbs positioned close behind the gills. The Greater siren can reach a length of 90cm (36in), making it one of the largest of the tailed amphibians, but the Lesser siren rarely exceeds 25cm (10in). The tiny dwarf sirens can be very numerous, particularly when they live among water hyacinths, a water weed that has become very abundant since its introduction to North America. When grasped, sirens commonly emit a yelping sound.

Many of the ponds and ditches where sirens live dry up in the summer, but they are able to survive periods of drought by going into a state called estivation. As the sand or mud dries out, the mucous coat covering their skin hardens to form a parchment-like cocoon that covers the entire body except the mouth. They can survive in this condition for many weeks until their habitat becomes flooded once again.

Siren reproduction is something of a mystery, since mating has never been observed. The absence in males of the glands that secrete the spermatophores in many other tailed amphibians, and the absence in the female of a receptacle in which sperm could be stored, suggests that they practice external fertilization. However, the female lays her eggs singly, dispersing them widely on aquatic vegetation, suggesting that they might be fertilized before they are laid. Either sirens show a form of internal fertilization quite unlike anything seen in other tailed amphibians or the male follows the female around during egg-laying, fertilizing each egg individually as it is deposited. This question requires further detailed study.

Mole Salamanders

FAMILY AMBYSTOMATIDAE

The mole salamanders are so named because they live in burrows for much of their lives. They are rarely seen except in the breeding season, when they migrate to ponds to mate and deposit their eggs. Found only in North America, most species are terrestrial, but all have aquatic larvae. Heavily built, they have broad heads and smooth skin. Many have bold, brightly-colored markings.

During early spring, the Spotted salamander shows spectacular breeding migrations, converging in very large numbers on ponds where, over 2–3 days, it engages in mass mating. Females lay up to 200 eggs in the water and then return to their terrestrial life, leaving the eggs to hatch into larvae that metamorphose into the adult form 2–4 months after hatching, leaving the water in late summer or fall. By contrast, the Marbled salamander lays its eggs in the fall in dry pond beds, the female coiling herself around the eggs and guarding them until the winter rains, when they hatch into larvae.

In some parts of North America, there are populations of various species of mole salamanders in which adults are paedomorphic, remaining in the larval form even when they reach sexual maturity. The most celebrated example of this is the Mexican axolotl, which in nature exists only in the larval form. It will metamorphose into the terrestrial form if it is injected with thyroid extract.

▷ **Right** Representative species of seven families of salamanders and newts: **1** Bolitoglossa schizodactyla; Plethodontidae, **2** Red salamander (Pseudotriton ruber); Plethodontidae. **3** Tiger salamander (Ambystoma tigrinum); Ambystomatidae. **4** Olm (Proteus anguinus); Proteidae. **5** Mudpuppy (Necturus maculosus); Proteidae. **6** Batrachuperus pinchonii; Hynobiidae. **7** Onychodactylus japonicus; Hynobiidae. **8** Tylototriton taliangensis; Salamandridae. **9** Eastern newt, red eft stage (Notophthalmus viridescens); Salamandridae. **10** Smooth newt (Lissotriton vulgaris); Salamandridae. **11** Greater siren (Siren lacertina); Sirenidae. **12** Two-toed amphiuma (Amphiuma means); Amphiumidae.

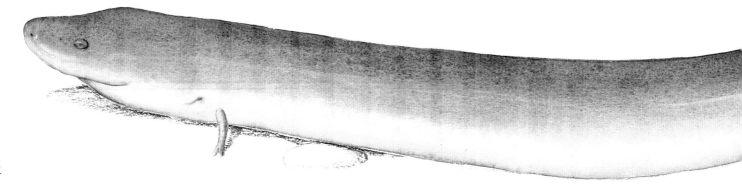

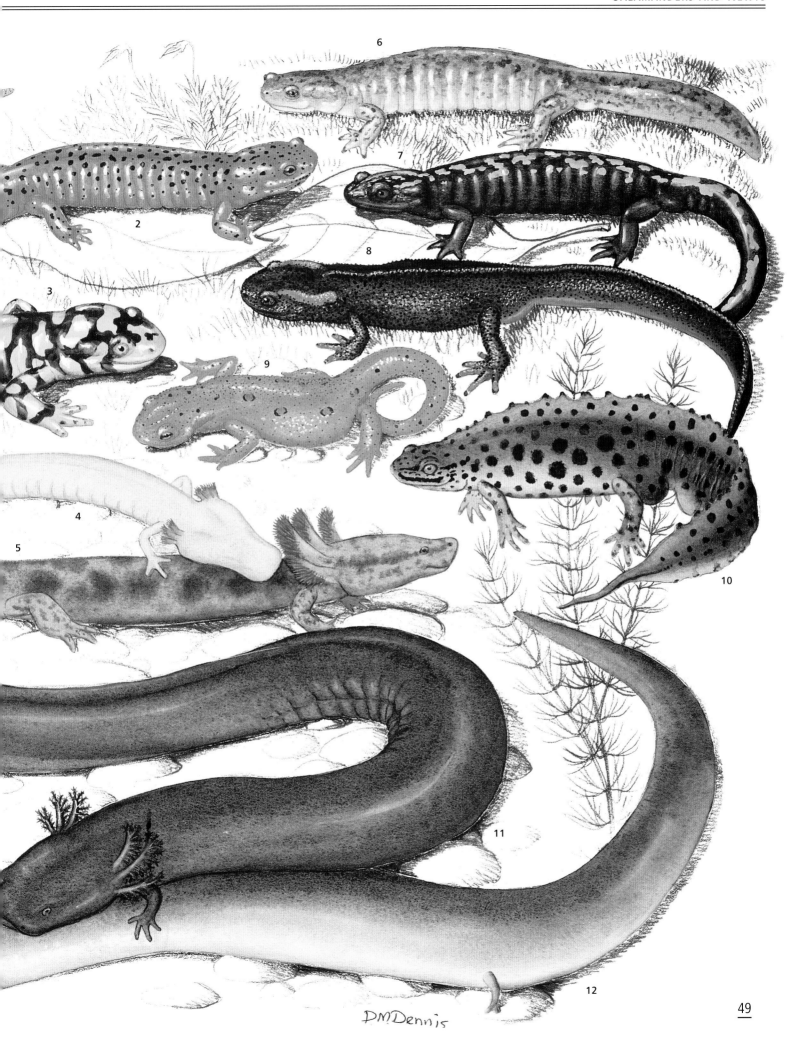

In the Mole salamander some larvae become cannibalistic, feeding on smaller larvae of their own species and growing to a very large size. There is a cost to this behavior, however; because cannibalistic larvae eat their own kind, they are more likely to become infected by parasites.

The larvae of all mole salamanders are easy prey for fish, and introduced fish have wiped out many populations. Others have been lost through habitat destruction, and some species, such as the California tiger salamander, are seriously threatened.

There are four species of Pacific mole salamanders (genus *Dicamptodon*), which were once considered to belong to a separate family (Dicamptodontidae). They live in or near mountain streams near the North American west coast and have lungs that are reduced in size, preventing them being buoyant and thus likely to be swept away. They can reach a large size, being the largest terrestrial salamanders, and big specimens can deliver a nasty bite. The most widespread species, the Pacific giant salamander, spends many years as an aquatic larva. In some parts of its range, a proportion of the population is paedomorphic, becoming sexually mature while still in the larval form. For example, Cope's giant salamander is an entirely paedomorphic species.

Pacific mole salamanders are not as abundant as they used to be, and very large, older specimens have become rare. They live in forests that are exploited for timber, and forestry affects them adversely by altering their terrestrial habitat and by silting up the streams in which they breed.

Newts and European Salamanders
FAMILY SALAMANDRIDAE

Of the nine families of tailed amphibians, the Salamandridae, including newts and European salamanders, have the largest range, covering North America, Europe, and parts of Asia. In terms of life history they are a very diverse family, and there is considerable variation between species in the proportions of their life spent in water and on land.

Fertilization is internal, sperm being transferred from male to female in a spermatophore, typically after a prolonged and elaborate courtship (see Courtship in Salamanders and Newts). In most species, the female lays her eggs in water, where they develop into larvae, but in the European Fire salamander the female retains the fertilized eggs in her body until they become fully-formed larvae. She then enters a pond or stream in order to give birth to her young, the only time that an adult returns to water.

◖ **Left** *The Marbled newt of southwestern Europe (Triturus marmoratus) is clearly set apart from other more somberly colored European members of the family Salamandridae by its vivid, mottled black and green skin. The distinctive orange-red stripe along the line of the backbone is a hallmark of juveniles (shown here) and females of this species.*

FINDING BREEDING PONDS

All newts and many salamanders return to water in the spring of each year in order to breed. They show remarkable powers of orientation and homing. It is not known how far they disperse from their breeding ponds, but newts in their terrestrial stage are commonly found several kilometers from the nearest suitable pond. Removal experiments with the Red-bellied newt in California have shown that individuals are able to find their way back to the exact stretch of stream where they were caught, despite having been taken to streams several kilometers away and separated from their home stream by mountainous territory.

A number of senses, some rather unusual, appear to be involved in this migration. Vision and smell are important, although even blinded animals with their olfactory system destroyed are able to find their way to water. The pineal body, an outgrowth of the brain lying just beneath the bones of the skull, is also sensitive to light, particularly polarized light. It has been shown that some salamanders can perceive polarized light and can use it to determine the position of the sun in the sky, even when the sun itself is hidden by a cloud, so that they can use the sun as a "compass" to direct their movements. More recently, it has been demonstrated that some North American species, such as the Cave salamander and Eastern newt (BELOW; in eft phase), can detect the earth's magnetic field, and use this cue as a directional and positional reference system.

Newts and salamanders that breed in water often similarly show very strong fidelity to the pond or stream in which they grew up, returning there to breed over several successive years, even if they have moved a long way from their breeding site in the intervening months or years. This fact suggests that their nervous systems may be able to develop and store detailed "maps" of their environment.

Many members of the family are brightly colored and release noxious secretions from glands in their skin. The American Red-spotted newt is unusual in that the eft stage is especially toxic and is colored bright red. Californian newts have skin secretions that are among the most toxic substances known.

Whether they live on land or in water, newts and European salamanders typically feed on small invertebrates, including worms, slugs, insects, and crustaceans. While in the aquatic phase, some species, such as the Smooth newt, are voracious predators of frog tadpoles. Prey detection is primarily visual, and it is usually movement by a prey item that allows detection. Moving prey can also be detected through lateral-line organs.

Vision, smell, and touch are also important in courtship. In newts, the male performs a variety of displays that send visual, olfactory, and tactile stimuli to the female. The European newts are unusual among tailed amphibians in that, during courtship, the male does not capture and hold onto the female. Associated with this kind of courtship, males and females in these species are much more strikingly different in appearance during the breeding season than in any other kind of tailed amphibian. The male develops a large crest that runs along the length of his body and tail, and the Smooth newt has many large black spots on his body and tail.

In some species, adults show a very strong tendency to return to the same breeding site in successive years. Newts in the terrestrial eft phase also show a strong affinity for their natal ponds, but some do move to other ponds, and it is probably at this stage of the life cycle that newts disperse from one breeding site to another.

Many species in this family have declined dramatically in the last 50 years, often because changes in agricultural practice have involved the destruction of their breeding ponds. They are also threatened by pollution and the loss of their terrestrial habitat.

Torrent Salamanders
FAMILY RHYACOTRITONIDAE

The family of torrent salamanders consists of four species belonging to a single genus. Confined to the Pacific Northwest region of the USA, they are distinguished from all other salamanders by the presence, in males, of large, square-shaped glandular swellings behind and to each side of the cloaca; the function of these is uncertain. Torrent salamanders have heavily-built bodies and tails and are semi-aquatic, living in or close to streams and seepages in cool, humid conifer forests. Their larvae take 3–5 years to complete their development, probably because they live in very cold water. Little is known of their habits and behavior, but they are threatened by forestry activity, which degrades their habitat and causes the silting up of their breeding streams.

Most European newts, by contrast, spend nearly half of each year in the water once they become adult. It is in the water that they court, lay their eggs, and then build up the fat reserves necessary to survive the winter on land. In some populations of crested newts, however, adults remain in the water all year round.

When, in the early spring, newts return to ponds, they undergo a number of physical changes representing a partial metamorphosis back to the larval condition. The skin becomes thin, smooth, and permeable to oxygen, the tail becomes a deep, flattened structure that enables them to swim powerfully, and their eyes become a slightly different shape, for focusing under water. Within the skin, a large number of lateral-line organs develop; these are sensitive to water vibrations and are important for detecting prey. Male Palmate newts acquire webbing between the toes of their hindlimbs during the breeding season, a feature that enables them to swim a little faster than females during their energetic courtship.

Newt and European salamander larvae have external, feathery gills and, initially, no limbs. They grow rapidly during their first few months of life until, in the fall, they lose their gills and leave the water as tiny replicas of their parents. They enter a terrestrial phase, sometimes called the eft stage, which lasts from 1–7 years, depending on both species and locality. In the Smooth newt, for example, the eft stage is much longer in Scandinavia than it is in southern Europe.

Amphiumas
FAMILY AMPHIUMIDAE

Because of their elongated body shape, the amphiumas were formerly referred to as "Congo eels." This designation was doubly misleading, since these animals bear only a superficial resemblance to eels, which are fish, and they are found in North America, not in Africa. There are only three species in the family, all of which are found in the swamps of the southeastern United States. They have a long, thin cylindrical body, smooth, slippery skin, and limbs that are so small that they cannot be of any use in locomotion. In the larval stage, however, the limbs are larger, relative to the size of the body, and are used for walking. The Three-toed amphiuma can reach a length of 90cm (36in), and large individuals are capable of delivering a painful bite when caught. Adults have lungs and no gills, although they do possess the larval feature of one open gill slit.

Amphiumas lead a totally aquatic life, although they may make brief excursions onto the land following very heavy rain. For much of the time they live in burrows and are active by night, feeding on a variety of prey including frogs, snails, fish, and crayfish. Consequently, they are considered a pest by fisherman, who catch and kill them.

In the breeding season females sometimes outnumber males, and several females may be seen rubbing a male with their snouts to attract his attention. During mating, male and female coil around one another and sperm is transferred directly into the female's cloaca. She lays as many as 200 eggs, joined together in a long string, and guards them, her body coiled around them, until they hatch after about 20 weeks. Egg-laying often occurs when the water level is high and, as it falls, the female and her eggs may be left in a damp hollow beneath a log. When they hatch, the young have to find their way back to water.

◑ **Left** *With its elongated body, rudimentary limbs, and vestigial eyes hidden beneath a membrane of skin, the olm (Proteus anguinus) is perfectly adapted to the dark coastal caves of northeastern Italy, Slovenia, and Croatia. This subterranean habitat makes it one of the least well-studied of all salamanders; indeed, its very existence only came to light in the 18th century when underground flooding washed an olm to the surface.*

◐ **Below** *A Two-toed amphiuma (Amphiuma means) observed, uncharacteristically, during the day on a lily-pad in the Florida Everglades. This and the other two species of amphiuma are overwhelmingly nocturnal, hunting a wide range of fish, freshwater crustaceans, and frogs. During rainstorms, when the ground is wet, it can sometimes be seen moving overland from one body of water to another.*

Lungless Salamanders

FAMILY PLETHODONTIDAE

Much the largest family of tailed amphibians, the lungless salamanders are arguably the most successful of all amphibian groups in terms of number of individuals. In the northeastern United States they are extremely numerous.

This success is paradoxical in view of the fact that, during their evolution, they have lost one of the most basic features of all terrestrial vertebrates, their lungs, and can absorb oxygen only through their skins and mouth lining. This limitation potentially imposes severe constraints on where they can live, how active they can be, and the size to which they can grow. Large animals have a small surface area in proportion to their volume, and so have greater difficulty than small animals in supplying all their tissues with oxygen if they are dependent on their skin for respiration. Despite this fact, some lungless salamanders are more than 20cm (8in) long.

The most important factor in using the skin for respiration is that it must be moist at all times for oxygen to be taken up by the blood in capillaries beneath the skin. For this reason, lungless salamanders living in temperate habitats are confined for most of their lives to damp hiding places, and they can only emerge from these in wet weather, typically at night, to mate or to find food.

The life of a lungless salamander thus consists of brief periods of activity, interspersed with periods of inactivity that are often very long. They are able to survive the inactivity because, having a very low metabolic rate, they have very low energy requirements. They do not need to feed very often and, when they do feed, they are able to store much of what they eat as fat. Some species are territorial, defending an area around their retreat where they feed and mate. The Red-backed salamander defends its territory by marking its borders with fecal pellets.

Lungless salamanders typically have slender bodies, long tails, and prominent eyes. A distinctive feature of the family is a shallow groove, the nasolabial groove, running from each nostril to the upper lip; its function is to carry waterborne odors from the ground to the nasal cavity.

The lungless woodland salamanders are wholly terrestrial and have no aquatic larval stage. Their eggs, laid in moss or rotting logs, hatch directly into tiny replicas of their parents. Hiding by day in burrows or under logs or stones, they emerge on damp nights to feed on a wide variety of invertebrate prey, including slugs, worms, and beetles. Jordan's salamander occurs in a wide range of color patterns, varying from one locality to another. The basic body color is black, but while some populations are totally black, others have several different combinations of red

▶ **Right** *Within the large and diverse family of lungless salamanders, one particularly striking example of specialization to habitat is found among certain species in the genus* Bolitoglossa, *which inhabits the tropical rainforests of Central and South America. Fully arboreal species such as* Bolitoglossa mexicana *have evolved a strong prehensile tail, which allows them to move with great agility around the forest canopy.*

◗ **Below** *In a feeding technique similar to that of chameleons, certain plethodontids can capture prey by means of a lightning-fast protrusion of their tongue. In absolute terms, as well as relative to body size, salamanders of the genus* Hydromantes *have the longest tongues of all, extending for some 5cm (2in), or more than half the length of their body (excluding the tail). The tongue skeleton (hyobranchial apparatus) that enables the tongue to be used as a projectile is a highly sophisticated apparatus, consisting of no less than seven cartilaginous elements linked by flexible joints. Most plethodontids are native to North, central, and South America, but the species shown here,* Hydromantes supramontis, *inhabits the mountains of Sardinia.*

legs, cheeks, and dorsal stripes. When they are handled, woodland salamanders produce a sticky slime that is very difficult to remove from the hands. Predators such as snakes may find their jaws glued together.

The dusky salamanders are typically found near streams, where they lay their eggs. They wander far afield, however, and sometimes climb up trees and shrubs in search of food. The Mountain dusky salamander is very variable in appearance, and much of this variation is due to the fact that, in some localities, it mimics the local form of Jordan's salamander; such mimetic forms are called "imitator" forms. The Pigmy salamander, whose maximum length is about 5cm (2in), lives at high altitudes, is totally terrestrial, and goes through its larval stage within the egg. It is sometimes found high above the ground among foliage.

Many lungless salamanders are specialized for life in specific kinds of habitat. The Spring salamander, a brightly-colored animal with a red, salmon pink, or orange skin spotted with black or brown, lives in mountain streams, where its wedge-shaped head enables it to push itself between rocks. The Red salamander, also bright red, burrows into the mud near springs and streams. The Cave salamander lives inside the entrances to caves, where its long, prehensile (grasping) tail enables it to climb among rocks with great agility.

Perhaps the most bizarre lungless salamanders are those that live in deep caves and underground bodies of water. The Texas blind salamander of the southwestern United States bears a remarkable resemblance to the European olm, a member of another family, the Proteidae. It has thin, feeble legs, a flattened snout, vestigial eyes, pink external gills, and an otherwise white body. The Grotto salamander, found in the Ozarks region of the United States, is pale pink to white in color and has vestigial eyes, but has no external gills. Its life history is unique. As a larva, it lives in ordinary streams and has a typical salamander form with fully functional eyes, external gills, a large tail fin, and gray or brown coloration. At metamorphosis, however, it retreats into caves and loses the tail fin, gills, and skin pigmentation. Its eyes cease to grow and become covered by skin.

There is much variation among lungless salamanders in terms of the timing of their breeding activities. Species in eastern North America are active in the summer, typically mate in spring or fall, and lay their eggs in early summer. In the west, they are inactive during the hot, dry summer months and mate throughout the winter and spring. In tropical Central and South America, they are active all year round, and some species can breed at any time of year. Several species of lungless salamander are threatened by habitat destruction, especially those that have very restricted ranges or very specific habitat requirements, such as caves.   Updated by CM

Salamander and Newt Families

THE ORDER CAUDATA IS DIVIDED INTO NINE families on the basis of a wide variety of characters, some of which relate to details of the positions of the skull bones relative to one another, others to the detailed distribution of teeth in the skull bones. The families Cryptobranchidae and Hynobiidae are the most ancient families and are sometimes grouped together in the superfamily Cryptobranchoidea. These families and the family Sirenidae have external fertilization; the other six families have internal fertilization.

Cryptobranchidae

Hynobiidae

Proteidae

Sirenidae

Ambystomatidae

Salamandridae

Rhyacotritonidae

Amphiumidae

Plethodontidae

Giant salamanders
Family: Cryptobranchidae

Tropic of Cancer

◗ Right Cladogram showing the relationships among the nine families of salamanders and newts. The families are grouped under three superfamilies.

◗ Below This is a male Alpine newt, a member of the wide-ranging family Salamandridae.

3 species in 2 genera. E and C USA, E China, Japan. Wholly aquatic, living in clear, flowing rivers and streams. Species: **Chinese giant salamander** (*Andrias davidianus*), **Japanese giant salamander** (*A. japonicus*), **hellbender** (*Cryptobranchus alleganiensis*).
SIZE: Length 60–160cm
COLOR: Brown or gray
FORM: Body and head massive and squat; 4 short limbs; tail short, heavy, laterally-compressed; folds of skin along flanks; eyelids lacking; lungs present but vestigial; single pair of gill slits; 4 fingers and 5 toes.
BREEDING: Fertilization external. Eggs laid in nest in stream bed, guarded by male.
CONSERVATION STATUS: The **Chinese giant salamander** is classed as Critically Endangered.

Asiatic salamanders
Family Hynobiidae

Tropic of Cancer

50 species in 10 genera. C and E Asia, including Japan and Taiwan. Most species are terrestrial, but migrate to ponds or streams to breed. Species include: **Siberian salamander** (*Salamandrella keyserlingii*), **Semirechensk salamander** (*Ranodon sibiricus*), **Dusky Oriental salamander** (*Hynobius nebulosus*).
SIZE: Length 10–21cm
COLOR: Various
FORM: Heavy-bodied; tail long and thick; eyelids movable; lungs small or absent; costal grooves on body; 4 fingers and 4–5 toes.
BREEDING: Fertilization external; eggs laid in gelatinous masses.
CONSERVATION STATUS: 5 species are Critically Endangered; 10 are Endangered.

Olm, mudpuppies, and waterdogs
Family Proteidae

Tropic of Cancer

6 species in 2 genera. Balkans and N Italy (olm), E and C N America. Wholly aquatic, living in mud or stagnant water; olm lives in caves. Species include: **Gulf Coast waterdog** (*Necturus beyeri*), **mudpuppy** (*N. maculosus*), **olm** (*Proteus anguinus*).
SIZE: Length 11–35cm
COLOR: Gray or brown above, gray below; the olm is creamy white.
FORM: Head large and flat, body stout, tail short and laterally compressed, limbs small; olm has elongated body, pointed head; eyelids lacking. All species are paedomorphic; lungs present but small; external gills and 2 gill slits; 4 fingers and 4 toes, or 3 fingers and 2 toes (olm); costal grooves along the side of the body.
BREEDING: Fertilization internal
CONSERVATION STATUS: The **Alabama waterdog** is listed as Endandered and the **olm** as Vulnerable.

NOTES	
	Length = length from snout to vent.
	Approximate nonmetric equivalents: 10cm = 4in 1kg = 2.2lb

Sirens
Family Sirenidae

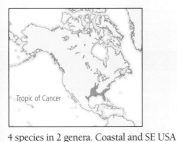

Tropic of Cancer

4 species in 2 genera. Coastal and SE USA and Mississippi Valley, NE Mexico. Live permanently in heavily vegetated lakes, marshes, and swamps. Species: **Northern dwarf siren** (*Pseudobranchus striatus*), **Southern dwarf siren** (*P. axanthus*), **Greater siren** (*Siren lacertina*), **Lesser siren** (*S. intermedia*).
SIZE: Length 10–90cm
COLOR: Olive green or dark gray, pale spots.
FORM: Eel-like, with external gills and one or three pairs of gill slits; horny beak instead of premaxillary teeth; tiny fore-limbs close to gills; eyelids lacking; eyes very small; lungs present but small; 3–4 fingers; no hindlimbs; costal grooves on skin above the ribs.
BREEDING: Fertilization mode uncertain, but presumed to be external. Eggs deposited singly or in small clusters on vegetation.

Mole salamanders
Family Ambystomatidae

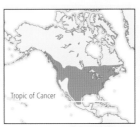

Tropic of Cancer

35 species in 2 genera. N America. Most species terrestrial, migrating to water to breed in winter or spring. Species include: **Mexican axolotl** (*Ambystoma mexicanum*), **Mole salamander** (*A. talpoideum*), **Marbled salamander** (*A. opacum*), **Santa Cruz long-toed salamander** (*A. macrodactylum croceum*), **Spotted salamander** (*A. maculatum*), **Tiger salamander** (*A. tigrinum*), **California tiger salamander** (*A. californiense*), **Coastal giant salamander** (*Dicamptodon tenebrosus*), **Californian giant salamander** (*D. ensatus*).
SIZE: Length 8–22cm
COLOR: Brown or black, spotted, mottled, or striped with white, yellow, green, pink, orange, red, or blue.
FORM: Sturdy, head broad, tail laterally compressed; eyelids movable; lungs present; 4 fingers and 5 toes; costal grooves

⬥ Above *The Tiger salamander is one of the most widely distributed of the mole salamanders. This subspecies, the Barred tiger salamander* (A. t. mavorticum) *is found in the US Midwest, from Nebraska and Wyoming south to New Mexico.*

along the side of the body. Axolotl is paedomorphic, as are some populations of other species, e.g. Mole salamander.
BREEDING: Fertilization internal
CONSERVATION STATUS: 9 species in the genus *Ambystoma* are listed as Critically Endangered, and 2 others are Endangered.

Newts and European salamanders
Family Salamandridae

Equator

74 species in 20 genera. W and E North America, Europe, Mediterranean Africa, W Asia, China, Taiwan, SE Asia, Japan. Most

species terrestrial, migrating to ponds or streams in spring to breed. Some species mate on land, e.g. Fire salamander. Species include: **Sardinian brook salamander** (*Euproctus platycephalus*), **Alpine newt** (*Mesotriton alpestris*), **Great crested newt** (*T. cristatus*), **Palmate newt** (*Lissotriton helveticus*), **Smooth newt** (*L. vulgaris*), **Fire salamander** (*Salamandra salamandra*), **Gold-striped salamander** (*Chioglossa lusitanica*), **Red-bellied newt** (*Taricha rivularis*), **California newt** (*T. torosa*), **Red-spotted newt** (*Notophthalmus viridescens*), **Sharp-ribbed newt** (*Pleurodeles waltl*), **Spiny newt** (*Echinotriton andersoni*).
SIZE: Length 7–35cm
COLOR: Brown, black, or green above, yellow, orange, or red below; often with dark spots.
FORM: Slender with long tail; rough-skinned except when in water; eyelids movable; lungs present; 4 fingers and 4–5 toes; differ from other members of Salamandroidea in lacking costal grooves on the side of the body. Some populations of some newt species are paedomorphic, e.g. Smooth newt.
BREEDING: Fertilization internal
CONSERVATION STATUS: **Yunnan Lake newt** (*Cynops wolterstorffi*) is thought to be extinct; of the 7 species of Lycian

salamanders (genus *Lyciasalamandra*), 1 is Critically Endangered, 5 are Endangered, and 1 is Vulnerable; 10 other species in the family are Critically Endangered or Endangered.

Torrent salamanders
Family Rhyacotritonidae

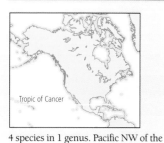

4 species in 1 genus. Pacific NW of the USA. Semi-aquatic residents of humid conifer forest. Larvae and adults found in rocky rubble in cold, well-aerated forest streams and springs; adults forage on forest floor during heavy rain. Species include: **Cascade torrent salamander** (*Rhyacotriton cascadae*), **Olympic torrent salamander** (*R. olympicus*).
SIZE: Length 9–12cm
COLOR: Brown, reddish-brown, or greenish-brown above, yellow below.
FORM: Sturdy body and tail; 4 short, well-developed limbs; eyelids movable; lungs present; 4 fingers and 5 toes; costal grooves present above the ribs; distinctive swollen glands around the cloaca.
BREEDING: Fertilization internal

Amphiumas
Family Amphiumidae

3 species in 1 genus. SE USA, including S part of Mississippi Valley. Wholly aquatic, living in stagnant waters and swamps. Species: **One-toed amphiuma** (*Amphiuma pholeter*), **Two-toed amphiuma** (*A. means*), **Three-toed amphiuma** (*A. tridactylum*).
SIZE: Length 22–76cm, occasionally up to 116cm.
COLOR: Dark brown or black above, gray below.
FORM: Heavily-built, long, eel-like, and flexible; limbs very small; eyelids lacking; lungs present; 1–3 fingers and toes; costal grooves along the side of the body.
BREEDING: Fertilization internal

Lungless salamanders
Family Plethodontidae

377 species in 24 genera. Much the largest family in the Caudata. N, C, and S America, Europe (SE France, Italy, Sardinia), and SW Korea. The most evolutionarily advanced salamanders, living in woodland, caves, or mountain streams. Adults of some species aquatic, but many are terrestrial and some are very small, tree-living forms. Species include: **Cave salamander** (*Eurycea lucifuga*), **Texas blind salamander** (*E. rathbuni*), **Northern two-lined salamander** (*E. bislineata*), **Northern dusky salamander** (*Desmognathus fuscus*), **Allegheny Mountain dusky salamander** (*D. ochrophaeus*), **Pigmy salamander** (*D. wrighti*), **Grotto salamander** (*Eurycea spelea*), **Jordan's salamander** (*Plethodon jordani*), **Shenandoah salamander** (*P. shenandoah*), **Red-backed salamander** (*P. cinereus*), **Slimy salamander** (*P. glutinosus*), **Korean crevice salamander** (*Karsenia koreana*), **Red salamander** (*Pseudotriton ruber*); and a number of miniature salamanders in the genus *Thorius*.
SIZE: Length 2.5–21cm
COLOR: Brown, black, red, yellow, or green; many spotted, mottled, or striped with red, yellow, or white. Cave species, such as some *Eurycea*, and all *Haideotriton* and *Typhlotriton*, white.
FORM: Long and slender, head narrow, tail long and cylindrical; eyelids movable; lungs absent; typically 4 fingers and 5 toes; costal grooves along the side of the body.
BREEDING: Fertilization internal. Eggs laid in water or on land.
CONSERVATION STATUS: **Ainsworth's salamander** (*Plethodon ainsworthi*) is listed as Extinct. Another 116 species are Critically Endangered or Endangered, mainly because of very small ranges and the danger of habitat destruction.

◗ **Right** *The Mud salamander lives at low elevations along the eastern seaboard of the United States. All its subspecies (here, the Midland mud salamander;* Pseudotriton montanus diastictus*) are vividly colored.*

NOTES   Length = length from snout to vent.

Approximate nonmetric equivalents: 10cm = 4in 1kg = 2.2lb

COURTSHIP AND MATING IN SALAMANDERS AND NEWTS
The twists and turns of an elaborate reproductive ritual

THE OVERWHELMING MAJORITY OF NEWTS AND salamanders exhibit an extremely unorthodox form of mating. Fertilization is internal, with the sperm meeting the eggs inside the female's reproductive tract, but the male has no penis to introduce the sperm into her. Instead, the sperm is transferred from male to female in a roughly cone-shaped capsule known as a spermatophore, which is made up of two parts: a broad, gelatinous base surmounted by a sperm-filled cap. In salamanders that mate on land, the base is rigid and hard; by contrast, in newts, all of which mate in water, it is a soft, fluid-filled sac. In both these situations, the function of the base of the spermatophore is to hold the sperm cap above the ground or the floor of the pond, where it can be picked up by the female's cloaca. The base is secreted by a variety of glands that open into the

male's cloaca, which characteristically becomes greatly swollen during the breeding season.

The transfer of sperm involves complex and elaborate forms of behavior. In many species of salamander, the male captures the female and clasps her in a firm grip called amplexus. Species vary as to precisely which limbs they use for this purpose; various combinations of the limbs and tail may be involved. In salamanders such as the European Fire salamander and the Mountain salamanders of Corsica, Sardinia, and the Pyrenees, amplexus is maintained throughout mating, and the spermatophore is transferred directly to the female. On the other hand, in species like the North American Red-spotted and Red-bellied newts, the male clasps the female initially but releases her just before the transfer of the spermatophore. In the European Smooth

Below *Application of sexual stimulants in salamanders and newts.* **1a** *The Two-lined salamander, a semi-aquatic North American species. The male has protruding teeth* **1b** *with which he lacerates the female's skin, after first covering it with secretions from his chin gland.* **2** *Jordan's salamander, a terrestrial woodland species from eastern North America. The male leads the female, repeatedly turning to slap a gland under his chin against the female's snout.* **3** *The Redbelly newt, an aquatic species from western North America. The male clasps the female with both fore- and hindlimbs and rubs a gland under his chin over her nostrils.* **4** *The Smooth newt. The male fans his tail to generate a water current that carries odor from numerous skin glands to the female's nose.*

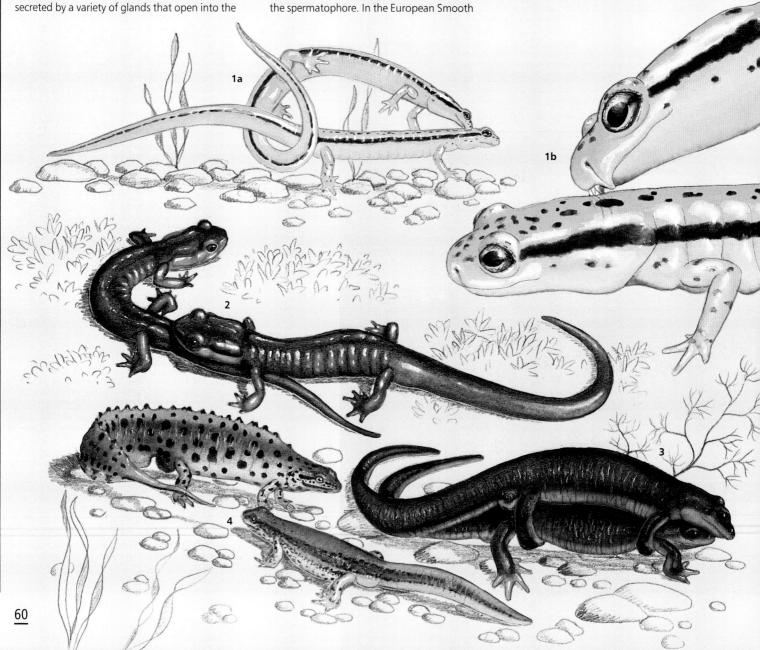

⚫ **Above** *Amplexus in three salamanders and newts. 1 The Fire salamander, a terrestrial European species. The male clasps the female from below; he deposits a spermatophore on the ground and then flicks his body and tail to one side so that the female's cloaca falls onto it. 2 An aquatic European brook salamander. The male captures the female with his tail and the spermatophore is transferred from the male's cloaca directly to the female's. 3 The Red-spotted newt, an aquatic North American species. The male clasps the female's neck with his powerful hindlimbs; in this position he rubs large glands on his cheek over her nostrils.*

newt and in the North American woodland salamanders, amplexus does not take place and the female is entirely free to move about at all stages of courtship.

Because female newts are free to move about, they are also free to choose their mates in a way in which salamanders with amplexus are not. In the European newts, female choice requires males to compete with one another to have the most conspicuous and stimulating appearance and behavior. Uniquely among tailed amphibians, male European newts develop striking color patterns and a large crest along the back during the breeding season. Females are more receptive to males with larger crests.

Amplexus fulfills two functions; the male can stimulate the female with secretions from various glands while holding her in his grasp, and can also prevent other males reaching and attempting to court her. In species in which the female is free to move during mating, she must be stimulated by the male before she will behave appropriately. Male newts and salamanders possess a variety of glands producing secretions for this purpose, applied to the female in a variety of ways. Male European newts develop a huge gland in the abdomen that opens into their cloaca; this secretes a pheromone during courtship that the male wafts towards the female's snout by rapidly fanning his tail to create a current of water. Some male salamanders secrete courtship pheromone from a gland under the chin that they rub over the female's head and body. They also scratch the female's skin with special teeth, so that the pheromone enters her blood stream.

In contrast to the variety of behaviors that different species exhibit in the initial phases of courtship, many show very similar behavior during the transfer of the spermatophore. The male creeps in front of the female and then stops, quivering his tail. The female touches the tail, and the male responds by depositing a spermatophore. He then creeps away from it, moving through an arc to take up a position at right angles to his previous direction. He then stops and prevents the female, who has followed him, from moving any further. She nudges his tail once more, and is now positioned with her cloaca directly above the spermatophore. The sperm mass is taken up by her cloacal lips, leaving the spermatophore base behind.

This process is inherently imprecise, and in many salamander and newt species a high proportion of spermatophores are missed by the female and so are wasted. To compensate for their loss, males are able to produce large numbers of spermatophores, and mating often involves several repetitions of the elaborate transfer sequence.

After mating, the sperm is stored by the female in special receptacles, called spermathecae, until she is ready to lay her eggs. Sperm storage means that mating and egg-laying can be separated in both time and space in ways that they cannot be in frogs, which have external fertilization. Many newt and salamander females deposit their fertilized eggs in safe places, such as deep underground. European newts laboriously wrap each egg individually in a folded leaf; this covering protects them both from predators and from ultraviolet radiation.   TRH

REPELLENT DEFENDERS

The antipredator arsenal of salamanders and newts

AS VERTEBRATES GO, SALAMANDERS ARE SMALL, slow, and weak; animals with these characteristics could be expected to mature quickly and be short-lived. Yet salamanders are typically long-lived, with the record going to a Fire salamander, *Salamandra salamandra*, that survived 50 years in captivity. In the wild, salamanders and newts are often attacked by shrews, birds, snakes, other salamanders, and even beetles, centipedes, and spiders. Heavy pressure from these predators has led to the evolution of antipredator mechanisms that combine repulsive or toxic skin-gland secretions with other defensive devices.

A number of newts and salamanders have evolved deadly toxins for their protection, but in each case that has been examined, one or more snake species has evolved resistance to the toxin and continues to feed on the salamanders. For example, the Rough-skinned newts (*Taricha*) have a neurotoxin, tetrodotoxin, concentrated in their skin at extreme levels. One newt can carry enough toxin to kill 25,000 mice, but can still be eaten by garter snakes. Some members of the genus *Bolitoglossa* have an undescribed neurotoxin that can kill some snake species after a single bite to their tails – yet several sympatric snake species eat these very toxic salamanders.

Snakes are thus the most dangerous predators for most salamanders because many have evolved resistance to the repulsive skin secretions. Most salamanders respond to the flick of a snake's tongue by running away or quickly striking a defensive posture. In contrast, populations of salamanders where there are no snake predators (for instance, at high elevations in Central America) do not respond to snake tongue flicks. Also, some tropical salamanders only respond with a defensive posture when temperatures are high enough for snakes to be active.

The Fire salamander has evolved a unique mechanism to employ the neurotoxic substance samandarine and related toxins present in giant glands along its mid-dorsal line. The animals are able to pressurize these glands and spray the toxins in a controlled and directional manner for as far as 4m (13ft). The spray causes burning and temporary blindness in humans, and presumably also in would-be predators. Spraying defensive toxins adds a powerful weapon to the antipredator arsenal of this species.

Salamanders typically employ defensive behavior patterns that maximize the efficacy of their chemical defenses. Certain species have concentrations of glands at the back of the head (parotoid glands) that produce unpalatable secretions. A number of these species, such as the Spotted salamander (*Ambystoma maculatum*), bend their heads down or hold them flat against the ground when attacked, thus presenting a predator with the most distasteful part of their bodies. More complex is the head-butting behavior of some other *Ambystoma* mole salamanders and newts, including the Sharp-ribbed newt (*Pleurodeles*) of Spain, Portugal, and Morocco. These species hold their bodies high off the ground; the head is bent down and the back of the head, bearing well-developed glands, is swung or lunged into the predator, an effective way of repulsing shrews. Most species vocalize while head-butting, and in several the glands are brightly colored with spots of yellow or orange as a warning to experienced

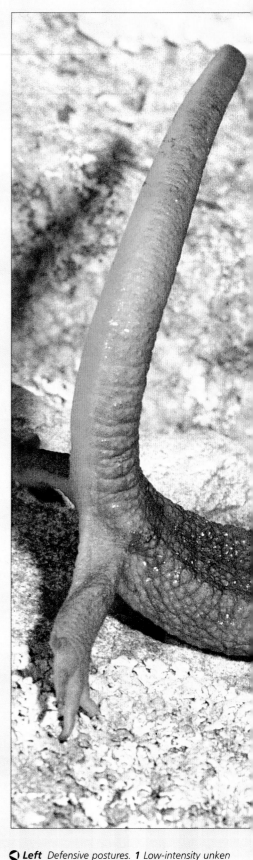

◁ **Left** Defensive postures. **1** Low-intensity unken reflex in the Spiny newt. **2** High-intensity unken reflex in the Red-bellied newt. **3** Tail lashing in the Ensatina (Ensatina eschscholtzii). **4** Undulating tail posture in the Cave salamander. **5** Head-butting in the Mole salamander (Ambystoma talpoideum).

attackers. Predators without color vision, such as shrews, probably learn to recognize the distinctive odors or vocalizations of these salamanders.

Tail lashing is characteristic of species with well-developed tail muscles and concentrations of poison glands on the upper surface of the tail. The Tiger salamander (*Ambystoma tigrinum*) and the Sharp-ribbed newt are examples of species that forcefully slap a secretion-laden tail into the approaching predator (typically, a shrew). This behavior draws the attention of the attacker to a repellent and expendable part of the anatomy. There is sometimes warning coloration on the upper surface of the salamander to establish unpleasant associations for would-be predators.

Many species have concentrations of skin glands on the upper surfaces of long, slender tails that are not strong enough to lash at attackers. Some of these species undulate their tails, usually in a vertical position, while holding their bodies still. This is most common among the lungless salamanders (Plethodontidae), which are also often capable of shedding their tails. The distastefulness of the object to which the attack is directed deters future attacks, and the active thrashings of the tail after it breaks off distract the predator while the salamander escapes to grow a new one. However, losing the tail does have a cost; not only is tissue lost but the salamander is more susceptible to future attack, because it loses much of its chemical repellent and cannot run as fast.

Some newts are also tail undulators, but under intense attack shift to the "unken reflex" that is more characteristic of their family: they hold themselves rigidly immobile with tail and chin elevated, showing brightly-colored undersides. Birds quickly learn to avoid species such as *Taricha* and *Triturus*, inedible animals with an upper surface abundantly supplied with poison glands and a deep yellow-to-red belly. Some newts (*Cynops* and *Paramesotriton*) with brightly colored bellies even flip onto their backs when harassed by predators.

Imitators in the genera *Desmognathus*, *Gyrinophilus*, *Pseudotriton*, *Plethodon*, and *Eurycea* take advantage of the defensive secretions of more distasteful species by mimicking their warning colors. The immobility of a salamander showing the unken reflex also has the effect of inhibiting the attack reflexes of predatory birds, reducing the likelihood of a serious wound from a bird that has not yet learned that the salamander, or a species it is mimicking, is inedible.

Perhaps the most remarkable antipredator mechanisms belong to the Sharp-ribbed (*Pleurodeles*) and Spiny (*Echinotriton*) newts, which in addition to their other defenses have sharp, elongated ribs whose tips protrude through the skin when the animals are grasped. The rib tips of Spiny newts pass trough enlarged glands on the side of the body, and painful skin secretions are injected into the mouths of would-be predators.

As a result of pressure from persistent and resilient predators, salamanders have evolved an arsenal of defensive chemicals and patterns of behavior that use these compounds effectively against predators.   EDB

◑ **Above** *The unken reflex in action. A Rough-skinned newt* (Taricha granulosa) *becomes immobile and lifts its head and tail to reveal its ventral surface.*

◐ **Right** *Poison-rib defense. The Spiny newt* (Echinotriton andersoni), *a Japanese species, has long, sharp-pointed ribs. If it is grasped by a predator, the ribs push out through poison glands in the skin.*

Frogs and Toads

FROGS AND TOADS ARE THE MOST NUMEROUS *and diverse of the amphibians – about 5,453 species occupy habitats ranging from deserts and savannas to high mountains and tropical rain forests. Most live both in water and on land for at least parts of their lives, but a few are entirely aquatic and many others are entirely terrestrial or arboreal.*

Although the richest variety of anurans (more than 80 percent of all species) is found in the tropics and subtropics, a great many live in the temperate zones. They are found on most islands and on all continents except Antarctica. The European common frog (*Rana temporaria*) and the Wood frog (*Lithobates sylvatica*) of North America have ranges extending north of the Arctic Circle.

Built for Leaping
FORM AND FUNCTION

The most obvious features that distinguish anurans from other amphibians are the long hind limbs, short trunk, and lack of a tail. Two "ankle," or tarsal, bones are elongated, thereby increasing the length of the hind limb. The anterior, or front, part of the backbone is short and usually contains eight or fewer vertebrae; the posterior vertebrae that compose the tails of other amphibians are fused to form a long, bony rod (the urostyle) that forms the posterior end of the backbone in anurans. Thus, the backbone consists of usually 12 or fewer bones in anurans, whereas there are 30–100 individual bones in newts and salamanders, and as many as 250 in caecilians, both of which lack the pelvic and hindlimb specializations of anurans. The modified tarsal bones of anurans add an additional segment to their hind limbs, and the pelvis is modified to form a leverlike joint with the fore part of the backbone. From a sitting position, the animal can rapidly and forcefully unfold each joint in its hind limbs in sequence, thereby providing a propulsive force to project the forepart of its body forward in a leap.

"Frogs" are frequently distinguished from "toads," but in fact both are anurans, and the distinction between the two can be misleading since definitions vary in different parts of the world. For example, Europeans and North Americans usually refer to smooth-skinned, toothed water frogs and tree frogs as "frogs," and the rough-skinned, toothless spadefoot or parsley frogs and so-called "true toads" as "toads." On the other hand, Africans refer to native smooth-skinned, aquatic frogs as "clawed toads." The distinction between frogs and toads is in fact meaningless in the greater scheme of anuran diversity and evolution.

Depending on their lifestyle, anurans have one of several different bodyplans, each characterized by different sets of physical characteristics. Anurans adapted to similar lifestyles can share many physical similarities despite not being closely related. For example, semiaquatic anurans such as the Edible frog, American bullfrog, and Spotted grass frog frequent the edges of ponds and lakes. These frog species all have pointed heads, smooth skin, streamlined bodies, exceedingly long hindlimbs, and long, webbed toes. Normally, they sit at the edge of the water; when disturbed, they escape by leaping into the water and swimming away. Their body form adapts them to move rapidly and easily between and within both terrestrial and aquatic habitats.

In contrast, anurans that spend most of their time out of the water (for instance the European common toad, Gold frog, Darwin's frog, Brazilian horned toad, and the poison frogs) usually have blunt heads, may have rough skin, stout bodies, and short hind limbs, and have little toe-webbing. These anurans are adapted to hopping about on the ground. To escape danger they either sit still and rely on their cryptic colors and shapes to hide them, or hop quickly away and confuse predators by changing direction frequently. Several different kinds of anurans that live in highly seasonal or arid climates are burrowers (for example the Burrowing toad, spadefoot toads, Australian shovelfoots, and narrow-mouthed frogs). Typically, these anurans are not large and have blunt snouts, broad, high heads, robust, globular bodies, short and stout limbs, and unwebbed toes. Most burrow backward into loose soil with their hind limbs, and several species have "spades" on their feet to assist in digging.

Most anurans are adapted to perching on various kinds of vegetation, such as swamp reeds and grasses (sedge and bush frogs), bushes (*Eleutherodactylus* species), and branches and leaves of trees (glass frogs, leaf frogs, and most tree frogs). Their bodies tend to be flattened and their hind limbs long; their toes (and sometimes their fingers) are partially webbed. In many instances, the ends of the fingers and toes are expanded; species with these terminal discs are thereby aided in gaining purchase on plant surfaces and in climbing on vegetation.

◁ **Left** *Red-eyed leaf frogs* (Agalychnis callidryas) *are superbly well adapted to their arboreal habitat in the rain forests of Central America, with slender bodies, plus long hind limbs and adhesive disks on all their digits for greater traction on wet stems and branches.*

◑ **Right** *The prominent eyes common to all frogs and toads are especially evident on this Western spadefoot toad* (Pelobates cultripes). *Vertical pupils enable aquatic frogs to see above the water while keeping the rest of the body submerged.*

There are surprisingly few anurans that one would term fully aquatic in the sense that they spend most of their time in the water; among those that do are the clawed frogs and Surinam toads, pseudid frogs, and the Lake Titicaca frog. Whereas each has special adaptations to living in the water, they all look vastly different from one another and are not closely related. The most fully aquatic are the tongueless pipids (clawed frogs and Surinam toads), which have flattened bodies and heads, small, dorsally-placed eyes, laterally sprawled limbs, and fully webbed toes. Pseudid frogs have pointed heads, dorsal eyes, a streamlined body, powerful hind limbs with fully webbed toes, and elongated fingers and toes. In contrast to both of these, the Lake Titicaca frog is one of the larger frogs and has a robust, chunky body with powerful hind limbs, as well as distinctive large folds of loose skin along its flanks that help it breathe in the extremely cold waters of the high-altitude lake it inhabits.

The heads of anurans are flush with the trunk, so the animals cannot move their heads from side to side. Adults have lungs for breathing air on land, but they obtain most of their oxygen directly through the skin. Some frogs living in cold water at high elevations, which is very low in dissolved oxygen, have extremely baggy skin, or hairlike projections enriched with capillaries; these modifications increase the area of the respiratory surface and maximize oxygen uptake.

The eyes are large in most anurans, as one would expect of organisms that locate their food by sight. All anurans, except one small group, have a tongue. In most, the tongue is padlike and attached only in the front of the mouth at the jaw; this arrangement allows the anuran to flip the tongue out and over the lower jaw to pick up food with the sticky upper surface of the tongue pad. The eyes have special glands to keep them moist, and they are protected from dust and soil by movable lids. Just behind the eye of most species is a

FACTFILE

FROGS

Order: Anura (Salientia)

About 5,453 species in 374 genera and 45 families

DISTRIBUTION All continents except Antarctica

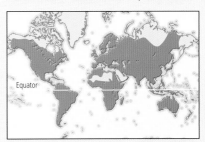
Equator

HABITAT Nearly all habitat types used; absent only from the polar regions and most very dry deserts. Adults arboreal, terrestrial, amphibious, or aquatic.

SIZE Length from snout to vent 1–35cm (0.4–14in), but most are 2–12cm (0.8–4.7in).

COLOR Extremely variable including green, brown, black, red, orange, yellow, and even blue and white.

REPRODUCTION Fertilization mostly external; eggs may be laid in water or on moist ground, or develop in or on one of the parents. Embryos of most species pass through a free-swimming and feeding tadpole stage, but some develop within the egg capsule and hatch as froglets.

LONGEVITY In captivity, adults commonly live 1–10 years, though some reported to live more than 35 years; very little is known about longevity in the wild.

CONSERVATION STATUS 14 recent species, including the Vegas Valley leopard frog, both mouth-brooding frogs from Australia, and the Golden toad from Central America, are listed as Extinct (the true number is probably far greater); in addition, 387 species are Critically Endangered, 628 Endangered, and 542 Vulnerable. In total, these represent one-quarter of all frog species.

See species table ▷

large, conspicuous eardrum (tympanum). Frogs are the most primitive vertebrates to have a middle ear cavity for transferring sound vibrations from the eardrum to the inner ear. Correlated with the development of the ear and with jumping locomotion is the appearance of a true voice box (larynx) and a large, expandable vocal sac(s), making possible a wide variety of vocalizations.

Coping with Heat and Cold

TEMPERATURE AND WATER REGULATION

Frogs are "cold-blooded" animals (ectotherms), which means that although they produce some internal heat by metabolism, they depend primarily on environmental sources of heat to regulate body temperature. Their body temperature is usually close to that of their surroundings and ranges from 3–36°C (37–97°F) depending on climatic conditions. At low temperatures during temperate-zone winters, anurans cannot maintain activity, and their only option is to enter torpor. Most species can survive temperatures between 0–9°C (32–48°F) for long periods during torpor, and some, such as the European common frog and the Wood frog in North America, can survive temperatures as low as −6°C (21°F) by producing glycerol, which acts as an antifreeze in their tissues. Some frogs increase their body temperatures during the day by basking in the sun with their bodies and legs outstretched, but this results in water loss; thus, basking is restricted to species living near permanent water. Frogs living in hot, arid climates avoid water loss by burrowing during the hottest seasons or parts of the day, and emerging only when the rains start or at night.

Anurans always face desiccation because of their highly permeable skin; hence, most are nocturnal and active when temperatures are lower and atmospheric humidity is higher. When submerged in water, anurans can absorb water and salts through the skin and the surfaces lining the

mouth cavity and lungs, but most have no physiological ability to control evaporation of body water on land; rather, they rely on behavioral adaptations – primarily posturing their bodies to expose more or less surface area, depending on prevailing conditions. Species from arid regions best tolerate water loss. The Western spadefoot toad of the dry plains and deserts of North America can lose up to 60 percent of its body water, whereas the aquatic Pig frog can tolerate only a 40-percent loss. Arid-adapted species also can rehydrate faster than species from wetter regions. Some toads, such as true and spadefoot toads, absorb water simply by sitting on an area of damp soil. The skin on the lower surface of the body is thin and rich in blood vessels; they have an absorbent "seat-patch."

Many burrowing anurans store water in the bladder; this feature enables them to remain underground for long periods without drying out. When water is needed, it diffuses through the bladder walls into the body. The Australian Water-holding frog has large, baggy lymph glands beneath the skin and appears bloated when fully hydrated with water, which accounts for half its weight. This species and other burrowers secrete a mucous membrane that hardens and acts as a water-retaining cocoon. During dry periods, burrowing frogs may remain inactive for many months, even several years. In spadefoot toads, emergence from below ground after a dry spell is triggered by decreased barometric pressure associated with oncoming storms.

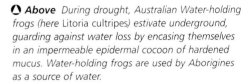

Above *During drought, Australian Water-holding frogs (here* Litoria cultripes) *estivate underground, guarding against water loss by encasing themselves in an impermeable epidermal cocoon of hardened mucus. Water-holding frogs are used by Aborigines as a source of water.*

Left *A spadefoot toadlet (genus* Scaphiopus) *emerges from its hatching pool to seek shade and avoid desiccation. In the arid environment of the southwestern USA and Mexico, development must be fast to exploit the ephemeral pools left after rainstorms, so very rapid hatching and larval periods (3–4 days and 6–8 days, respectively) have evolved among these toads.*

Right *Male Common toads (*Bufo bufo) *clinging to a much larger female during amplexus. In the relatively short breeding season of this species – a feature it shares with most other anurans from temperate zones – males arrive at breeding ponds well in advance of the females, which they heavily outnumber.*

Some frogs living in dry regions combine unusual anatomies with remarkable behaviors to avoid water loss. The casque-headed tree frogs have expanded skulls covered by skin that is fused with the bone below. They fill cavities and block holes with their peculiar heads (a behavior known as phragmosis) to keep moisture from escaping from their refuges. Some leaf frogs secrete lipids and spread the secretion over their skin to form a nearly impermeable barrier to water loss. A few frogs are known to excrete uric acid in a semisolid form rather than as urea, which is water-soluble and involves a much greater water loss.

Continuous or Explosive
REPRODUCTION AND DEVELOPMENT

Frogs and toads display two basic reproductive patterns. In most temperate-zone species, the timing of reproduction is dependent on a combination of temperature and rainfall. Especially in temperate and dry tropical regions, anurans aggregate in large numbers for breeding, and choruses of calling males produce an impressive noise that can be heard from great distances. Such large breeding aggregations usually last only a few nights, so these species are referred to as "explosive" breeders. However, in most species that inhabit humid regions in the tropics and subtropics, anurans can reproduce throughout the year; the major factor controlling timing of reproduction is rainfall. These tropical frogs are referred to as "continuous" or "opportunistic" breeders.

Migrations to breeding sites are often highly synchronized among species and may involve large numbers of animals. This is especially true in arid zones, simply because there are not many occasions when conditions are suitable for breeding. In the humid tropics and subtropics, on the other hand, anurans commonly live near their breeding sites, such as extensive swamps and mountain streams.

Some species breed in permanent ponds or lakes and show remarkable site-fidelity, returning to exactly the same pond year after year. Even if individuals are transported to other suitable ponds in the area, they will attempt to return to the pond they visited before. Several environmental cues are thought to be involved in migration to breeding ponds, including smell, humidity gradients, landmarks, the position of celestial bodies, and the calls of other frogs. For a long time it was believed that European common frogs found their "home" pond solely by smell, but this cannot be the only cue used, since there are frequent and well-documented reports of frogs returning to sites where ponds have been drained, filled in, or built over.

Anurans have extremely diverse reproductive modes. Although nearly all anurans in temperate and arid regions (as well as several tropical species) deposit their eggs in ponds where free-swimming tadpoles develop, many species in the tropics deposit their eggs on vegetation, on the ground, or in excavations. The egg masses of frogs deposited in water vary enormously in size and shape. In cold water, egg masses tend to be globular in shape, whereas in warmer water where oxygen levels are low, the eggs form a thin film on the surface, so that each embryo is able to obtain sufficient oxygen. The eggs of most true toads (Bufonidae) are in strings that are deposited as the mated pair moves through the water.

Eggs laid openly in water are subject to desiccation if the water level drops; also, the eggs are in danger of being eaten by fish and various aquatic insects. Many kinds of anurans have overcome this fragile part of their life history by depositing their eggs out of ponds and streams. For example, several species of South American Leptodactylidae lay their eggs in foam nests floating on water, while some Australian ground frogs (Limnodynastidae) deposit eggs in foam nests either in water or

in terrestrial burrows. The foam is a mixture of water, air, sperm, and eggs. This combination is whipped up by the female's energetic kicking; the outer part that is exposed to the air hardens like a meringue, whereas the interior remains moist and provides a protected environment in which the eggs develop.

Many species of Amero-Australian tree frogs (Hylidae), some reed frogs (Hyperoliidae), and all glass frogs deposit eggs on vegetation above water, or in or above water in tree holes or bromeliads. Upon hatching, the tadpoles either drop into water below the nests or complete their development in tree holes or bromeliads. Male gladiator frogs (Hylidae) excavate and defend shallow basins in sand or mud adjacent to ponds or streams; tadpoles pass their early development in these basins.

Many species in various families deposit their eggs on land. In many of these the tadpole stage is eliminated and the embryos develop directly into froglets. In contrast to the clutches numbering hundreds or thousands of small eggs that are deposited in water, clutches deposited on vegetation or land are much smaller. The terrestrial eggs are much larger because each contains enough yolk to sustain the young through development. In many cases, terrestrial eggs are attended by a parent, usually the female, who may sit on the eggs to keep them moist and to prevent them from being eaten by predators.

Many more elaborate kinds of parental care have also evolved in anurans. Males of the midwife toads carry the eggs entwined on their hind legs. Upon hatching in terrestrial nests, the tadpoles of poison frogs and Seychelles frogs are transported to water on the backs of parents; in some poison frogs, females deposit infertile eggs in bromeliads to feed developing tadpoles. The hatchling tadpoles of the Australian Pouched frog (*Assa darlingtoni*) wriggle into pouches on the sides of the males; here they complete development. As the terrestrial eggs of Darwin's frog (*Rhinoderma darwinii*) hatch, the male picks them up in his mouth; the tadpoles develop into froglets inside the vocal sacs. In the South American marsupial frogs and their allies, eggs develop on or in pouches in the backs of females, where they develop into froglets, but in some cases the eggs hatch as tadpoles and the female deposits them in ponds.

The ultimate in parental care has evolved independently in West African live-bearing toads and the Puerto Rican live-bearing frog. Females produce only two young per litter. The provision of nourishment for the developing embryos within the oviducts corresponds to the condition in placental mammals.

Unlike the larvae of other amphibians, tadpoles have short, almost spherical bodies. Their mainly vegetarian diet requires a long gut; it has extensive absorbing surfaces and is coiled into a tight ball.

Most are herbivorous grazers or browsers that rasp vegetable matter from algae and other aquatic plants. Tadpoles of a few species, such as the South American bullfrog, are truly carnivorous and have much shorter guts than herbivores. Most tadpoles are also capable of filter-feeding and can survive for months without visible food, as they feed on algae and other small particles in the water. Water is taken in through the mouth, passed over specialized structures there that filter out plankton, and finally expelled to the outside through a tube—the spiracle. Tadpoles have internal gills that extract oxygen from the water.

The mouthparts of most tadpoles consist of a pair of fleshy lips with rows of horny "teeth" arranged upon them like the teeth of a comb. The number and appearance of the teeth (which are

◑ **Left** *Many Afro-Asian tree frogs (Rhacophoridae) construct foam nests on vegetation above water. In some of these species, one or more males assist in whipping up the foam. When the tadpoles hatch, the foam nest dissolves and the tadpoles drop into the water below. .*

◐ **Below** *In common with other members of the True toad family (Bufonidae), the Green toad (Pseudepidalea viridis) of Europe and Asia lays its clutches of small, pigmented eggs in long strings underwater. These eggs hatch into free-living tadpoles.*

ANURAN TOXINS AND POISONS

All frogs and toads produce poisons from special glands in their skin. Most are not harmful to humans, but a few species, such as the poison frogs of Central and South America, produce some of the most poisonous biological toxins known. One of the most lethal is the batrachotoxin in the skin secretions of the Koikoi poison frog of Colombia and Panama; a mere 0.00001g (0.0000004oz) is enough to kill an average-sized man. The Choco Indians of the Darien jungle poison the tips of as many as 50 arrows with secretions extracted from one tiny frog. Today they use these arrows for hunting game such as deer, monkeys, and birds, but in earlier times, their blowguns and poison darts are said to have been used with deadly effect against neighboring hostile tribes.

Many poisonous frogs are brightly colored to warn off potential predators, and birds that normally eat frogs quickly learn to avoid certain species.

When provoked, many leaf frogs quickly extend their legs to reveal bright patches on their sides and along the insides of their thighs. Fire-bellied toads arch their backs and twist their legs to expose their bright orange and black undersides. The False-eyed frog of South America, which has a large pair of eyespots on its rump, turns away from an attacker, inflates its lungs, and lifts its backside in the air to display the markings. This species also produces an unpleasant secretion from glands around the eyespots.

The Cane toad can squirt toxic secretions from its parotoid glands up to 1m (3.3ft) into the eyes or mouth of a predator. Dogs attempting to eat these toads have later suffered extreme pain. Human fatalities have even occurred in Fiji and the Philippines, when people used to eating frogs were not warned of the dangers of eating this particular species on its introduction there. Peruvian peasants have died merely from eating soup prepared from its eggs.

○ *Above* A somewhat different defensive strategy to signal toxicity and deter would-be predators has evolved in the False-eyed frog (Physalaemus nattereri) of Brazil, in the form of unnerving posterior eyespots.

◗ *Right* The Blue poison frog (a form of Dendrobates tinctorius) has striking electric-blue coloration, a warning of its extreme toxicity.

unlike the jaw teeth of adults) vary and are important features in the identification of species. Tadpoles of clawed frogs and Surinam toads have fleshy, tubular, sensory extensions of skin (barbels or tentacles). For detecting pressure changes and vibrations in the water, tadpoles possess lateral-line organs—a series of special sensory cells (neuromasts) arranged in orderly rows on the head and body. In some totally aquatic frogs (for instance the clawed frogs), the lateral-line system is retained in adults. The tadpole stage lasts from a few days in some species to more than three years in those in cold temperate regions.

The Human Impact

CONSERVATION AND ENVIRONMENT

Frog legs are eaten in many parts of the world, but most often as a delicacy for gourmets rather than as a major source of nutrition. In Europe, the Edi-

ble frog, a hybrid of the Marsh and Pool frogs, is the main table species, but now most frogs intended for consumption are imported from the developing countries. Most species are in fact edible, but it is only economical to market the larger ones (such as the American and Asian bullfrogs). Unfortunately, when these species are introduced for farming, they tend to eliminate local frog populations.

Frogs have been, and still are, widely used for teaching and research; studies of anatomy, development, and physiology in frogs have contributed greatly to our understanding of vertebrate evolution. As experimental animals, anurans are important biomedically. Anuran studies have elucidated suitable methods for organ transplants, and they were used widely for human pregnancy tests until better methods were devised. Techniques have also been developed to extract certain alkaloids

from toxic species for use in therapeutic drugs.

Despite the benefits man has reaped from the study of frogs, the effect of human activity on these animals generally is negative. The single most harmful impact comes from habitat destruction, followed closely by global warming resulting from human activities. More than three-quarters of all species occur in tropical rain forests, which are being destroyed at an ever-accelerating rate. Clearing land for buildings, drainage of swamps and marshes, damming rivers to form lakes, acid rain, and the heating of water as part of the cooling process of nuclear power plants—all of these activities have had adverse effects on frog populations. The use of pesticides and herbicides on crops also has contributed to the demise of some populations; adults are affected when they eat contaminated arthropods, and run-off makes ponds, streams, and lakes unsuitable for develop-

1

ing tadpoles. Most recently, a chytrid fungus has been found to be responsible for the disappearance of many populations of frogs around the world. The fungus affects the skin of adult frogs and the mouthparts of developing tadpoles; the reasons for its sudden appearance are unknown.

New Zealand Frogs
FAMILY LEIOPELMATIDAE

The six species of New Zealand frogs share most of the same primitive morphological features, but the four species that live in New Zealand lack the tail-like cloacal appendage of those living in North America, the tailed frogs. New Zealand frogs live on mountainsides where streams and pools of water are scarce. Fertilization is external, and females lay small clutches of 1–22 unpigmented, heavily-yolked eggs in damp crevices under rocks and logs. The male provides parental care by remaining with the eggs, and in Hamilton's frog the tadpoles climb onto his back, where they are kept moist.

The two species of tailed frogs occur in northwestern North America in cold, fast-moving mountain streams with cobbled bottoms. They are among the most primitive of living anurans, having nine presacral vertebrae, a prepubic bone, free ribs, and vestigial tail-wagging muscles in adults. Tailed frogs are nocturnal and exhibit several interesting adaptations to life in the torrential streams. In males, the cloaca is modified to form an intromittent organ (the so-called "tail"), which is used to transfer sperm to the female; this internal fertilization ensures that the sperm are not swept away in the current. Males also have greatly enlarged forearms during the breeding season to enable them to hang on to females during mating. Fertilized eggs are deposited in strings under rocks in the stream. The tadpoles develop slowly, taking between 1 and 4 years to metamorphose, and young frogs do not mature sexually before they are 7 or 8 years old. The tadpoles have large, suctorial oral disks armed with many rows of denticular teeth; the disks allow them to cling to the undersides of rocks in the stream and scrape food from the rock surface.

Clawed Frogs and Surinam Toads
FAMILY PIPIDAE

The clawed frogs and Surinam toads are highly specialized aquatic anurans; indeed, adults from this family rarely venture onto land. Pipids have flattened bodies and limbs that extend laterally from the body; they have long hind legs, fully webbed feet, and long fingers that are rotated medially. The eyes are small and directed upward. African clawed frogs have keratinized tips on the toes, and the Surinam toad and its relatives have star-shaped sensory structures on the ends of the fingers. Pipids are unique among anurans in that they lack tongues and have a unique vocal apparatus, consisting of a partially ossified box containing two bony rods, the rattling of which produces clicks that are transmitted through the water; vocal cords are absent.

Most pipids inhabit ponds and rivers, where they forage on the bottom; aquatic arthropods and small fish are shoved into the mouth with the fingers. The laterally-positioned limbs provide no support on land; when the frogs do cross land, they essentially "swim" across the substrate. African clawed frogs often live in stagnant pools where oxygen concentrations are low; they have lungs that are proportionately large for their body sizes, and regularly come to the surface to gulp air.

During mating, which takes place in water, males clasp females around the waist. African clawed frogs deposit their eggs on aquatic plants. Amplectant pairs of other African pipids go through vertically circular maneuvers that culminate in them depositing their eggs on the surface of the water. Similar maneuvers are characteristic of the Surinam toad and its relatives in the genus *Pipa*, but in their case the eggs are swept by the male's hind feet onto the back of the female, where they adhere and become embedded in the skin. In some *Pipa* species, the eggs hatch as tadpoles; after about four weeks the tadpoles emerge from the skin pockets and develop as free-swimming tadpoles. However, the eggs of the Surinam toad and some other *Pipa* species develop directly into toadlets; after 3–4 months, the small toads partially emerge from the pockets and leave the mother. Pipid tadpoles have long barbels (tentacle-like projections) bordering the mouth. These midwater filter–feeders are characteristically oriented head down at about 45°.

6

Mexican Burrowing Toad
FAMILY RHINOPHRYNIDAE

The Mexican burrowing toad shares several anatomical features with the African clawed frogs

○ **Above** *Representative species of five families of frogs and toads: 1 Syrian spadefoot toad (Pelobates syriacus) of eastern Europe and Southwest Asia, jumping; Pelobatidae. 2 Oriental fire-bellied toad (Bombina orientalis); Bombinatoridae. 3 Tailed frog (Ascaphus truei); Leiopelmatidae. 4 Male Iberian midwife toad (Alytes cisternasii), carrying eggs; Discoglossidae. 5 Surinam toad (Pipa pipa), with young emerging from pits on the female's back; Pipidae. 6 Couch's spadefoot (Scaphiopus couchii), in amplexus; Pelobatidae.*

and Surinam toads, but it is terrestrial. This globular, short-legged, small-headed anuran has hind feet that are specialized for digging; a large spade-like tubercle is present on the edge of the hind foot. It spends most of its time underground, where it feeds mainly on termites. It does not flip the tongue out in the usual frog fashion. The front of the tongue is not attached to the floor of the mouth and, when feeding, the tongue is protruded forward. Breeding occurs in temporary ponds after heavy rains. Males have internal vocal sacs and call while floating on the surface of the pond. Eggs develop rapidly, and the tadpoles, which have sensory barbels, are midwater filter-feeders.

Midwife Toads and Painted Frogs
FAMILY ALYTIDAE

The midwife toads and painted frogs are primarily European. Painted frogs resemble true frogs (see below), except that they have round or triangular pupils rather than horizontal ones. The eardrum is inconspicuous and the tongue is disk-shaped. Males grasp the female around the waist during amplexus and in the breeding season have large nuptial pads on the fingers, chin, and belly. These anurans are active by day and night and often sit with their heads just above the surface of shallow water, in which they breed. The male's call is best described as a quiet, rolling laugh.

Midwife toads are squat and rough-skinned. The Common midwife toad only occurs in western Europe, in woodlands, gardens, quarries, and rocky areas at elevations up to 2,000m (6,500ft). It is mainly nocturnal, and by day hides in crevices under logs or digs shallow burrows with its front legs. The midwife toads are highly terrestrial, even mating away from water. The males of all species in this family grasp the female around the waist during amplexus and, except for midwife toads, possess large nuptial pads during the breeding season. The females extrude strings of large eggs that the males catch with their feet and twist around their hind legs. They carry them like this for several weeks, keeping them moist by occasionally visiting pools of water. When the embryos are about to hatch, the males deposit them in shallow water. The burrowing Iberian midwife toad is found in sandy locations in central Spain and Portugal, and the Mallorcan midwife toad is restricted to Mallorca in the Balearic Islands. The latter species is Critically Endangered.

Fire-bellied Toads and Barbourulas
FAMILY BOMBINATORIDAE

The fire-bellied toads and barbourulas inhabit Eurasia. Fire-bellied toads are small and warty-skinned, with a rather flattened body shape. Bright black and red-to-yellow markings on their undersides warn potential predators against their distasteful and slightly toxic skin secretions. Usually they are found in shallow water at the edges of rivers, streams, marshes, drainage ditches, and tem-

⟲ Above *A Tyrrhenian painted frog (Discoglossus sardus) in an advanced phase of metamorphosis. At this stage, several weeks after hatching, both the fore and hind limbs are well developed and the eyes are* prominent, but the tail has not yet been resorbed. The discoglossids are confined to southwestern Europe and the coastal fringe of North Africa; this species is able to tolerate brackish water.

porary bodies of water such as ruts and small puddles; they have subdued voices, and breed in water two or three times a year. Males of all species in this family grasp the female around the waist during amplexus and possess large nuptial pads during the breeding season that are present on the fingers but also occur on other parts of the body (chin, belly, and toe-webs of painted frogs; forearms of fire-bellied toads). Barbourulas are completely aquatic in streams on the islands of Borneo and Palawan in the Philippines. The fire-bellied toads and barbourulas are largely aquatic and breed in water.

▷ **Right** *The Horned frog (Megophrys nasuta) of Malaysia uses its highly effective camouflage to sit and wait among leaf litter on the rainforest floor for its prey, which includes invertebrates and smaller frogs.*

▽ **Below** *When harassed by a snake or other predator, the defensive posture of the Yellow-bellied toad (Bombina variegata) involves lifting all four limbs to expose the bright coloration on its belly and soles. This posture is known as the "Unken reflex," from the German vernacular name for this genus of toad.*

Parsley Frogs
FAMILY PELODYTIDAE

The three species of parsley frogs, inhabiting western Europe and southwestern Asia, are terrestrial and nocturnal, except that breeding takes place by night and day in the spring and autumn. Strings of pigmented eggs deposited in ponds or slow-moving streams hatch into free-swimming tadpoles. Tadpoles in some populations can withstand brackish water.

American Spadefoot Toads
FAMILY SCAPHIOPODIDAE

American spadefoot toads live in arid regions, including deserts, where they spend most of the year in underground chambers, protected from the extremes of heat. Breeding takes place in spring and summer, with the onset of heavy rain following several warm days, when toads may appear on the surface in huge numbers. In two days, the females finish laying eggs and the toads disappear underground again. The voice is a loud, harsh clucking that can be heard from more than 2km (1.2mi) when many males are calling together. Sometimes males fight for females, which may be wounded by the sharp spades on the struggling males' hind feet. Eggs are deposited in temporary ponds and hatch within a few days; tadpoles of North American spadefoot toads complete their development in 1–3 weeks. The feeding tadpoles commonly swim about in huge aggregations and feed on suspended organic material. Some tadpoles of desert-dwelling spadefoots develop different jaws and teeth from plankton-feeding tadpoles of the same species and become cannibals; these grow to larger sizes, often reaching 10cm (4in). With two types of tadpole in a pond, the desert spadefoots are prepared for different eventualities.

If more rain falls, plankton-feeders will thrive; if there is no more rain, the cannibals will feast on stranded herbivores.

Eurasian Spadefoot Toads
FAMILY PELOBATIDAE

Spadefoot toads are usually confined to dry areas with sandy soils in Europe and western Asia. A keratinized, sharp-edged tubercle ("spade") on each hind foot is used for burrowing. Adults are strictly nocturnal; during the day and during long dry periods they hide in deep burrows, where they can tolerate high levels of water loss. On warm, moist nights during the summer, they emerge to feed, and eat almost any kind of terrestrial arthropod. Outside the breeding season, adults do not move very much but sit and wait for prey to come to them. The home range of the Eastern spadefoot is estimated to be no more than 9sq m (100sq ft).

Breeding occurs in the spring, often in relatively deep pools, and both males and females may be active by day at this time of the year. Males call from under the surface. The eggs are laid in wide strings and are usually entangled around aquatic plants. the tadpoles are not especially fast-growing, unlike their North American counterparts, and grow very large before metamorphosing; some may overwinter as tadpoles.

Asian Toadfrogs
FAMILY MEGOPHRYIDAE

The Asian toadfrogs occur in Southeast Asia and associated islands. Some of these frogs are streamside inhabitants, and have stream-adapted tadpoles with large, suctorial mouths and low fins. The Malaysian horned toad and its relatives have fleshy "horns" on each upper eyelid and on the

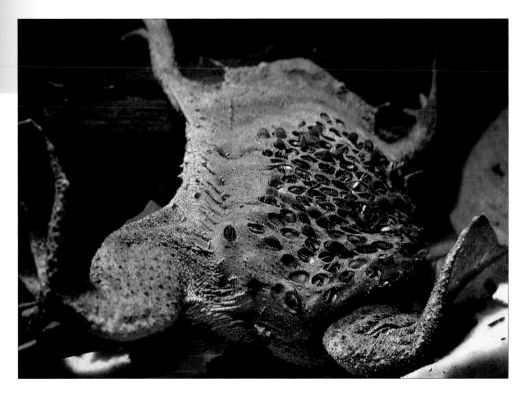

○ **Above** *Toadlets emerging from the distinctive "pockets" on the back of the female Surinam toad. In this species, development from egg to toadlet is direct, whereas in other* Pipa *species, larvae emerge and complete their development as tadpoles.*

○ **Below** *The Crucifix toad (*Notaden bennettii*) – an Australian limnodynastid named for the cross-shaped pattern of dark coloration that appears on its back – inflates its body and raises itself on its arms and legs when it is threatened.*

snout. Their bodies are mottled in various shades of brown, and the skin is ribbed so that the frogs appear to be dead leaves. This cryptic coloration makes them almost impossible to see when they are sitting on the forest floor. Males are conspicuous during heavy rain when they sit in shallow streams and produce very loud, mechanical-sounding "clanks." The tadpoles of many horned frogs have large, upturned mouths, and feed at the surface of the water.

Ghost Frogs
FAMILY HELEOPHRYNIDAE

The ghost frogs of South Africa live in and along fastflowing mountain streams. Adults have a glandular fold behind the eye, no webbing between the fingers, and toes that are nearly fully-webbed. The tips of the fingers and toes are expanded and spatulate. Breeding males are slightly smaller than females and have loose folds of skin and horny spines on various surfaces of the body, especially the hands and forelimbs. Large, unpigmented eggs are deposited either in streams under rocks or in shallow pools, or else under rocks on wet gravel at the edges of streams. The tadpoles are adapted to live in torrents; they have long, muscular tails and large, suckerlike mouths. The tadpoles feed by grazing over algae-covered rocks; at night they use their mouths to climb wet, vertical rocks. The tadpoles require more than a year until metamorphosis.

Seychelles Frog and Purple Frog
FAMILY SOOGLOSSIDAE

Four species of Seychelles frogs occur on granitic islands in the Seychelles group in the Indian Ocean. They are terrestrial and nocturnal. Eggs are laid in small clumps on moist ground and are attended by females. Upon hatching and after two weeks of incubation, the tadpoles, which already have rudimentary hind limbs, wriggle onto the female's back; they respire through their skin and lack gills and spiracles. After metamorphosis, the froglets stay on their mother's back for a short time. The eggs of Gardiner's frog (*Sooglossus gardineri*) require longer to develop and hatch as froglets. The recently described Purple frog occurs in the Western Ghats of southern India. It is plump, with a pointed snout and is purplish-gray in color, and therefore very unlike the Seychelles frogs with which it has been classified. The biology of the Purple frog is unknown.

Central Chilean Frogs
FAMILY CALYPTOCEPHALELLIDAE

There are four species, which were previously placed in the family Leptodactylidae. They live in the Valdivian forests of Chile. These frogs are aquatic species and are never found away from water. Three species in the genus *Telmatobufo* are extremely rare and have very small ranges. They live under stones on the bottom of streams and rivers. The Helmeted frog (genus *Caudiverbera*) lives in lakes and ponds. Its range is also shrinking as a result of development and pollution, but its numbers are more stable away from towns.

Australian Ground Frogs
FAMILY LIMNODYNASTIDAE

The Australian ground frogs, also found in New Guinea, are terrestrial or fossorial inhabitants of rain forests, grasslands, and deserts; *Philoria* inhabit moist, cool mountains. Many species of *Limnodynastes* inhabit marshes and swamps, but some burrowing forms live in deserts. The latter emerge only after heavy rains; they store water in their bodies, and can survive for long periods under ground. All have free-swimming tadpoles. *Neobatrachus* and *Notaden* mate in water, whereas other genera have foam nests.

The shovelfoots or spadefoots inhabit Australian deserts. These medium-sized, globular limnodynastids have warty skin and short limbs. They possess sharp, spadelike tubercles on their feet for burrowing into loose soil. The Desert shovelfoot exploits the Bulldog ant (*Myrmecia regularis*) by living in the nests of its insect host. When molested, the Australian shovelfoots inflate their bodies to absurd sizes and rise on all four limbs to face intruders. All species exude a thick, white, smelly poison when aroused; this dries to form a strong, elastic substance.

Australian Toadlets and Water Frogs
FAMILY MYOBATRACHIDAE

Australian toadlets and water frogs have restricted distributions in Australia and southern New Guinea. They inhabit dry grassland, scrub, savannas, marshes, stream and lake shores, and rain forest. Most myobatrachids have aquatic or terrestrial eggs that hatch into free-swimming tadpoles. Among the smallest are the toadlike members of the genus *Pseudophryne*, which are rarely more than 3cm (1.2in) long and often have brightly colored bellies. Some species also have bright colors on the back. Bibron's toadlet lays its eggs in damp cavities under stones, and males are often found sitting next to several clutches of eggs. Breeding takes place at any time during the 6–7 months of the year when rain is sufficient to wet the ground. If a drought occurs after egg-laying, the eggs may not hatch for over three months, but the tadpoles eventually must swim to, and complete their metamorphosis, in standing water.

The Sandhill frog (*Arenophryne rotunda*) and Turtle frog (*Myobatrachus gouldii*) burrow headfirst into sandy soils, where each deposits a few large eggs that develop directly into burrowing toadlets. The small Hip-pocket frog (*Assa darlingtoni*) deposits a few eggs on moist ground. As the eggs hatch, the male allows the tadpoles to wiggle onto its body, whereupon they move into pouches on either side. Here the tadpoles complete their development and emerge after two months as

CANE TOADS IN AUSTRALIA

In the 1930s, Australian sugarcane growers enthusiastically greeted reports of a toad from Central and South America that had been introduced to Puerto Rican canefields in 1920. Up to 23cm (9in) long, it was variously called the Marine, Giant, or Mexican toad. Word had it that in Puerto Rico it had eaten large numbers of sugarcane pests, raising hopes that it might similarly control the Grayback cane beetle (*Dermolepida albohirtum*) that was menacing Australian sugarcane.

Although biologists warned that, without natural enemies, the toad, which breeds year-round, would soon pose as much of a pest as introduced rabbits and cacti had already done, 100 adults were imported in 1935. Of more than 1.5 million eggs laid by these new arrivals, 62,000 reached the young-adult stage and were released into selected areas of Queensland, where the species became known as the Cane toad.

Despite their voracious appetites, the toads disappointed farmers. Queensland cane fields provided inadequate cover for daytime hiding, so the toads quickly moved into the surrounding countryside and gardens. Here, populations rapidly increased to near-plague proportions (LEFT). Today they are so abundant in some places that nighttime gardens become a slowly moving, dark, shuffling sea of toads, and in the morning roads are littered with squashed bodies run over in the night.

More troubling is the evidence that these toads eat just as many creatures that are beneficial to agriculture as they do pests (BELOW). Naturalists worry that the toads may be numerous enough to have a damaging effect on the populations of native frogs, which contribute to their diet.

One unexpected benefit of the Cane toad's introduction, however, is its usefulness as a laboratory animal in schools, universities, and hospitals. In addition, raw Cane toad skins are tanned for leather.

froglets. Females of the stream-dwelling gastric-brooding frogs (*Rheobatrachus* species, probably now extinct) are the only frogs known in which the young develop in the stomach. The female swallows the eggs or the tadpoles. The digestive properties of the stomach shut down, and froglets emerge from the mouth after 6–7 weeks.

Casque-headed Frogs
FAMILY HEMIPHRACTIDAE

The six species of casque-headed frogs are strange-looking, semi-arboreal rain forest amphibians from Central and South America. They are rarely seen, but some species have large ranges. Their shape and coloration is cryptic and they are strictly nocturnal, usually seen on low vegetation. They have bony, triangular heads and enormous gapes; if threatened they open their mouths widely and lunge at their attacker. Females carry their eggs on their backs; the eggs hatch directly into tadpoles.

Three-toed Toadlets
FAMILY BRACHYCEPHALIDAE

Previously restricted to a handful of small toadlets from South America, this family has been greatly expanded to include several hundred species

75

previously placed in the Leptodactylidae. Nearly 500 species of *Eleutherodactylus* species live in moist habitats from Honduras to Brazil and several, notably the coqui (*E. coqui*), have been introduced accidentally to other subtropical regions, including Florida and Hawaii. The Puerto Rican live-bearing frog (*E. jasperi*) has internal fertilization and gives birth to small froglets; the few large eggs complete their development in the oviducts. This species has not been seen since 1981, however, and may be extinct. Members of the other 16 genera include the three-toed toadlets from Brazil, one of which, the Gold frog (*Brachycephalus ephippium*) has a bright golden-yellow body. All members of the family lay terrestrial eggs in leaf litter or bromeliad plants, and the eggs develop directly into miniatures of the adults.

Backpack Frogs
FAMILY CRYPTOBATRACHIDAE
These are small rain forest frogs that live in humid forests, often near waterfalls and streams. They were previously classified with the tree frogs (Hylidae) and are rare and poorly known. Their fertilized eggs stick to the female's back and remain there until they are fully developed.

Marsupial Frogs
FAMILY AMPHIGNATHODONTIDAE
Members of this family are arboreal and terrestrial frogs that carry their eggs in pouches on the female's back. Amplexus is axillary and, as the eggs emerge, the male fertilises them and then manoeuvers them into the female's pouch, which opens to the posterior. The eggs and tadpoles

remain in the pouches for varying periods, depending on the species. *Flectonotus* species are small and live in montane forests, where four of the five species exclusively use bromeliad plants in which to deposit their tadpoles when they hatch; *F. ohausi* from Brazil uses broken stems of hollow bamboo. Their tadpoles are well-developed at the time of release and complete their development without feeding in a few days. The remaining species (genus *Gastrotheca*) are more variable and may live in lowland or montane forests, or above the tree-line in parts of the Andes. Lowland species tend to retain their tadpoles until they are fully-formed whereas those from higher and cooler environments release them as tadpoles, using their long hind toes to scrape them out of the pouch and into small bodies of water, where development continues for several months.

● **Right** *Representative species of nine families of frogs and toads:* **1** *Japanese tree frog* (Rhacophorus arboreus); *Rhacophoridae.* **2** *South African bullfrog* (Pyxicephalus adspersus); *Pyxicephalidae.* **3** *Seychelles frog* (Sooglossus seychellensis); *Sooglossidae.* **4** *Asiatic painted frog* (Kaloula pulchra); *Microhylidae.* **5** *A leaf frog* (Phyllomedusa bicolor); *Hylidae.* **6** *Ornate horned frog* (Ceratophrys ornata); *Ceratophryidae.* **7** *Marine toad* (Chaunus marinus); *Bufonidae.* **8** *Koikoi poison frog* (Phyllobates bicolor); *Dendrobatidae.* **9** *Common frog* (Rana temporaria); *Ranidae.*

Amero-Australian Tree Frogs
FAMILY HYLIDAE

The Amero-Australian tree frogs are most diverse in the Americas, Australia, and New Guinea; a few occur in Eurasia. Most have enlarged, adhesive pads on the tips of the digits; these are used for climbing and holding onto vegetation, where they are active at night. Most Australian and New Guinean species (subfamily Pelodryadinae) deposit their eggs in water, but a few place their eggs on vegetation above water; all have free-swimming tadpoles. The same is true for the widespread American subfamily Hylinae, but a few of its species deposit their eggs in bromeliads or tree holes; in some of these, such as the Central American *Anotheca spinosa* and the Amazonian *Osteocephalus oophagus*, the tadpoles feed on additional eggs provided by the female.

The leaf frogs (subfamily Phyllomedusinae) are mostly large, brightly colored animals of lowland rain forest and montane cloud forest in Central and South America. Eggs are deposited on leaves over water; the hatchlings drop into the water and complete their development as free-swimming tadpoles. During oviposition, some *Phyllomedusa* species wrap the eggs in a leaf and deposit water-filled capsules on top of them; the water in the capsules is transmitted to the embryos in the course of their development.

The aquatic paradox frogs of South America are well adapted for their existence. They have eyes on top of the head, muscular hind limbs, and long, slender, fully-webbed toes. The largest and best-known species is the eponymous Paradox frog (*Pseudis paradoxa*), from whose extraordinary development the group derives its vernacular name. Paradox frog tadpoles often reach an enormous size, sometimes over 22cm (9in) long, but metamorphose into adults of only 6.5cm (2.5in).

Glass Frogs
FAMILY CENTROLENIDAE

The majority of glass frogs are small, green anurans about 3cm (1.25in) in length that live principally in montane rain forests of Central and South America. However, two species of *Centrolene* are robust, attaining lengths of more than 7cm (2.5in). In some species, the skin of the abdomen is transparent, so that their organs are visible. These nocturnal, arboreal frogs have expanded tips to their digits and breed on vegetation overhanging streams. The territorial males defend their calling sites (the upper or lower surfaces of leaves). Eggs are deposited on the surfaces of

⊳ **Right** *During the day, Green treefrogs (Hyla cinerea), which are native to the southern USA, rest under large leaves or in other moist places.*

leaves; in some species they are attended by males, who may perch on the eggs. Hatchling tadpoles drop into streams.

South American Water Frogs
FAMILY LEPTODACTYLIDAE

Many genera that were previously placed in this genus have been removed to other families, leaving a less variable set of 91 species in four genera. They are found in a range of habitats, closely paralleling the ranid frogs of the northern hemisphere but most are closely associated with water. Many species build foam nests that float on water, either in ponds or flooded forests, and have free-living tadpoles. The Puerto Rican white-lipped frog (*Leptodactylus albilabris*) has an unusual method of communication. As well as uttering audible chirp-like calls, the males produce seismic waves by pounding the vocal sac on the ground as they call. Neighboring males have acute sensitivity to ground-borne vibrations and, when they detect them, respond by producing a range of calls at a faster rate. The waves are probably used by males to assess their distance from rival callers.

South American Horned Frogs
FAMILY CERATOPHRYIDAE

This family includes large voracious species such as the horned frogs (genus *Ceratophrys*) from forests of tropical South America and Budgett's frogs (genus *Lepidobatrachus*) from arid pampas. These are sit-and-wait predators that have huge gapes and will tackle small mammals as well as invertebrates and other frogs. The horned frogs get their name from fleshy horns above their eyes; their tadpoles are carnivorous. The Budgett's frogs aestivate by burrowing down into the mud and the bottom of their pools during the dry season, where they form a cocoon in the hard-baked clay, to re-emerge when the next rainy season begins. The aquatic *Telmatobius* species occur in the Andes. The Lake Titicaca frog (*T. culeus*) is adapted to life in a cold lake, where there is little dissolved oxygen; it has extremely baggy skin that acts as a respiratory organ. Nine species of *Atelognathus* live on the Patagonian steppe of Argentina and Chile where they inhabit the margins of permanent ponds and lagoons.

Smooth Horned Frogs, Darwin's Frogs
FAMILY CYCLORAMPHIDAE

A diverse group of South American frogs the button frogs (genus *Cycloramphus*) from the Atlantic forests of Brazil. Many live in fast-flowing streams where they, and their tadpoles, cling to wet water-splashed rocks, but a few live in drier habitats and their eggs are thought to develop directly into froglets. The spiny-chested frogs (genus *Alsodes*)

also live in or near streams, mostly in the mountains of Argentina and Chile; they have free-swimming larvae. The smooth horned frogs (genus *Proceratophrys*) occur in the forests of Brazil and neighbouring countries and are well-camouflaged leaf-litter frogs that breed in streams and rivers, and have free-swimming larvae.

The two species of mouth-brooding frogs of the humid temperate forests of southern South America have a unique type of parental care. Females lay some 20 eggs on land; each egg clutch is attended by a male. The eggs hatch in about 20 days and the male picks the tadpoles up in his mouth. In one species (*Rhinoderma rufum*) the male releases the tadpoles into the water, where they develop in the normal way. In Darwin's frog (*R. darwinii*), the tadpoles move into the male's vocal pouch, where they metamorphose in about 50 days and then emerge from the male's mouth.

Tree Toads and Spiny-thumb Frogs
FAMILY HYLODIDAE

Three genera of poorly-known forest-dwellers from Brazil. Most, perhaps all, live along forest

streams, cascades and waterfalls and the tadpoles live in flowing water. All are rare; those from higher elevations seem to be disappearing from parts of their ranges. The *Crossodactylus* species are also known as spiny-thumb toads and the *Megaelosia* species as big-toothed frogs. Formerly classified within the Leptodactylidae and sometimes placed in the Cycloramphidae.

Rocket Frogs
FAMILY AROMOBATIDAE

Closely related to the poison frogs, the rocket frogs lack bright coloration and the strong skin toxins of the dendrobatids. Most are brown with cream lines on their flanks, or speckled beneath. Eggs are laid on the ground and attended by the male. As they hatch, the tadpoles wriggle onto his back and he carries them to a stream, pond, or small body of water in a leaf axil, bromeliad, or similar. The eggs of some species remain in a terrestrial nest where they complete their development without feeding, while the breeding habits of many species are unknown. The skunk frog (*Aromobates nocturnus*) is larger than most species and has an unusual defence; it secretes non-toxic but very pungent substances from its skin.

Poison Frogs
FAMILY DENDROBATIDAE

The poison frogs are mostly diurnal, terrestrial inhabitants of tropical forests in Central and South America. Nearly all anurans have at least a trace of poison in their skin glands (see Anuran Toxins and Poisons), but toxins are developed to a high degree in some of the poison frogs. They have a pair of platelike scutes on the upper surface of each toe and fingertip. The nontoxic genera (*Aromobates*, *Colostethus*, and *Mannophryne*) are cryptically colored, whereas the toxic genera (*Dendrobates*, *Epipedobates*, *Minyobates*, and *Phyllobates*) have vibrant warning colorations of bright green, blue, red, or gold with darker spots and stripes.

Small clutches of large eggs are laid in moist, terrestrial situations. The eggs of some *Colostethus* species hatch as nonfeeding tadpoles that complete their development in the nest, but the eggs of most dendrobatids hatch as tadpoles that wiggle onto the back of an attending parent, who carries them to water – streams, or constrained bodies of water in bromeliads, husks, or logs. In some species of *Dendrobates*, the tadpoles are carried by females, who deposit unfertilized eggs for the tadpoles to eat. Many dendrobatids are known for their aggressive behavior including calling and color changes (in males), postural displays, chases, attacks, and wrestling (in both sexes). Prolonged fights usually occur between frogs of the same species and sex.

Dwarf Frogs, Foam Frogs, and Relatives
FAMILY LEIUPERIDAE

Seven genera of small frogs from Central and South America make up this family. Many species are common and ubiquitous, found around villages, fields and along roadsides, breeding in ditches, wheel ruts and flooded fields, but others are scarce and have restricted ranges. The Tungara frog is small and toad-like, and calls from flooded open spaces with a voice that carries many hundreds of metres. Its foam nest is about the size of a golf ball and the tadpoles hatch and disperse in about three days. This, and related species, are preyed on by frog-eating bats of the genus *Trachops*, which home in on their calls.

True Toads, Harlequin Frogs, and Relatives
FAMILY BUFONIDAE

True toads, harlequin frogs, and related species are nearly worldwide in their distribution except for Madagascar; Australia has only one introduced species. These toads are mainly terrestrial, but a few are arboreal or aquatic. Species of the largest genus, *Bufo*, move by short hops. Some run, for example the Natterjack toad. None can escape from predators by leaping, but skin poisons compensate. Protective skin secretions are concentrat-

ed in the prominent parotoid glands just behind the head, and some large species also have poison glands on the legs.

Species of toads living in subhumid regions have short breeding seasons (in other words, they are "explosive" breeders); the males call from shallow water, where they will clasp any small object that moves and fight vigorously for females. In species with prolonged breeding seasons such as the Natterjack toad, males call loudly at the edge of temporary pools to attract females.

Typically, toads deposit many small eggs (over 10,000 per clutch in the Marine toad) in rosary-like strings usually wrapped around vegetation in water. The eggs hatch into free-swimming tadpoles that metamorphose in 2–10 weeks. The dwarf toads of southeast Asia lay a few, large-yolked eggs that develop in small, rain-filled depressions on the forest floor and hatch into nonfeeding tadpoles. Many harlequin frogs (genus *Atelopus*) are brilliantly colored, rivaling some of the poison frogs in beauty and toxicity. One, *Atelopus oxyrhynchus*, has an unusually long period of amplexus; one record-holding pair remained in the state for 125 days. Harlequin frogs, which live in Central and South America, and slender toads (genus *Ansonia*) in Southeast Asia deposit eggs in strings in swift-flowing streams; the tadpoles have large, suckerlike mouths and adhere to rocks in the streams. A few bufonids have terrestrial eggs that undergo direct development into toadlets.

A remarkable transition from external fertilization and terrestrial eggs to internal fertilization and live-bearing takes place in the African genus *Nimbaphrynoides*.

○ **Above** *One member of the relatively speciose Glass frog family is the Reticulated or La Palma glass frog (Centrolenella valerioi).* It is a highly adept climber, which lives in the cloud forests of the Panamanian isthmus.

Narrow-mouthed Frogs
FAMILY MICROHYLIDAE

The narrow-mouthed frogs, which are widely distributed throughout the Old and New World tropics, include both terrestrial and arboreal species. Many are small, stout-bodied burrowers with tiny heads and short legs. Many of the tree-dwellers have discs on their toes to assist climbing. Most terrestrial species remain underground until breeding begins after heavy rains. These species characteristically have aquatic eggs that hatch into tadpoles that lack keratinized mouthparts. Other terrestrial species lay a few, large-yolked terrestrial eggs that undergo direct development. The eggs of the globular African rain frogs (genus *Breviceps*) develop into froglets in underground chambers. Some arboreal genera in Madagascar deposit their large eggs in tree holes or leaf axils; these hatch into nonfeeding tadpoles that are attended by the male.

African Rain Frogs and Relatives
FAMILY BREVICIPITIDAE

Rain frogs (genus Breviceps) are plump, almost globular frogs with short heads. They spend most of the year underground, only emerging during rain in order to breed. Amplexus is a difficult affair because of the great girth of the females and the short limbs of the males and they solve this problem by producing a sticky skin secretion that holds them together for the duration of mating. The eggs are laid in underground chambers where they undergo direct development. Related species, from East Africa and Ethiopia are poorly known but probably also have direct development.

Shovel-Nosed Frogs
FAMILY HEMISOTIDAE

Shovel-nosed frogs occur in mostly subhumid areas of sub-Saharan Africa. These frogs have pointed, hardened snouts and burrow headfirst. Males call from the ground and adhere to the female while she burrows into a subterranean incubation chamber near a pond. The eggs are laid and fertilized in the chamber, and the male digs his way out of the chamber, while the female remains with the eggs. When the eggs hatch, the female digs an escape tunnel and guides or carries the tadpoles to water.

Reed and Sedge Frogs
FAMILY HYPEROLIIDAE

The reed and sedge frogs are mainly small to medium-sized climbers found on reeds, shrubs, and trees near water in Africa (also in Madagascar and the Seychelles); a few species live on the ground or are completely aquatic. The ground-dwelling genus *Kassina* runs instead of hopping. Several sedge frogs (*Hyperolius* species) display remarkable variation in coloration; in addition to geographical variation, colors change with temperature, humidity, light intensity, and stress. Most arboreal species lay eggs in a gelatinous mass on vegetation over water. To prevent desiccation of the eggs, the females of some species fold leaves around the eggs and glue them together with sticky secretions from the oviducts. In nearly all species, the hatchling tadpoles drop into water below the nest, but the eggs of *H. obstetricans* develop directly into froglets.

Squeakers and Cricket Frogs
FAMILY ARTHROLEPTIDAE

Squeakers and cricket frogs are small to medium-sized frogs that live in large areas of sub-Saharan Africa. Most are terrestrial, and some live along mountain streams. Several species deposit terrestrial eggs that hatch into froglets. Others have aquatic eggs and tadpoles. Male Hairy frogs (*Trichobatrachus robustus*) have hairlike dermal villi that aid cutaneous respiration when the frogs attend eggs on the bottom of streams. Bush frogs of the genus *Leptopelis* bury their eggs near water. The tadpoles of this genus hatch out during heavy rain and wriggle across damp ground to standing water.

African Grass Frogs
FAMILY PTYCHADENIDAE

Mostly active terrestrial and semi-aquatic frogs from Africa that have enormous hind legs and make prodigious leaps; ornate burrowing frogs (genus Hildebrandtia, with three species), however, are stocky with short limbs. Breeding takes place in ponds and streams and development is typical. Tadpoles of the ornate frogs, *Hildebrandtia ornata*, are apparently carnivorous.

Ground Frogs, Forest Frogs, Leaf Frogs
FAMILY CERATOBATRACHIDAE

A varied grouping of frogs from Asia, formerly included within the Ranidae. Most are stocky, with short legs and live on forest floors among the leaf litter where they are well camouflaged. The Asian leaf frog (*Ceratobatrachus guentheri*) is an excellent example of a leaf-mimic, paralleling the unrelated Malaysian horned frog. Where known, reproduction in this family involves clusters of terrestrial eggs that undergo direct development into froglets. In some species, details of reproduction remain unknown.

Indian Torrent Frogs
FAMILY MICRIXALIDAE

Small frogs from India that are often associated with streams in forests and foothills. Breeding, where known, is in streams and the tadpoles are free-swimming. *Micrixalus elegans* is known from a single specimen collected in 1937 and others are little studied.

Indian Frogs
FAMILY RANIXALIDAE

Small frogs from the Western Ghats of India and surrounding hills. They live alongside forest streams and breed by laying their eggs among wet rocks. The tadpoles wriggle about in leaf litter and rocks, presumably grazing algae, and do not enter the water. Habitat change and pollution has caused the decline of at least five species.

Puddle Frogs
FAMILY PHRYNOBATRACHIDAE

Seventy-two puddle frogs are all placed in the genus *Phrynobatrachus*. They were previously placed in the family Ranidae. They are small frogs that typically live in and around shallow pools and ditches in forest clearings and grasslands. They breed in water and have free-swimming tadpoles.

African Water Frogs
FAMILY PETROPEDETIDAE

Medium-sized to very large frogs, this family includes the Goliath frog (*Conraua goliath*), the world's largest species, which can grow up to 30cm in length and weighing over 3 kilograms. It lives in cascades and torrents with clear water and a covering of the aquatic plant *Dicraea warmingii* on the rocks; the tadpoles feed exclusively on this plant. The *Petropedetes* species live in closed canopy forests and are associated with waterfalls and cascades but they attach their eggs to rocks and leaves in the splash zone and the tadpoles develop out of water but in spray or pockets of high humidity.

◁ **Left** In swampland, amid the abundant leaves of Water hyacinth, a large adult bullfrog (Lithobates catesbeiana) grapples with a Western ribbonsnake (Thamnophis proximus). Other prey items of this wide-spread North American species include crayfish, minnows, insects, and small frogs.

▷ **Right** Some Southeast Asian rhacophorid treefrogs, such as Wallace's flying frog (Rhacophorus nigropalmatus; seen here) of Malaysia and Borneo, or the Malaysian flying frog (R. reinwardtii), glide between trees by using the enormous areas of webbing between their fingers and toes as a parachute.

◁ **Below right** In the wild, the vividly colored Tomato frog (Dyscophus antongilii), a microhylid, is confined to the northwest of Madagascar, where it has been classified as Vulnerable due to habitat loss. This species has, however, been widely bred in captivity.

Pyxie Frogs, Dainty Frogs, and Relatives
FAMILY PYXICEPHALIDAE

A varied group of African frogs that includes small as well as very large species, many of them occurring in regions that are seasonally arid while others are more dependent on water. The African bullfrog (*Pyxicephalus adspersus*) lays its eggs in shallow pans and the male guards the tadpoles. If the pool in which they live begins to dry out he may excavate a deeper area or even create a channel to a larger body of water. The pyxie frogs (genus *Tomopterna*) also breed in response to heavy rain and burrow into the ground to avoid desiccation at other times, in a similar fashion to the spadefoot toads of Europe and North America, which they resemble superficially.

Stream Frogs, Wart Frogs, and Relatives
FAMILY DICROGLOSSIDAE

A large group of frogs from Africa, the Middle East and Africa, formerly placed in the Ranidae. Many are common species, sometimes associated with paddy fields, irrigation ditches and other disturbed habitats. Others, however, do not adapt as well and are endangered owing to habitat changes. Most species breed in shallow pools, flooded fields, etc. in response to rain. The breeding season may extend for many months if conditions are suitable. Males of the Bornean species *Limnonectes finchi* attend terrestrial clutches of eggs; the tadpoles are transported on the male's back to water.

Mantellas and Relatives
FAMILY MANTELLIDAE

The mantellas demonstrate a remarkable degree of convergence with the Neotropical poison frogs (Dendrobatidae); small, brightly coloured, toxic, diurnal species. The green-backed mantella (*Mantella laevigata*) even lays its eggs in tree holes or broken bamboo stems and returns to the tadpoles regularly to provide them with an infertile food egg, like members of the dendrobatid genus *Oophaga*. Bright-eyed frogs (genus *Boophis*) are mostly small and green, with large eyes but some are larger and dull in colour. They are arboreal, males call from bushes and trees along water courses or from the edges of ponds. They have free-swimming tadpoles. Members of the genus *Guibemantis* are

strongly associated with large pandanus plants and live on the leaves, laying their eggs in the water-filled leaf axils, and rarely, if ever, leaving the plant. Their tadpoles are elongated and can move over wet surfaces by flipping their whip-like tail. Some *Mantidactylus* species have dispensed with water altogether and produce tadpoles that have no mouth parts but which complete their development in a terrestrial nest without feeding.

Afro-Asian Tree Frogs
FAMILY RHACOPHORIDAE

The Afro-Asian tree frogs vary from the luridly colored red, yellow, or orange mantellas of Madagascar to large tree-dwelling species in Africa and southern Asia. Most are arboreal foam-nesters. In the southern African Gray tree frog (*Chiromantis xerampelina*; see Conscientious Parents), several males cling to a female and beat the foam, apparently competing to fertilize the eggs as the female deposits them into the nest. Several South-East Asian species of *Rhacophorus* have heavily webbed feet that they use to glide from high branches to the forest floor, in a similar fashion to the American leaf frogs, *Agalychnis* and *Phyllomedusa*, to which they are only distantly related.

Wart Frogs and Night Frogs
FAMILY NYCTIBATRACHIDAE

Poorly known frogs from the Western Ghats of southern India (genus *Nyctibatrachus*), and Sri Lanka (*Lankanectes corrugatus*). They occur along fast and slow moving streams in forests, and have free-swimming tadpoles.

True Frogs
FAMILY RANIDAE

The so-called "true" frogs have the widest range of all the families of frogs – nearly all areas of the world except the polar regions and most of Australia and South America. Many species have long, muscular legs, usually with webbed hind feet, and their streamlined bodies are ideal for jumping and swimming; the skin of amphibious species is usually smooth and brown or green. Others are terrestrial and some burrow; a few are arboreal. A number of species are able to tolerate brackish water, and the Crab-eating frog (*Rana cancrivora*) inhabits saline mangrove swamps.

Most species lay aquatic eggs, and the adults are rarely found far from water. Some *Rana* species breed for several months during the year, and males establish mating territories. However, many of those in temperate latitudes breed for just a few days, early in the year after ice has melted from the surface of ponds. During the breeding season, males of many species develop rough, swollen pads on their thumbs for gripping females. The European common frog and its North American equivalent, the Wood frog, deposit globular egg masses in communal clusters, which can contain hundreds or thousands of separate clutches. It is likely that communal spawning is an adaptation that helps to prevent the eggs from freezing during the spring. The small, black embryos absorb heat from the sun, and the thick jelly mass serves to insulate them. Temperatures within the cluster may be as much as 6°C (11°F) higher than the surrounding water.

Eggs of the Bornean genus *Meristogenys* are deposited in mountain streams, and the tadpoles have an abdominal sucker for adhering to rocks in fast-flowing water. Parental care is limited but diverse in Ranidae.

Frog and Toad Families

THE ORDER ANURA HAS 45 FAMILIES, which together contain nearly 5,453 species. Families are distinguished on the basis of the internal anatomy of adults, features of tadpoles, and reproductive biology. Evolutionarily basal anurans are usually referred to as archaeobatrachians, an informal group containing nine families and 209 species; the remaining 36 families of anurans are neobatrachians.

ARCHAEOBATRACHIANS

New Zealand frogs
Family: Leiopelmatidae

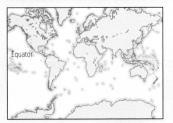

6 species in 2 genera. New Zealand and NW N America. Aquatic (*Ascapus*) or terrestrial in humid forest, and nocturnal (*Leiopelma*). Species include: **Tailed frog** (*A. truei*); **Hamilton's frog** (*L. hamiltoni*).
LENGTH: 3–5cm
COLOR: Mainly brown
BODY FORM: Stout; head wider than long; snout rounded; fingers long, slender and unwebbed; toes webbed; pupil vertical elliptical. Male *Ascaphus* have short, tail-like cloacal appendage.
BREEDING: Amplexus inguinal; fertilization internal or external; small clutches or strings of unpigmented eggs laid in streams (*Ascaphus*) or in cavities on land. Male *Leiopelma* attend eggs, which hatch as late-stage tadpoles that complete their development in terrestrial nests. *Ascaphus* tadpoles are free-swimming with large suctorial oral disks and feed on undersides of rocks in streams. Larval stage in *Ascaphus* lasts 1–4 years.
CONSERVATION STATUS: **Hamilton's frog** is Endangered and **Archey's frog** (*L. archeyi*) Critically Endangered. Two other *Leiopelma* species Vulnerable.

Clawed frogs and Surinam toad
Family: Pipidae

31 species in 5 genera. Tropical S America and sub-Saharan Africa. Aquatic; noctur-

nal and diurnal. Species include: **African clawed frog** (*Xenopus laevis*), **Surinam toad** (*Pipa pipa*).
LENGTH: 3–17cm
COLOR: Brown, gray, green; venter white.
BODY FORM: Broad, depressed; snout rounded; fingers long, unwebbed; toes long, nearly fully webbed; keratinized claws on toes of African clawed frogs; pupil round.
BREEDING: Amplexus inguinal; fertilization external. Mating occurs in water and consists of elaborate maneuvers resulting in the eggs being deposited on the surface of the water (clawed frogs) or on the back of the female (Surinam toads). In some species of Surinam toads the eggs hatch as froglets; in all others the free-swimming tadpoles lack keratinized mouthparts and are suspension feeders.
CONSERVATION STATUS: **Cape clawed toad** (*Xenopus gilli*) and **Myers' Surinam toad** (*Pipa myersi*) are Endangered.

Mexican burrowing toad
Family: Rhinophrynidae

1 species: *Rhinophrynus dorsalis*. Rio Grande valley, Texas, USA, to Costa Rica. Terrestrial, burrowing; nocturnal.
LENGTH: 8cm
COLOR: Dark gray with orange-red markings.
BODY FORM: Globular; head small; snout pointed; limbs short; fingers short and toes long, both basally webbed; "spade" on hind feet; pupil vertically elliptical.
BREEDING: Amplexus inguinal; fertilization external. Eggs laid in temporary ponds; free-swimming tadpoles aggregate in schools.

Midwife toads and painted frogs
Family: Alytidae

11 species in 2 genera. W, C, and S Europe; NW Africa; Asia Minor. Terrestrial (midwife toads) or edges of rocky streams (painted frogs); nocturnal.

Species and genera include: **Common midwife toad** (*Alytes obstetricans*), **Iberian midwife toad** (*A. cisternasii*), **Mallorcan midwife toad** (*A. muletensis*), **European painted frog** (*Discoglossus pictus*), **Israel painted frog** (*D. nigriventer*).
LENGTH: 4–8cm
COLOR: Primarily brown
BODY FORM: Stout in midwife toads, slender in painted frogs; head as long as, or longer than, wide; snout blunt in midwife toads, acutely rounded in painted frogs; fingers short, unwebbed; toes long, basally webbed; pupil vertically elliptical.
BREEDING: Amplexus inguinal; fertilization external. Female midwife toads produce up to 100 eggs in strings, which are fertilized by the male and entwined around his legs; the male carries the eggs for 3–5 weeks before returning to water, whereupon the eggs hatch as free-swimming tadpoles. Painted frogs deposit

500–1,000 eggs in slow-moving streams, where the tadpoles develop for 3–8 weeks before metamorphosing.
CONSERVATION STATUS: The **Mallorcan midwife toad** is Vulnerable, one other *Alytes* species is Vulnerable, and four **painted frogs** are Near Threatened. The **Israel painted frog** is listed as extinct.

Fire-bellied toads and barbourulas
Family: Bombinatoridae

10 species in 2 genera. Europe, E and SE Asia, Borneo, Philippines (Palawan). Semiaquatic in marshes (fire-bellied

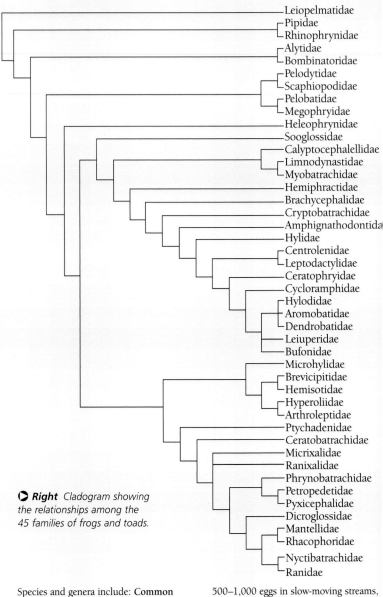

Right *Cladogram showing the relationships among the 45 families of frogs and toads.*

toads) or in montane streams (barbourulas); nocturnal. Species and genera include: **fire-bellied toads** (genus *Bombina*), **barbourulas** (genus *Barbourula*).
LENGTH: 4–10cm
COLOR: Brown, gray, or green; fire-bellied toads vividly marked below with red, orange, yellow, white, and black.
BODY FORM: Depressed; head as wide as long; snout blunt; fingers short and unwebbed; toes moderately long, webbed; pupil vertically elliptical. Dorsum warty in fire-bellied toads, nearly smooth in barbourulas.
BREEDING: Amplexus inguinal; fertilization external. Female fire-bellied toads deposit 60–200 pigmented eggs in numerous small clutches attached to vegetation in marshes. After 4–10 days, free-swimming tadpoles hatch and spend 35–45 days in water prior to metamorphosis. Female barbourulas produce 70–80 large ova; presumably the eggs are deposited in, and tadpoles develop in, streams.
CONSERVATION STATUS: Three species of *Bombina* (*fortinuptialis, lichuanensis, microdeladigitora*) Vulnerable.

Parsley frogs
Family: Pelodytidae

3 species in 1 genus. SW Europe, SW Asia. Terrestrial; nocturnal, except for breeding. Species include: **Parsley frog** (*Pelodytes punctatus*).
LENGTH: 4–6cm
COLOR: Green
BODY FORM: Stout, depressed; snout acutely rounded; fingers long, unwebbed; toes long, basally webbed; pupil vertically elliptical.
BREEDING: Amplexus inguinal; fertilization external. Aquatic eggs (up to 1,600) in ponds hatch as free-swimming tadpoles, which metamorphose in 75–80 days.

American spadefoot toads
Family: Scaphiopodidae

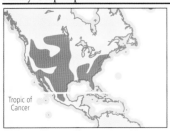

7 species in 2 genera. Terrestrial and fossorial; nocturnal. Species include:

Eastern spadefoot (*Scaphiopus holbrookii*) **Plains spadefoot** (*Spea bombifrons*).
LENGTH: 5–8cm
COLOR: Brown, gray, green.
BODY FORM: Stout; snout blunt; limbs short; fingers short, unwebbed; toes long, basally webbed; prominent sharp-edged "spade" on hind feet; pupil vertically elliptical.
BREEDING: Amplexus inguinal; fertilization external. Aquatic eggs in shallow ponds hatch as free-swimming tadpoles that develop rapidly (6–23 days).

Eurasian spadefoot toads
Family: Pelobatidae

4 species in 1 genus. Terrestrial and fossorial; nocturnal. Species include: **Common spadefoot toad** (*Pelobates fuscus*).
LENGTH: 5–8cm
COLOR: Brown, gray, green.
BODY FORM: Stout; snout blunt; limbs short; fingers short, unwebbed; toes long, basally webbed; prominent, sharp-edged "spade" on hind feet; pupil vertically elliptical.
BREEDING: Amplexus inguinal; fertilization external. Aquatic eggs in shallow ponds hatch as free-swimming tadpoles, which may overwinter twice before metamorphosing.
CONSERVATION STATUS: **Moroccan spadefoot toad** (*Pelobates varaldii*) is Endangered.

Asian toadfrogs
Family: Megophryidae

136 species in 11 genera. SE Asia, Philippines, Borneo, Sumatra, Java. Primarily terrestrial in forests; many are streamside inhabitants; primarily nocturnal. Species and genera include: **Asian horned toad** (*Megophrys nasuta*) **Himalayan stream frog** (*Scutiger boulengeri*) **Pope's spiny toad** (*Vibrissophora liui*).
LENGTH: 1–12cm
COLOR: Green, brown, yellow, gray.
BODY FORM: Stout; limbs short to long; snout bluntly to acutely rounded; fingers long, unwebbed; toes long,

basally to nearly fully webbed; pupil vertically elliptical.
BREEDING: Amplexus inguinal; fertilization external. Eggs laid in water and hatch into free-swimming larvae; tadpoles of many streamdwellers have large suctorial mouths and low fins.

NEOBATRACHIANS

Ghost frogs
Family: Heleophrynidae

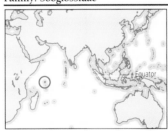

6 species in 1 genus. S Africa. Fast-flowing rocky streams; nocturnal. Species include: **Cape ghost frog** (*Heleophryne purcelli*).
LENGTH: 6.5cm
COLOR: Mottled green and brown.
BODY FORM: Slender, depressed; head flattened; snout blunt; limbs long; digits expanded distally; pupil vertically elliptical.
BREEDING: Amplexus inguinal; fertilization external. Unpigmented eggs attached to rocks in streams. Tadpoles with large, suctorial mouths cling to rocks in streams and require 1–2 years to metamorphose.
CONSERVATION STATUS: **Hewitt's ghost frog** (*Heleophryne hewitti*) and the **Table Mountain ghost frog** (*H. rosei*) both Critically Endangered.

Seychelles frogs and purple frog
Family: Sooglossidae

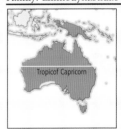

5 species in 2 genera, *Sooglossus* and *Nasikabatrachus*. Seychelles Islands and Western Ghats, India. Terrestrial, in moist forests.
LENGTH: 2–5cm
COLOR: Tan to brown (*Sooglossus*); purple (*Nasikabatrachus*).
BODY FORM: Moderately slender in *Sooglossus*, plump in *Nasikabatrachus*; tips of fingers pointed or blunt, pupil horizontally elliptical.
BREEDING: Amplexus inguinal; fertilization external; eggs deposited on land. Eggs hatch as froglets of non-feeding tadpoles that wriggle onto back of female. Breeding in *Nasikabatrachus* unknown.

CONSERVATION STATUS: All Seychelles frogs Vulnerable, **Purple frog** (*N. sahyadrensis*) Endangered.

Central Chilean frogs
Family: Calyptocephalellidae

4 species in 2 genera. Mountains and forests of Central Chile. Aquatic or terrestrial close to water, nocturnal. Species include: **Chilean helmeted frog** (*Caudiverbera caudiverbera*); ***Telmatobufo australis***.
LENGTH: 12–32cm
COLOR: Dull gray or brown.
BODY FORM: Heavily built; fingers unwebbed; toes basally webbed.
BREEDING: Chilean helmeted frog lays up to 10,000 eggs laid in shallow water. Tadpoles grow to 15cm and take up to 2 years to develop, but typically 5–12 months. Breeding in *Telmatobufo* unknown.
CONSERVATION STATUS: The three *Telmatobufo* species are listed as Vulnerable, Endangered, and Critically Endangered.

Australian ground frogs
Family: Limnodynastidae

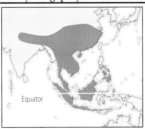

44 species in 8 genera. Australia and New Guinea. Terrestrial or fossorial, in rain forests and in dry forests and grasslands; nocturnal. Species and genera include: **Baw-baw frog** (*Philoria frosti*), **bullfrog** (*Limnodynastes dumerilii*), **Desert spadefoot toad** (*Notaden nichollsi*), **Trilling frog** (*Neobatrachus centralis*), **sphagnum frogs** (genus *Kyarranus*).
LENGTH: 2–10cm
COLOR: Mostly brown or tan, with darker markings dorsally; some with brightly patterned venters.
BODY FORM: Most are robust, with relatively short legs (long in *Mixophyes*), blunt heads, and little webbing; pupil horizontally or vertically elliptical.
BREEDING: Amplexus inguinal (axillary in *Mixophyes*); fertilization external. Eggs deposited as clumps at edges of streams (*Mixophyes*), as strings in ponds (*Neobatrachus* and *Notaden*), or as foam

nests in burrows or on water; in those deposited as foam nests in burrows, the tadpoles develop in the nests, whereas others have free-swimming tadpoles. CONSERVATION STATUS: Several **sphagnum frogs** (*Philoria*) are listed as Endangered.

Australian toadlets and water frogs
Family: Myobatrachidae

82 species in 11 genera. Australia and S New Guinea. Terrestrial or fossorial, in rain forests, grasslands, and deserts; nocturnal, except for *Taudactylus*. Species and genera include: **Pouched frog** (*Assa darlingtoni*), **Sandhill frog** (*Arenophryne rotunda*), **toadlets** (genera *Pseudophryne* and *Uperoleia*).
LENGTH: 3.5cm
COLOR: Highly variable, but most are shades of brown or gray.
BODY FORM: Most are robust, with relatively short legs, blunt heads, and little webbing on the feet; pupil horizontally elliptical.
BREEDING: Amplexus inguinal; fertilization external. Aquatic eggs and tadpoles in most species. Burrowers (*Arenophryne* and *Myobatrachus*) deposit large, unpigmented eggs in underground chambers, where the eggs undergo direct development. *Assa* deposits eggs on the ground; hatchling tadpoles wriggle into pouches in the male and undergo direct development. *Taudactylus* and *Rheobatrachus* inhabit mountain streams; the former has free-living tadpoles, whereas the tadpoles of the latter are swallowed by the female and develop into froglets in the stomach. CONSERVATION STATUS: Several **toadlets** (*Pseudophryne*) are listed as Endangered or Critically Endangered. All six torrent frogs are listed, including the **Mount Glorious Torrent frog**, which is probably extinct. Both **gastric brooding frogs** (*Rheobatrachus silus* and *R. vitellinus*) are also extinct.

Casque-headed frogs
Family: Hemiphractidae

6 species in 1 genus. C and S America. Terrestrial or semi-arboreal in rain forests. Species include **Spix's casqued tree frog** (*Hemiphractus scutatus*).

LENNGTH: 4–8cm
COLOR: Various shades of brown.
BODY FORM: Slender body with a massive, triangular head with fleshy horns over the eyes and on the snout. Tips of digits slightly expanded. Pupil horizontally elliptical.
BREEDING: On land. Females carry the eggs on their backs and they hatch directly into tadpoles.
CONSERVATION STATUS: **Johnson's casqued frog** (*Hemiphractus johnsoni*) is listed as Endangered.

Three-toed toadlets, litter frogs, and their relatives
Family: Brachycephalidae

803 species in 17 genera. Southern North America, Central America, South America and Caribbean region. Variable; terrestrial or arboreal in rain forest, mostly nocturnal but sometimes diurnal.
SPECIES INCLUDE: **Bold frog** (*Brachycephalus ephippium*); **Puerto Rico live-bearing frog** (*Eleutherodactylus jasperi*).
LENGTH: 1–4cm
COLOR: Variable; sometimes orange or green, usually shades of brown.
BODY FORM: Variable. Slightly expanded toe pads or without pads. Pupil horizontally elliptical. Toadlets (*Brachycephalus*) have only 2 or 3 digits on forefeet.
BREEDING: Fertilization usually external, except in a few *Eleutherodactylus* species, of which *E. jasperi* gives birth to live young. Otherwise, terrestrial eggs hatch directly into froglets.
CONSERVATION STATUS: Several species listed as Vulnerable, Endangered or Critically endangered.

Backpack frogs
Family: Cryptobatrachidae

21 species in 2 genera. Northern South America. Terrestrial in mist forests, usually associated rocks in or near streams and waterfalls. Species include: **Fuhrmann's backpack frog** (*Cryptobatrachus*

> **NOTES**
> Length = length from snout to vent.
> Approximate nonmetric equivalents:
> 10cm = 4in 1kg = 2.2lb

fuhrmanni), *Stefania scalae*.
LENGTH: 2–4cm
COLOR: Shades of brown.
BODY FORM: Bluntly rounded snout. Horizontally elliptical pupils. Toes slightly expanded.
BREEDING: Amplexus axillary; fertilization external; eggs carried on female's back hatch directly into froglets.
CONSERVATION STATUS: *Cryptobatrachus boulengeri* is listed as Endangered. Some other species are known from only a few specimens.

Marsupial frogs
Family: Amphignathodontidae

58 species in 2 genera. Central and South America. Terrestrial and arboreal in rain forests and montane grasslands; mostly nocturnal. Species include **Marsupial frog** (*Gastrotheca marsupiata*).
LENGTH: 3–6cm
COLOR: Variable, mostly brown but some are green or mottled brown and green.
BODY FORM: Stocky, with relatively short legs, little or no webbing on feet and expanded discs on toes. Females have an enclosed pouch on back.
BREEDING: Amplexus axillary; fertilization external; eggs maneuvered into pouch on female's back may be released as well-grown tadpoles or retained until they metamorphose, depending on species.
CONSERVATION STATUS: Several *Gastrotheca* species, and **Fitzgerald's marsupial frog** (*Flectonotus fitzgeraldi*) are listed as Endangered.

Tree frogs
Family: Hylidae

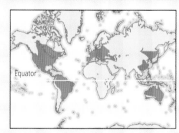

830 species in 46 genera. N to S America, W Indies, Eurasia, N Africa, Australia, New Guinea. Primarily arboreal and nocturnal; some terrestrial or fossorial. Examples include: **Egg-eating frog** (*Osteocephalus oophagus*); **Fringe-limbed tree frog** (*Ecnomiohyla miliaria*); **Gladiator frog** (*Hypsiboas rosenbergi*); **Australasian tree frogs** (*Philoria species*); **chorus frogs** (genus *Pseudacris*); **cricket frogs** (genus *Acris*); **leaf frogs** (genera *Agalychnis*,

Phyllomedusa, and relatives).
LENGTH: 1.2–14cm
COLOR: Mostly green, yellow, or brown; many with bright flash colors on flanks or limbs.
BODY FORM: Usually slender and slightly depressed, with long limbs and webbed feet; tips of digits with expanded adhesive disks; some genera have modified skulls, with skin adherent to underlying bones; *Cyclorana* and *Pternohyla* robust with short limbs, little webbing on feet, and small disks; pupil horizontally elliptical in most, vertically elliptical in *Phyllomedusa* and *Nyctimystes*.
BREEDING: Amplexus axillary; fertilization external. Most species deposit eggs in water, but others deposit eggs on vegetation above ponds or streams, and some lay eggs in bromeliads or water in cavities in trees; all of these have free-swimming tadpoles. Male gladiator frogs, such as *Hypsiboas boans*, construct and defend basin-like nests in which tadpoles develop. The tadpoles of *Anotheca spinosa*, *Osteopilus brunneus*, and some *Osteocephalus* species develop in tree holes or bromeliads and feed on unfertilized eggs deposited by females.
CONSERVATION STATUS: Very many, especially from Central America and Australia, are listed as Vulnerable, Endangered, or Critically Endangered.

Glass frogs
Family: Centrolenidae

144 species in 4 genera. C and S America. Arboreal in humid forests (mostly in mountains); nocturnal.
Length: 2–8cm
COLOR: Mostly green, with or without darker or paler flecks; skin on venter nearly transparent.
BODY FORM: Slender; head large with prominent eyes; fingers and toes partially webbed; expanded adhesive discs on digits; pupil horizontally elliptical.
BREEDING: Amplexus axillary; fertilization external. Small clutches of eggs deposited on vegetation above streams; hatchlings drop into water and develop in streams. *Allophryne ruthveni* apparently lays its eggs in water.

South American water frogs
Family: Leptodactylidae

91 species in 4 genera. Extreme southern N America, C America, S America, and W Indies. Terrestrial, burrowing, aquatic, and arboreal; mostly nocturnal. Species include: **South American bullfrog** (*Leptodactylus pentadactylus*), **White-lipped frog** (*L. albilabris*).
LENGTH: 1–18cm
COLOR: Highly variable.
BODY FORM: Moderately slender bodies with long limbs and little or no webbing; pupil horizontally elliptical.
BREEDING: Amplexus axillary; fertilization external; eggs laid in water or aquatic foam nests, hatching into free-swimming tadpoles.
CONSERVATION STATUS: **Dominican ditch frog** (*Leptodactylus dominicensis*) Endangered.

South American horned frogs
Family: Ceratophryidae

86 species in 8 genera. South America, in rain forests, grasslands and semi-desert regions; terrestrial and burrowing; nocturnal. Species include: **horned frogs** (genus *Ceratophrys*); **Budgett's frog** (*Lepidobatrachus laevis*).
LENGTH: 4–12 cm
COLOUR: Brown or gray, sometimes marked with green and, occasionally, red.
BODY FORM: Rotund or globular with short heads and wide mouths; short limbs with webbing on hind feet; eyes small and positioned dorsally; fleshy projections over eyes in Ceratophrys. Skin granular or smooth.
BREEDING: Explosive breeders; amplexus axillary; fertilization internal. Eggs laid in water develop into free-swimming tadpoles that may grow very large before metamorphosing.
CONSERVATION STATUS: At least 20 species of **water frogs**, *Telmatobius*, are Endangered, and another 8 are Critically Endangered.

Smooth horned frogs, Darwin's frogs, and relatives
Family: Cycloramphidae

96 species in 13 genera. Southern S America in subtropical and temperate forests; terrestrial. Species and genera include **Darwin's frog** (*Rhinoderma darwinii*); **button frogs** (genus *Cycloramphus*).
LENGTH: 3–6cm
COLOR: Mostly brown or tan; sometimes green.
BODY FORM: Varied. Darwin's frog is stocky with a pointed snout; other species stocky with toad-like bodies and short heads; toes with little or no webbing; pupils horizontally elliptical.
BREEDING: Amplexus axillary; fertilization external. Terrestrial eggs are attended by male for about 20 days; males gather hatching tadpoles in their mouths where they continue development until metamorphosed. Other species breed aquatically and have free-swimming tadpoles.
CONSERVATION STATUS: **Darwin's frog** listed as Vulnerable; *Rhinoderma rufum* is Critically Endangered and may be extinct.

Tree toads, spiny-thumb frogs, and big-tooth frogs
Family: Hylodidae

37 species in 3 genera. NW to S Brazil and N Argentina, in forests, near streams. Terrestrial, on rocks and forest floor; males, and possibly females, of most are diurnal. *Genera* are spiny-thumb frogs (genus *Crossodactylus*), big-tooth frogs (genus *Megaelosia*), and tree toads (genus *Hylodes*)
LENGTH: 3–cm
COLOR: Brown, usually with darker marking markings.
BODY FORM: Slender or moderately stocky; toes webbed; pupils horizontal.
BREEDING: Virtually unknown. Males call from streamsides, tadpoles have been found in running water.
CONSERVATION STATUS: Mostly Data Deficient. Several species are disappearing from parts of their ranges.

Poison dart frogs
Family: Dendrobatidae

163 species in 11 genera. C America and tropical S America, in rain forests. Terrestrial and arboreal; diurnal. Species include **Strawberry frog** (*Oophaga pumilio*), **Green and black poison dart frog** (*Dendrobates auratus*).
LENGTH: 1.2–6cm
COLOR: Red, yellow, blue, or green, usually with black markings.
BODY FORM: Slender, with short heads and slender limbs; fingers unwebbed; toes with or without webbing; tips of digits slightly expanded, with pair of scutes on dorsal surface; pupil horizontally elliptical. Brightly colored genera (*Dendrobates, Epipedobates, Minyobates,* and *Phyllobates*) produce strong skin toxins.
BREEDING: Amplexus absent; fertilization external. Eggs deposited on ground and attended by male or female; in most species, hatchling tadpoles wriggle onto back of parent and are transported to streams or water in bromeliads or other water-filled structures where they are free-swimming. In some species of *Dendrobates*, female provides infertile eggs as food for tadpoles in bromeliads.
CONSERVATION STATUS: **Yellow poison dart frog** (*Phyllobates terribilis*) and many others listed as Endangered.

Dwarf frogs, foam frogs, and their relatives
Family: Leiuperidae

76 species in 7 genera. S Mexico to C Chile and C Argentina, in a variety of habitats, including disturbed and polluted places. Terrestrial and nocturnal. Species and genera include **Chile four-eyed frog** (*Pleurodema thaul*), **Tungara frog** (*Engystomops pustulosus*), **foam frogs** (genus *Physalaemus*).
LENGTH: 3–6cm
COLOR: Mostly brown, sometimes with green markings.
BODY FORM: Slender, with a pointed snout and streamlined body, to short and toad-like. Little or no webbing on toes; pupil horizontal.

BREEDING: Extended breeding season, often in temporary pools. Eggs laid in gelatinous strings or in a foam nest.
CONSERVATION STATUS: Many Endangered.

Rocket frogs
Family: Aromobatidae

91 species in 6 genera. Lower C America, S America. Terrestrial and partially arboreal in tropical forests, often near rivers and streams. Diurnal. Species include: **Trinidad rocket frog** (*Mannophryne trinitatus*); **Dunn's rocket frog** (*Prostherapis dunni*).
LENGTH: 1.5–4cm
COLOR: Various shades of brown, black, and cream.
BODY FORM: Slender, with slender limbs; no webbing on feet or hands; small toe-pads; pupil horizontally elliptical.
BREEDING: Amplexus axillary; fertilization external. Terrestrial eggs are attended by male or female; hatching tadpoles wriggle onto back of parent and are carried to small bodies of water, often in bromeliad plants. Non-feeding tadpoles develop in terrestrial nest in some species.

True toads, harlequin frogs, and relatives
Family: Bufonidae

495 species in 47 genera. All continents except Australia and Antarctica; Marine toad introduced on many islands and in W Australia. Mostly terrestrial, but a few aquatic or arboreal; nocturnal and diurnal. Species and genera include: **American toad** (*Anaxyrus americanus*); **European common toad** (*Bufo bufo*); **Golden toad** (*Ollotis pereglenes*); **Marine toad** (*Chaunus marinus*); **harlequin frogs** (genus *Atelopus*); **slender toads** (genus *Ansonia*).
LENGTH: 2–23cm
COLOR: Most true toads are shades of brown; many diurnal species of *Atelopus* and *Melanophryniscus* are brightly-colored combinations of yellow, red, and black.
BODY FORM: Robust and short-legged with warty skin in most true toads; smooth skin, slender bodies, and long limbs in many harlequin frogs; pupil horizontally elliptical.
BREEDING: *Amplexus* usually axillary;

fertilization external except in *Mertensophryne, Nimbaphrynoides* and some *Nectophrynoides* species. Eggs typically deposited as strings in water, hatching into free-swimming tadpoles; others (for example *Osornophryne*) have terrestrial eggs that hatch as toadlets. Those species with internal fertilization give birth to a few living young. CONSERVATION STATUS: Golden toad, *Atelopus ignescens,* and *A. longirostris* are thought to be Extinct and most other **harlequin frogs** are Extinct, Critically Endangered, or Endangered. Two species of **livebearing toad** from Mount Nimba (genus *Nimbaphrynoides*) are Critically Endangered, and there are many more Endangered species.

Narrow-mouthed frogs
Family: Microhylidae

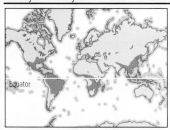

419 species in 48 genera. SE N America, C and S America, sub-Saharan Africa, Madagascar, SE Asia, Indonesia, New Guinea, N Australia. Mostly terrestrial or fossorial, in rain forest, savannas, and temperate and montane forests; some arboreal; nocturnal. Species and genera include: **Tomato frog** (*Dyscophus antongilii*), **narrow-mouth toads** (genera *Gastrophryne* and *Microhyla*).
LENGTH: 0.9–8.8cm
COLOR: Mostly gray or brown, but some with contrasting patterns of black and red or yellow.
BODY FORM: Terrestrial forms have stout bodies and small heads, with pointed or blunt snouts; toes with little webbing; pupil horizontally elliptical.
BREEDING: Amplexus axillary or inguinal; fertilization external. Eggs laid on ground; hatch into froglets or nonfeeding tadpoles; other genera deposit eggs in water and have free-swimming tadpoles.
CONSERVATION STATUS: **Black microhylid** (*Melanobatrachus indicus*) is Endangered; several others Vulnerable or Endangered.

African rain frogs
Family: Brevicipitidae

26 species in 5 genera. Sub-Saharan Africa. Fossorial and nocturnal in grasslands,

---

forests, and semidesert regions; nocturnal.
LENGTH: 2.5–6cm
COLOR: Mostly brown or tan.
BODY FORM: Globular, with short heads and limbs. No webbing on hands or feet; pupils horizontally elliptical.
BREEDING: Amplexus inguinal; fertilization external. Male "glues" himself to female's back. Eggs laid in underground chambers hatch directly into froglets.

Shovel-nosed frogs
Family: Hemisotidae

9 species in 1 genus. Sub-Saharan Africa. Terrestrial burrowers, mostly in savannas and scrub forest; nocturnal. Genera include: **African shovel-nosed frogs** (genus *Hemisus*).
LENGTH: 3–8cm
COLOR: Olive green or brown, usually with yellow or orange markings.
BODY FORM: Robust, with pointed snout; tympanum absent; limbs short;

Reed frogs
Family: Hyperoliidae

199 species in 17 genera. Sub-Saharan Africa, Madagascar, Seychelles Islands. Mostly arboreal (some terrestrial) in rain forests and savannas; nocturnal. Species and genera include: **Golden leaf-folding frog** (*Afrixalus brachycnemis*), **Marbled reed frog** (*Hyperolius marmoratus*), **Seychelle Islands tree frog** (*Tachycnemis seychellensis*), **sedge frogs** (genus *Hyperolius*).
LENGTH: 1.7–8.7cm
COLOR: Highly variable, in shades of green or brown; many have brightly-colored (yellow, red, orange) markings.
BODY FORM: Usually slender, with moderately long limbs and expanded disks on digits; pupil horizontally elliptical.
BREEDING: Amplexus axillary; fertilization external. Eggs are deposited in water, on leaves, in foam nests on vegetation, or in cavities on land adjacent to water; all have free-swimming tadpoles.
CONSERVATION STATUS: **Knysna banana frog** (*Afrixalus knysnae*) and several *Hyperolius* species listed as Endangered.

Squeakers, bush frogs, and relatives
Family: Arthroleptidae

129 species in 8 genera. Sub-Saharan Africa, in forests and grasslands. Terrestrial or arboreal; some fossorial. Most nocturnal. Examples include: **West African squeaker** (*Arthroleptis poecilonotus*); **Hairy frog** (*Trichobatrachus robustus*); **bush frogs** (genus *Leptopelis*).
LENGTH: 2–6.5cm
COLOR: Mostly shades of brown; some bush frogs are green or green and brown.
BODY FORM: Varied; stout bodies; moderately long limbs; expanded toe disks in arboreal species (*Leptopelis*).
BREEDING: Amplexus axillary; fertilization external. Terrestrial eggs may develop directly into froglets or hatch and wriggle to water. Hairy frog breeds in streams, eggs laid on bottom and hatch into carnivorous tadpoles.
CONSERVATION STATUS: **Long-toed bush frog** (*Leptopelis xenodactylus*) and two other **bush frogs** Endangered; several **squeakers** Endangered.

African grass frogs
Family: Ptychadenidae

52 species in 3 genera. Africa, mostly south of the Sahara and Madagascar. Introduced onto several Indian Ocean islands. Terrestrial in grasslands and forests; diurnal and nocturnal.
LENGTH: 3–6cm
COLOR: Mostly brown, gray, or olive.
BODY FORM: Streamlined; narrow heads and pointed snouts; long legs; long partially webbed toes; some (*Hildebrandtia*) stocky, with shorter limbs; pupil horizontally elliptical; folded ridges of skin on back.
BREEDING: Axillary amplexus; fertilization external; aquatic eggs hatch into free-swimming tadpoles.
CONSERVATION STATUS: **Broadley's grass frog** (*Ptychadena broadleyi*) and **Newton's grass frog** (*P. newtoni*) listed as Endangered.

Ground frogs, forest frogs, and leaf frogs
Family: Ceratobatrachidae

73 species in 6 genera. W China, SE Asia, New Guinea and neighbouring islands. Examples include: **Solomon's leaf frog** (*Ceratobatrachus guentheri*); **forest frogs** (genus *Platymantis*).
LENGTH: 2–8cm
COLOR: Mostly shades of brown.
BODY FORM: Varied; mostly stocky; short legs; limited webbing between toes; pupils horizontally elliptical. **Solomon's leaf frog** has a triangular leaf-like body form.
BREEDING: Amplexus axillary; fertilization external; terrestrial eggs develop directly into froglets.
CONSERVATION STATUS: Several **forest frogs** (*Platymantis*) listed as Endangered.

Indian torrent frogs
Family: Micrixalidae

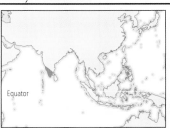

9 species in 1 genus. India. Terrestrial in hill forests, along streams and torrents. Species include **Brown torrent frog** (*Micrixalus fuscus*).
LENGTH: 3–5cm.
COLOR: Shades of brown.
BODY FORM: Moderately slender; pointed head; feet partially webbed; toes end in expanded digits; pupil horizontally elliptical.
BREEDING: Presumed amplexus axillary; fertilization external; aquatic eggs laid in streams; free-swimming tadpoles.
CONSERVATION STATUS: **Gadgil's torrent frog** (*Micrixalus gadgili*) Endangered.

Indian frogs
Family: Ranixalidae

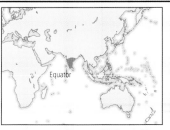

10 species in 1 genus. Central and SW India, especially in the Western Ghats

mountains, in forests near streams; terrestrial and nocturnal. Species include the **Toad-skinned frog** (*Indirana phrynoderma*).
LENGTH: 3–5cm
COLOR: Brown.
BODY FORM: Stocky; slightly warty back; rounded snout; horizontal pupils; webbing on toes.
BREEDING: Eggs are laid on wet rocks and the tadpoles live in the splash zone.
CONSERVATION STATUS: **Toad-skinned frog** and one other are Critically Endangered; another three are Endangered.

Puddle frogs
Family: Phrynobatrachidae

72 species in 1 genus. Sub-Saharan Africa. Terrestrial in forests and grasslands; mostly nocturnal. Examples include the **Natal puddle frog** (*Phrynobatrachus natalensis*).
LENGTH: 1.6–3.5cm
COLOR: Brown
BODY FORM: Moderately stout; variable amounts of webbing on hind feet; pupils horizontally elliptical.
BREEDING: Amplexus axillary; fertilization external; aquatic eggs laid in shallow pools or streams; free-swimming tadpoles.
CONSERVATION STATUS: *Phrynobatrachus ghanensis, P. irangi, P. kreffti,* and *P. pakenhami* are all listed as Endangered. Several others poorly known.

African water frogs
Family: Petropedetidae

16 species in 2 genera. Sub-Saharan Africa. Terrestrial and semi-aquatic in forests, near streams and waterfalls; nocturnal. Examples include **Goliath frog** (*Conraua goliath*), **Sierra Leone water frog** (*Petropedetes natator*).
LENGTH: 4–32cm
COLOR: Brown, olive.
BODY FORM: Mostly stocky; moderately long legs; varied degrees of webbing on back feet; toes expanded in some; pupils horizontally elliptical.
BREEDING: Amplexus axillary; fertilization external; aquatic eggs laid in rivers and streams where known; tadpoles free-swimming.

CONSERVATION STATUS: The **Goliath frog** is Endangered; *Conraua derooi* is Critically Endangered.

Pyxie frogs, dainty frogs, African bullfrogs, and relatives
Family: Pyxicephalidae

69 species in 13 genera. Africa, in a variety of habitats, including mountains and grasslands; nocturnal. Examples include **African bullfrog** (*Pyxicephalus adspersus*); **Cape dainty frog** (*Cacosternum capense*).
SIZE: 2.3–23cm.
COLOR: Brown, gray, reddish, olive, or green.
BODY FORM: Stout or slender; limbs moderately long to short; back covered with warts or ridges of raised skin; variable amounts of webbing on hind feet; pupils horizontally elliptical.
BREEDING: Amplexus axillary; fertilization external; aquatic eggs laid in shallow, often temporary, pools and pans; tadpoles free-swimming.
CONSERVATION STATUS: **Bale Mountain frog** (*Ericabatrachus baleensis*) listed as Endangered.

Stream frogs, wart frogs, and relatives
Family: Dicroglossidae

163 species in 15 genera. Africa, Middle East, India, Sri Lanka, China, South East Asia, Philippines, S Pacific islands, Japan, in a variety of moist habitats, usually near streams, and often in disturbed places; nocturnal or diurnal. Examples include **Paddy frog** (*Fejervarya limnocharis*); **wart frogs** (genus *Limnonectes*); **Oriental frogs** (genus *Occidozyga*).
SIZE: 3–12.5cm
COLOR: Brown or gray.
BODY FORM: Streamlined with pointed snout; most with warty back; hind feet partially webbed; pupils horizontally elliptical.
BREEDING: Amplexus axillary; fertilization external; aquatic eggs mostly laid in shallow water; tadpoles free-swimming.
CONSERVATION STATUS: Several species Endangered.

Mantellas, bright-eyed frogs, and Madagascan forest frogs
Family: Mantellidae

165 species in 12 genera. Madagascar and Comoros. Terrestrial and arboreal in forests; nocturnal (mantellas diurnal). Examples include **Golden mantella** (*Mantella aurantiaca*); **Green bright-eyed frog** (*Boophis viridis*).
SIZE: 1.5–6cm
COLOR: Brown, bright green, orange, multicoloured.
BODY FORM: Slight; long legs; toes end in expanded disks; hind feet partially webbed; eyes large in some, small in others; pupils horizontally elliptical. **Madagascan bullfrog** is stocky with short limbs.
BREEDING: Amplexus axillary (absent in mantellas), fertilization external. Otherwise variable: eggs laid on leaves, in water, or on the ground; tadpoles may drop or be washed into water, or live in leaf axils, treeholes, etc.; **Green-backed mantella** (*Mantella laevigata*) breeds in hollow bamboo stems and feeds tadpoles on infertile eggs.
CONSERVATION STATUS: *Boophis williamsi* and several **mantellas** (*Mantella aurantiaca milotympanum, expectata, viridis*) are Critically endangered. Other *Mantella* and several *Mantidactylus* species Endangered.

Afro-Asian tree frogs, bush frogs
Family: Rhacophoridae

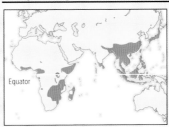

278 species in 10 genera. Tropical Africa, Asia, SE Asia, China, Japan. Terrestrial and arboreal, in forests, grasslands; nocturnal. Examples include **African foam-nest frog** (*Chiromantis xerampelina*), **Malaysian flying frog** (*Rhacophorus reinwardtii*), and **Asian foam-nest frog** (*Polypedates leucomystax*).
SIZE: 2–10cm.
COLOR: Gray, brown, olive, bright green.
BODY FORM: Mostly slender with long limbs and webbed feet, 'flying' species heavily so. Large eyes; pupils horizontally elliptical.
BREEDING: Amplexus axillary; fertilization external; arboreal species deposit eggs in arboreal foam nests; hatchlings drop into

water and develop as free-swimming tadpoles.
CONSERVATION STATUS: Up to 18 **bush frogs** (genus *Philautus*) believed to be Extinct. Many other species Critically Endangered or Endangered.

Wart frogs and night frogs
Family: Nyctibatrachidae

13 species in 2 genera. S India and Sri Lanka. Terrestrial, near streams in forests; nocturnal. Examples include **Sri Lanka warty frog** (*Lankanectes corrugatus*), **night frogs** (genus *Nyctibatrachus*).
LENGTH: 4–7 cm
COLOR: Brown, gray.
BODY FORM: Stocky, with shortish limbs; webbed hind feet; wrinkled skin; pupils horizontally elliptical.
BREEDING: Amplexus axillary; fertilization external; eggs apparently laid in streams; tadpoles free-swimming.
CONSERVATION STATUS: About half the night frogs are Endangered.

'True' frogs
Family: Ranidae

319 species in 19 genera. Worldwide except for the polar regions, S South America, Madagascar, New Zealand, and most of Australia. Aquatic, aquatic margin, terrestrial, from rain forests and grasslands; mostly nocturnal. Examples include **American bullfrog** (*Lithobates catesbeiana*), **Edible frog** (*Pelophylax esculenta*), **European common frog** (*Rana temporaria*), **torrent frogs** (genera *Amolops* and *Meristogenys*).
LENGTH: 1.5–20cm.
COLOR: Usually brown or green with darker markings.
BODY FORM: Mostly streamlined, with pointed heads; long limbs, with webbing between digits; pupil horizontally elliptical.
BREEDING: Amplexus axillary; fertilization external. Most lay eggs in water and have free-swimming tadpoles.
CONSERVATION STATUS: **Ramsey Canyon leopard frog** (*Rana subaquavocalis*) and several other species from Mexico and SW N America are Critically Endangered; Endangered and Vulnerable species in many parts of the world.

LEAPS AND BOUNDS
The mechanics of anuran locomotion

IN COMMON PARLANCE, FROGS JUMP AND toads hop. Even though anuran locomotion is in fact much more diverse than just jumping and hopping, there is nonetheless a real difference between a leaping jump and a bouncing hop, just as there is a difference in the morphology and posture of leapers and hoppers. Long versus short distance travel registers the difference between these two levels of locomotor performance. A hop is a short-distance jump, measuring just a few times the body length of the anuran. A jump is a long-distance leap, measuring many multiples of body length.

American toads, Marine toads, and related species – the "true" toads – jump an average of three to five times their body length, no matter whether they are tiny like the Oak toad (*Anaxyrus quercicus*), measuring just 2–3cm (1in) from snout to vent, or giants such as the Marine toad (*Chaunus marinus*), a massive 10–15cm (5in). Furthermore, the range of individual hops or jumps

making up the average is narrow: seldom less than two or more than six times body length. In marked contrast, prodigious jumpers like the Australo-papuan rocket frog (*Litoria nasuta*) cover an average of 25 body lengths with each leap, and a single leap can be over 50 times body length. Few other frogs can match this Olympian performance, although the jumping performance of many true frogs (*Rana*) and treefrogs (*Hyla*, *Litoria*) regularly measures 10 to 20 times body length.

Of course, a variety of frogs, including some true frog and treefrog species, fill the performance continuum between the hoppers and the strong jumpers. Furthermore, a few frogs (a small fraction of the total) do not jump at all. Some species

regularly walk rather than jump or hop, and even leapers and hoppers will occasionally walk for short distances. Like all other four-limbed vertebrates, frogs walk with a diagonal sequence of limb movements: right forelimb followed by left hind, then left fore and right hind. This sequence of limb movement provides maximum stability, with the body's center of gravity always lying within a tripod of supporting limbs.

A few species are strictly aquatic. Their swimming behavior, like that of most other frogs when in the water, is unlike any other vertebrate, except for humans when they use the breast stroke. Vertebrates that use their limbs for swimming typically use the same sequence of limb movements

3

4

as when walking on all fours, with the hindlimbs moving alternately. In frogs, however, the hindlimbs move synchronously, just as in jumping; but rather than extending the legs at right angles to the body and throwing the body upward and forward, swimming frogs extend the hindlimbs backward parallel to the body, pushing it forward. The forelimbs are pressed against the side of the body.

As the hindlimbs push back, the large, webbed feet are forced open like parachutes by the water pressure. When the hindlimbs are fully extended, they are flexed; the hindfeet fold shut and are feathered forward like oars on a boat. Thus, the swimming movement of frogs is "fast forward" as the limbs extend, followed by "drift forward" as the limbs are flexed. Although this motion may seem inefficient, frogs in fact swim quickly. African

clawed frogs (*Xenopus*) are so adapted to swimming that they cannot shift their limbs to the vertical position needed to execute the standard frog jump. On land, they slip and bounce along the surface, never rising above it, with the hindlimbs propelling them forward with the swimming kick.

A less common form of aquatic locomotion is skittering, and the master of this mode is the Indian *Euphlyctis cyanophlyctis*. The frog literally skips or bounces across the water's surface in a series of rapid jumps. Skittering can begin either on the land or in the water. It requires the frog to maintain a high degree of buoyancy and land flat on the water surface. The hindlimbs provide the propulsion, as in terrestrial jumping, but the hindfeet are held vertically as in swimming, in order to push against the water and lift the entire frog up

◑ **Above** *A frog jumping:* **1** *Forelimbs elevate and aim the forepart of the frog; ankles lift the hind limbs off the ground;* **2** *Take-off: hindlimbs swing open from the hip joint; upper and lower legs extend, propelling the frog forward and upward; ankles and hindfeet roll off the ground;* **3** *The flight path follows a curve of approximately 45°; the eyes are shut and withdrawn downward into the mouth cavity;* **4** *Landing – the forelimbs break the fall, and the chest hits the ground, followed by the rest of the underside; the hindlimbs flex and press against the body, ready to leap forward again.*

and slightly off the surface. Skittering has been reported for fewer than a dozen species (other examples are found in North American *Acris* and Asian *Occidozyga* species), although it is possibly used by many others too.

Among arboreal species living in trees or low shrubs, jumping serves commonly for escape, and also occasionally to cross gaps. Frogs climbing in trees use the same sequence of limb movements as when walking on the ground, but movement is

Continued overleaf ▷

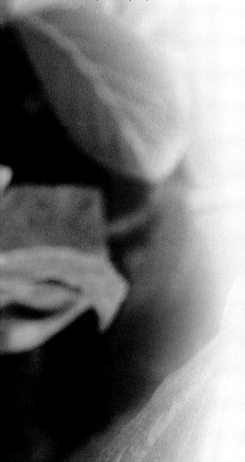

◑ **Left** *At full stretch, a Stripeless tree frog (Hyla meridionalis) launches itself from a branch with a powerful thrust of its hindlimbs. With their slim, agile build and adhesive pads on the ends of their digits, hylid frogs are extremely accomplished jumpers and climbers.*

◑ **Below** *The large, webbed hindfeet that characterize wholly or partially aquatic anuran species are clearly visible on this Common toad (Bufo bufo). The increased surface area of the hindfeet anchors the frog in the water, much as a drift anchor does a boat, and the body pushes forward from this anchor.*

slower and the placement of the feet is more methodical. With each step, the frog has to grasp the branch and ascertain that it will support its weight. When changing trees or branches, arboreal frogs use the standard jumping behavior, although arboreal jumping requires more precise targeting of the landing site to avoid a fall. From a frog's perspective, a fall is dangerous not so much because of the risk of injury from a crash landing as from the increased exposure to predators. In one Puerto Rican rainforest, a small frog, the coqui (*Eleutherodactylus coqui*) ascends trees at night. As day breaks and humidity drops, many coqui descend to the ground simply by jumping outward and falling; their spread-eagle posture keeps them stable and creates enough air resistance to prevent a hard landing.

Arboreal frogs usually have slender bodies and limbs, as well as digits tipped with "suctorial" disks. The clinging force of these disks derives from friction and wet adhesion; true suction, if present, is only a minor component. The enlarged toe pads have a specialized outer layer of skin with a fibrous, pile-like surface, divided by canals into regular blocks. This piling produces the major clinging force on rough, dry surfaces, through friction and the intermeshing of the fibers and the irregularity of the surface. On damp, smooth surfaces, the mucus covering of the toe pads creates adhesion through the force of surface tension.

The combination of friction and wet adhesion enables many tree frog species to cling to steeply inclined surfaces. It also enables some lightweight species weighing under 10g (0.35oz) to cling to, or even walk upside down on, smooth surfaces.

A number of unrelated species of tree frogs have heavily webbed feet, both fore and hind. Webbing, particularly of the forefeet, seems a peculiar morphology for nonaquatic frogs, but as it occurs in several unrelated groups of tropical treefrogs it must play a functional role in anuran arboreality. Possibly webbing improves an air-borne frog's steering; certainly it can serve as a parachute to slow the landing of a falling frog.

A few Asian *Rhacophorus* species are known as flying frogs, even though they certainly do not fly and are not even convincing gliders; their "flying" is really controlled falling at a steep angle rather than straight down, to move from one limb of a tree to another, or else to escape from predators. This form of locomotion required the evolution of a particular behavioral protocol to increase inflight stability; its features include keeping the body rigid and flat, the limbs outstretched, and toes and fingers widely spread to serve as parachutes and rudders.

❶ *Above* To break its fall when jumping from tree to tree in its tropical rainforest habitat, the Malaysian flying frog (Rhacophorus reinwardtii) of Southeast Asia spreads the heavily webbed toes on its front and hind feet.

❷ *Left* Together with other members of its genus, the African pig-nosed frog (Hemisus marmoratum) uses its pointed snout as an excavating tool. Cartilage serves to harden the snout against the rigors of tunneling.

❸ *Right* The large, well-muscled hindlimbs of the European species Pelophylax esculenta help propel it effectively through water. These same attributes have led to it being prized as a culinary delicacy, especially in France – its scientific name translates into English as "Edible frog."

Burrowing frogs are far less numerous than tree-dwellers, and most species burrow backward. Frogs dig with a sideways shuffle of their hindfeet. Limb movement is confined to the lower limb, from the knee to the heel, and each heel of most backward burrowers bears an enlarged, crescent-shaped metatarsal tubercle. This tubercle serves as a scraper and scoop. In digging, the heel is placed on the ground behind the middle of the body and then is pushed back and to the side. This curving sweep displaces soil to the side, and a rapid, side-to-side hindlimb shuffle sinks the frog into the soil.

Backward burrowing has the advantage of allowing the frog to survey its surroundings for predators and to jump forward to catch prey. All frogs, whether burrowers or not, possess a bony prehallux (an extra toe-bone in front of the first toe) beneath the metatarsal tubercle. Few frogs are headfirst burrowers, although members of one family, the Hemisotidae, dig exclusively in this fashion. These pig-nosed frogs use their snouts to dig tunnels, shoving them forward into the wall of the tunnel and lifting them up and down to compress the soil to the roof and floor.

the other or even both of these behaviors as the evolutionary force.

Two diametrically opposed schools of thought offer explanations of how the frog's jumping locomotion evolved. The older hypothesis is that the synchronized kick of the hindlimbs was an adaptation for aquatic movement, permitting frogs to lunge forward suddenly and quickly in the water, either to catch prey or to escape from predators. With the development and refinement of this kicking response, it is further postulated, frogs would have been preadapted for jumping when they moved onto land.

The opposing hypothesis suggests a terrestrial origin for jumping, viewing replacement of the tail with the hindlimbs for aquatic propulsion as too unlikely. It suggests instead that jumping evolved as a rapid means of escape, particularly to allow animals to move quickly from a resting site on land to a hiding place in the water.

No matter which hypothesis is correct (if in fact either is), both concur that the tail would need to have been shortened to permit an effective propulsive kick of the hindlimbs. As this type of kicking became more important, the tail's interference with limb movement and its frictional drag would have resulted in a tendency towards its reduction.

Anuran saltation is driven by the synchronous extension of the hindlimbs. From a resting, typically crouched posture, the frog elevates its head and forebody by a push-up extension of the forelimbs. Simultaneously the hindlimbs are tensed and the ankles are elevated toward the vertical. The movement of the ankle raises the thigh and crus (the tibiofibula), and the position of the ankles behind the midline of the body provides a focal point for the jumping force. This force comes with the thigh and crus snapping open, driving the frog forward and upward with the back and pelvic girdle also straightening. The force of the extending limbs literally hurls the frog off the ground and carries it forward. As the frog begins its downward trajectory, the forelimbs move forward to lessen the impact. Even so, the chest and chin of jumping frogs – although not of hoppers – smack the ground. The forelimbs also anchor the forebody, so the landing frog bends at the pelvic-sacral joint, cocking the frog for its next leap forward.

A jumping frog may not have the grace of a bounding kangaroo, but the anuran design has nonetheless proved to be an evolutionary success story, surviving for over 200 million years. It continues to serve the purposes of more than 4,000 species living today.					GRZ

Why Have Frogs Lost their Tails?

With few exceptions, all frogs can hop or jump, and this ability is reflected in their morphology. Anurans are designed for saltation (jumping) and even the oldest anuran, *Prosalirus bitis* of the Early Jurassic – there are earlier frogs (Salientia) – shows the major saltatory adaptation seen in living species. These adaptations include a shortened vertebral column; a robust pectoral girdle and forelimb skeleton; a unique pelvic girdle with a greatly reduced ischium and pubis and, exceptionally, lengthening of the ilium; elongation of the

hindlimb bones, especially the ankle bones; and an externally absent tail. The tail exists internally as a solid, bony rod (a series of caudal vertebrae in *Prosalirus*) between the ilia; these bones, the sacral vertebrae, and their muscles form an important element in postural orientation for saltation.

The factors that drove the evolution of jumping and the specialized anatomy that goes with it are unknown. Jumping is an effective means of catching prey from ambush, and also a quick method of distancing oneself from an approaching predator, but it is purely speculative to suggest one or

DECODING THE FROG CHORUS

Vocal communication in frogs and toads

IN MANY TROPICAL AND SEMI-TROPICAL AREAS, thousands of males of as many as two dozen species of frogs may call at the same time and place. These "choruses" are among the most impressive of biological phenomena and may be audible from more than 1km (up to 1mi) away. The calls are emitted by males, and their main function is to attract and stimulate females. Since females in the majority of species lay their eggs in water, choruses usually occur around either permanent or temporary bodies of water. Calling activity most commonly occurs at night, although in a number of species males call during the day.

There are two main types of frog choruses.

Males of "explosive breeding" species typically congregate and call for a short period of time, usually one or two nights, and females arrive synchronously. The result is that on any one night there are as many females as males present at the breeding site, with the consequence that most males mate. Explosive breeders are sometimes characterized by what is known as "scramble" competition, where males simply grasp anything that moves and calling is largely abandoned.

At the opposite extreme are species that fall into the category of "prolonged breeders." Males of these species congregate and call for periods of up to 6 months during the rainy season. Choruses occur on most nights during this period, usually starting at dusk and continuing until dawn. In these choruses individual male attendance at the breeding site varies; some males attend the chorus and call on every night of the season, while others may only call for a few nights. In prolonged breeders, female arrival is asynchronous and, as a consequence, a relatively small proportion of the female population is present in the chorus on any one night. Females generally lay 2 to 3 clutches a season and attend the chorus only when they are ready to mate and oviposit. On any one night, therefore, males usually greatly outnumber females, and there is intense competition to attract them; only a small percentage of males successfully mate. Another consequence is that individual males may vary considerably in the number of females they mate with over the breeding season. Although males are restricted to mating once per night, some males are successful in attracting females on a number of nights, while others never successfully attract a female. In spite of the competition, prolonged breeders rarely resort to actively pursuing females, but instead compete vocally in an attempt to attract a potential mate's attention.

After she has arrived at a multi-species chorus, a female faces a formidable task. She must identify and locate a suitable male of her own species by extracting the proper sound pattern from a background of intense and complex noise. The challenge is especially acute in prolonged breeding species, where males remain stationary and wait for females to approach them. Vision and odor may provide the female with some information, but playbacks of male vocalizations from loudspeakers have shown that the call alone is sufficient for a female to select a mate. This is possible because each species in a chorus produces calls with distinctive qualities that the female recognizes.

Once a female locates a male of the correct species she approaches him, pausing every now and then to reorientate to his call. Amplexus is usually initiated when the female physically touches the male, at which point he clasps her. The pair often remain at the male's calling site for several hours before moving to the water, where the eggs are fertilized by the male as the female releases them.

Frog calls not only attract females of the same species but may also be used in territorial and other disputes between males. In some species males call from all-purpose breeding territories containing oviposition and/or feeding sites, which they defend from intrusion by other males. In

Above *Frogs possess a variety of sizes and shapes of vocal sac. Most common is the single sac, set centrally below the throat (median subgular sac). However, several species have paired subgular sacs; shown here is the Edible frog (Pelophylax esculenta), an explosive breeder that inhabits temperate regions.*

Left *In the Brazilian rainforest, two male Red-legged treefrogs (Dendropsophus bipunctata) jockey for position to secure the best calling site from which to attract females. Competition for mating opportunities is fierce among such species of prolonged breeders in tropical zones.*

other species males simply defend an area around the calling site, even though it contains no resources required by the female for reproduction. Very often defense of territories or calling sites is facilitated by the same call as that used to attract females, which therefore serves to indicate the position and disposition of the male not only to potential mates but also to rival males; for this reason, what was formerly termed the "mating" call is now referred to instead as the "advertisement" call. In at least two species of territorial frogs (the

American bullfrog, *Lithobates catesbeiana*, and Green frogs, *Lithobates clamitans*), males are known to be able to recognize the advertisement calls of their neighbors, thereby reducing the costs associated with repeated aggressive interactions with non-threatening neighbors.

In case of a dispute between rival males, a frog may add extra notes to his advertisement call or may switch to an entirely different kind of call. These other calls have been variously termed "territorial" or "encounter" or "aggressive" calls. They appear to inform a rival of impending attack, or, in the early stages of conflict, they may serve mainly to disrupt the calling pattern of the competitor, making him less likely to attract females. In the species that have been tested, females are attracted to aggressive calls only when no male is making advertisement calls. Indeed, they may leave the vicinity of males that are sparring vocally to mate with a distant male producing advertisement calls.

There are two other kinds of call that are common in the repertoires of many kinds of frogs and toads. "Release" calls, usually given by a male clasped by another male, often have a similar

acoustic structure to the aggressive calls described above. Unreceptive females may also give a release call, often similar to but distinguishable from that of the male. A release call is usually sufficient to make the clasping male let go of the unwilling sexual partner.

"Distress" calls differ from other calls, because they are given with the mouth open. They are usually produced when a frog is grasped by a predator, and many sound like screams. Other frogs apparently ignore distress calls, so it is possible that their function is to startle the predator, which often drops the frog.

Choosing Mates by their Calls

While it is well established that calls allow a female to identify males of her own species, some researchers have suggested that they may also permit her to assess individual males in order to choose the fittest one. For example, in most species calls of the lowest frequencies are usually produced by the largest males, and females might

Continued overleaf ▷

in theory gain by mating with these individuals; large males may be older, for example, and their longevity may be due to genetically determined advantages that allow them to obtain food and avoid predators better than small, shorter-lived males. These advantages, because they are genetic and therefore heritable, could be passed on to the female's offspring.

This is a plausible hypothesis, but there is in fact little evidence that females prefer large males because they produce more attractive calls. In nearly every species in which large males have been found to mate more often than small ones, it has been shown that the large animals gain their advantage by excluding small males from territories in which females lay their eggs, or else by directly displacing them from females. Indeed, in several species there appears to be no correlation at all between call pitch and mating success.

In addition to call frequency, however, a female's choice of a mating partner may be influenced by other call characteristics such as the rate at which the call is produced or its duration. There is a fairly high degree of variability in these characteristics among the calls of conspecific males in a chorus. Most commonly, females appear to prefer calls that are produced at a faster rate and calls that are longer in duration. Both these factors could indicate that a male frog has "good genes," because calling is a highly energetic activity. It may be, therefore, that only those males that are efficient at sequestering and utilizing energy reserves can afford to call energetically – and, by mating with these males, females pass these characteristics on to their offspring.

Evidence has come to light recently that suggests that female preferences for males with longer calls may indeed indicate a preference for males of high genetic quality. In at least one species (the Gray treefrog, *Hyla versicolor*) in which females prefer to mate with males with long calls, offspring from such pairings have been shown to be of significantly higher phenotypic quality than offspring from pairings with males that emit relatively short calls.

Although there is a large body of evidence to suggest that females prefer males with certain call characteristics, numerous studies have also shown that these preferences are not always manifest in large natural choruses. It appears that the extent to which females are able to distinguish between male calls and so mate with preferred males depends to a large extent on social and environmental conditions. In some choruses males with preferred calls mate more often, while in others these males are no more successful in attracting

females than males with less attractive calls.

There is, however, one factor that seems to have a universal influence on determining how successful a male will be in attracting females: the number of nights he spends in the chorus. The more nights a male attends, the greater the number of times he mates over a breeding season. This relationship is not difficult to understand, since a male can only mate if he is present. However, it does raise the important question of why some males attend the chorus on more nights than

others. Because of the strong relationship with mating success, one would expect that all males would attend the chorus on every night of the season. It is likely that a major factor influencing a male's attendance is how efficient he is at meeting the energetic demands of calling activity.

In some frog species, males may be present in the chorus but do not vocalize to attract females. Instead, these "satellite" males take up positions near actively calling individuals and attempt to intercept approaching females. A number of

◁ **Left** *Inflating its impressively large vocal sac, the Great Plains toad* (Anaxyrus cognatus) *of North America emits a long, resonant trill, which can last for several minutes. This is the most prolonged vocalization of all anuran species.*

to the surrounding air, rather like the soundboard of a piano. The sac can also modify the frequency spectrum of calls that have harmonics (multiple frequency components) by emphasizing some and filtering out others. Frog calls range in duration from simple, brief clicks of 5–10 milliseconds, like those produced by the Spotted grass frog of Australia, to trills of several minutes' duration, as in the Great Plains toad of North America. In this case the long call is sustained by air being shunted back and forth across the larynx between the lungs and the vocal sac.

In terms of sound quality, frog calls range from the pure-tone whistles of the Spring peeper to the noisy, ducklike croaks of the Squirrel tree frog. In frogs that produce trills, there are two mechanisms. The airflow from the lungs may be pulsed by rapid contractions of the body wall musculature ("active" pulsing); or alternatively the larynx itself may, by a pressure-sensitive mechanism, break a steady flow of air from the lungs into a series of pulses ("passive" pulsing).

Most frogs and toads have conspicuous tympanic membranes (eardrums) on either side of the head, just behind the eyes. Sound waves displace each eardrum directly from the outside and indirectly via the opposite ear and a pathway connecting the two ears internally. Movements of the eardrum are transferred by the middle ear bones to a fluid-filled capsule (the otic capsule), which contains two hearing organs, the amphibian and basilar papillae.

The first of these organs is unique to amphibians. Sound waves in the inner ear organs cause the cilia of hair cells to bend, and this in turn results in the generation of nerve impulses that are sent to the brain via the auditory nerve. The frequency of the calls of a particular species roughly matches the frequency sensitivity of one or the other of the two inner ear organs. Indeed, in some species there are two emphasized frequencies in the call, one which best excites the amphibian papilla, the other the basilar. Thus, for each species, the sensitivity of the female's ear is matched to the frequencies contained in conspecific male calls. Recognition may, in part, be accomplished by this matching, but there are usually time cues, such as trill rates, that are also important and must be decoded in the female's auditory system.   MD/HCG

authors have suggested that satellites represent males with low mate-attractant abilities that become satellites as a means of making the "best of a bad job." Yet in most species where it occurs, satellite behavior is not a fixed strategy. Males switch between calling and satellite behavior, and which of these strategies a male adopts on any one night can depend on several factors including energy reserves and such variables as the size of the chorus, female arrival patterns, and the risk of predation.

How the Calls are Made

Frog calls are generated by vibrations of the vocal cords, located in the larynx; most calls are produced when the frog or toad exhales a large volume of air. Notable exceptions are the fire-bellied toads, which produce sounds either during inhalation or during both inhalation and exhalation.

As a frog vocalizes, it inflates one or more vocal sacs. The sac does not directly amplify (add energy to) the sound produced by the vocal cords, but instead helps to couple these vibrations effectively

FROM TADPOLE TO FROG

① *European common frogs (Rana temporaria) display amplexus, the clasping grip employed by males in mating. Fertlization is external: the female releases many hundreds of eggs and the male sheds sperm onto them. This pair is about to spawn onto a clump of eggs laid earlier by another couple. When many pairs do so, the resulting mass of eggs can form a heap within which the temperature is slightly higher than the surrounding water, speeding the embryos' development.*

② *Newly emerged tadpoles cluster around air bubbles. At this stage they have feathery external gills; sustained by the remains of the egg yolks, they move only occasionally, stirring the water around them to release its oxygen content. Their black coloration, due to melanin in the skin, protects their developing internal organs from the harmful effects of ultraviolet radiation.*

③ *For the most part tadpoles are herbivorous, surviving on a regimen of microscopic algae, but when the opportunity arises they occasionally scavenge off the carcasses of dead or dying animals, like the earthworm seen here. By this stage, Rana temporaria tadpoles have rounded, well-developed bodies, providing room for the long, coiled gut needed to digest their mostly vegetarian diet. The hindlimb buds are also starting to appear.*

4 The tadpole's hindlimbs develop before the forelimbs, growing in a membranous fold called the operculum that grows over the external gills as these shrivel. As they get larger, the legs are forced out of this protective sheath, dangling behind as the tadpole swims along, driven forward by beats of its large, powerful tail.

5 Once the front legs have also developed, the tadpole can emerge from the water, with its tail now much reduced in size. At this stage of its development it is very vulnerable to the weather and to predators; the remains of the tail hamper its movements as it seeks damp, safe places where it can sit out the danger time.

6 Froglets that have completed metamorphosis lose their tails, which are completely absorbed into the body, and become entirely reliant on their hindlimbs to hop about. Newly-metamorphosed frogs are carnivores, feeding on tiny insects and other animals, which they detect by means of their large, vigilant eyes. *TRH*

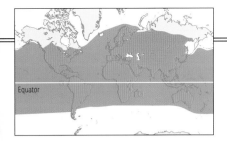

REPTILES

SOME PEOPLE REGARD SNAKES, CROCODILES, and lizards with revulsion. They are often thought of as cold, lurking creatures, ancient relics of a group on its way out, superseded by birds and mammals. Despite their low reputation, their extinct relatives, the dinosaurs and other "prehistoric monsters," are the best known of all past animals, with a firm place in popular fantasy.

However, reptiles are much more successful than is often thought, and the snakes have a recent evolutionary history of great diversification. In some habitats, notably deserts, reptiles are a dominant group. They have one huge advantage over birds and mammals; being less dependent on maintaining a constant body temperature, they can survive on a fraction of the vast food input that birds and mammals require. They are thus able to exploit environments where food supplies are sparse or sporadic.

The reptiles' most obvious feature is their covering of dry, horny scales, which amphibians do not possess. They resemble basal amphibians and birds in having only a single small bone in the ear, the columella or stapes, to conduct sound vibrations, and they have several bones in each side of the lower jaw. Mammals, on the other hand, have three small auditory bones and a single lower jaw bone, the mandible, which corresponds to the reptilian tooth-bearing element, the dentary. Reptiles also differ from amphibians and birds in having a single occipital condyle, the surface on the back of the skull that articulates with the vertebral column. Unlike birds and mammals, reptiles are mainly dependent upon external sources of heat, such as the sun's rays, for maintaining their body temperature (see Temperature Relations of Reptiles).

Reptiles reproduce by laying shelled eggs on land or by bearing their young alive. They do not pass through an aquatic larval stage like most amphibians. Their embryos, like those of birds and mammals, are provided with special membranes (amnion, chorion, and allantois) that are of great importance in terrestrial reproduction. The common possession of these membranes unites living reptiles, birds, and mammals together within the major tetrapod group, the Amniota.

How Reptiles Began
ORIGINS AND CLASSIFICATION

Amniotes arose from amphibianlike tetrapods, but the details of their early history are not clearly understood and current ideas are in a state of flux.

The early Carboniferous *Westlothiana* (340 million years ago) may lie close to the point of ancestry.

By a very early stage (early Upper Carboniferous, about 310 million years ago), basal amniotes had split into two main lineages, one leading eventually to mammals (Synapsida) and the other to living reptiles and birds (Reptilia). Remains of the earliest known reptiles have been found inside fossil tree stumps. They were small, terrestrial creatures superficially resembling lizards.

In combination with other characters, the "cheek" or temporal region of the skull behind the orbits (eye sockets) has long been used in amniote classification. In the most primitive amniotes, the temporal region presents an unbroken shell of bone without openings or "apses" (arched recesses). This is the anapsid condition, and it is retained today in a modified form in turtles and tortoises. In the line leading to mammals, a single opening appeared with a bar of bone beneath it, a feature that may have been associated with changes in the attachments of the jaw muscles. These amniotes constitute the Synapsida, a group of initially large carnivorous and herbivorous animals from which the first tiny mammals arose in the late Triassic period about 210 million years ago. In the Late Carboniferous (300 million years ago), another group of amniotes appeared with two temporal fenestrae on each side of the skull, each opening bounded below by a bony bar or "arch." These constitute the Diapsida.

Two main lines of diapsids (Lepidosauria and Archosauria) became numerous and diverse towards the end of the Triassic, when the large early synapsids were disappearing. Lepidosauria (the name means simply "scaly reptiles") contains the majority of living reptiles: the New Zealand tuatara – members of the largely extinct subgroup

Rhynchocephalia – and the Squamata – lizards, snakes, and the worm-lizards or amphisbaenians, a group of specialized burrowers. Archosauria ("ruling reptiles") includes the crocodiles, the pterosaurs, the dinosaurs, and their descendants, the birds.

Living tuatara have a skull structure similar to that of the early diapsids, with two distinct temporal openings. This trait used to be regarded as primitive, earning tuatara the description of "living fossils," but we now know that the condition has been reacquired. Early rhynchocephalians had a more lizardlike skull, in which the upper temporal opening was surrounded by bone but the bar beneath the lower fenestra had disappeared. Rhynchocephalians have been recovered from fossil deposits of Triassic (220 million years ago), Jurassic, and early Cretaceous age worldwide (in Europe, North America, South and Central America, India, Africa, Madagascar, and China). Current evidence suggests they became extinct in northern continents in the mid-Cretaceous, about 100 million years ago, but their survival history in southern regions remains largely unknown.

In contrast to that of rhynchocephalians, the early fossil record of squamates is relatively poor. Indirect evidence suggests that lizards arose and began to diversify before the end of the Triassic,

○ **Above** *A Panther chameleon (Furcifer pardalis) from the rain forests of Madagascar. This family of highly specialized arboreal lizards has a relatively recent geological history; the earliest fossil chameleon dates from the middle Miocene, c.20–15 million years before the present.*

3

○ **Left** *Early diapsids embraced a huge variety of forms:* **1** *Placodus belonged to a sister group (the Placodontia) of the marine plesiosaurs, and lived in the middle Triassic, 220 m.y.a.;* **2** *Archaeopteryx, the oldest known bird, from the late Jurassic, 147 m.y.a., was essentially a small theropod (meat-eating) dinosaur with feathers and some flying capability;* **3** *The terrestrial Gigantosaurus was an enormous saurischian dinosaur that flourished in the late Cretaceous, 70 m.y.a.*

but the earliest certain lizard fossils are Early to Middle Jurassic in age (around 185–165 million years ago). Snakes are known from the mid-Cretaceous; they are one of the most recently evolved of all the major groups of reptiles, and in terms of numbers of families and species they have been spectacularly successful. They arose from anguimorph lizards, but whether their ancestors were burrowers or swimmers is currently a major source of debate. Amphisbaenians (worm-lizards) are not recorded before the late Cretaceous, 75 million years ago.

The other main diapsid line, the Archosauria, became immensely successful during the Jurassic and Cretaceous periods (between about 205 and 65 million years ago), the peak of the so-called "Age of Reptiles." The most spectacular archosaurs were the dinosaurs, principally renowned for their very large size, although a few were little bigger than a pheasant. Experts recognize two major dinosaurian groups, the Saurischia, in which the hip bones were arranged in the typical reptilian pattern, and the Ornithischia, or "bird-hipped" forms.

The Saurischia contained all the carnivorous bipeds (theropods), culminating in *Tyrannosaurus*, with an overall length of 12m (40ft), and *Giganto-saurus*, up to 14.5m(48ft), but also in immense quadrupedal herbivores (sauropods) with tiny heads and long necks and tails. One of these,

AMNIOTE SKULLS

The arched recesses (apses) in the temporal region behind the eye sockets have been used to distinguish four major amniote skull types, and hence the four groups of reptiles: Anapsid, in which there are no apses, or if there is a hole in the temporal region, as in many turtles and tortoises, it is anatomically unlike those in other groups. Synapsid, in which there is a single apse with a bar of bone beneath it; this skull type characterizes pelycosaurs of the Lower Permian and their mammal-like descendants, the therapsids, from which mammals evolved at the end of the Triassic. Diapsid, in which there are two apses; the Diapsida encompasses two groups: the Lepidosauria (snakes, lizards, and their ancestors), and the Archosauria (dinosaurs, pterosaurs, and crocodilians). Although birds' skulls have only one temporal opening, they are also included in this latter group, since the single opening is thought to derive from the fusion of the two diapsid recesses. Euryapsid, or parapsid, a category previously used to describe the skull in the marine reptiles of the Mesozoic (i.e. plesiosaurs and ichthyosaurs); it has one apse high on the skull; these reptiles are now thought to be diapsid derivatives.

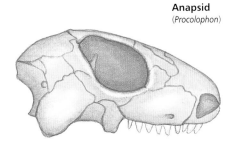

Anapsid
(*Procolophon*)

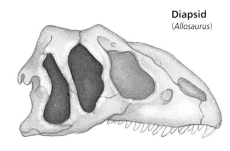

Diapsid
(*Allosaurus*)

◑ **Below** *Geological occurrence of extinct and living amniotes. The reptiliomorph* Westlothiana *was once thought to be the world's oldest amniote, but is now recognized as being close to their ancestry.*

Seismosaurus, with an estimated length of 40–50m (130–165ft), was among the longest, while *Argentinosaurus*, with a weight of around 70–100 tons, was one of the heaviest land animals that has ever existed. The earliest known true bird is *Archaeopteryx* from the Late Jurassic of Germany (147 million years ago), but bereft of its feathers this small animal is clearly dinosaurian in structure. Most workers currently agree that birds are descended from small theropod dinosaurs, and gracile, partially feathered theropods have recently been recovered from Cretaceous deposits in China.

The Ornithischia also contained both bipedal and quadrupedal types, but all were herbivorous and none reached the proportions of the largest sauropods. The bipeds include the well-known *Iguanodon*, with a spur on its thumb, and the duck-billed hadrosaurs, some of which possessed curious crests on the tops of their heads, almost certainly for trumpeting. Among the ornithischian quadrupeds were the lumbering *Stegosaurus*, with huge plates of bone down its back, and rhinoceros-like horned dinosaurs like *Triceratops*.

Other successful archosaurian groups were the flying reptiles (Pterosauria) and the Crocodylomorpha, which – like the dinosaurs – originated in the Triassic. The pterosaurs were similar to birds in certain features, such as the keel on the sternum for the attachment of the flight muscles and the presence of air spaces in many bones to reduce their weight. They were probably also warm-blooded, since some specimens show evidence of a hairlike covering. However, pterosaur wings were more like those of bats than birds, being membranous and supported by the elongated bones of the fourth finger.

Millions of Years Before Present								
443	417	354	295	248	205		144	65
PALEOZOIC ERA				MESOZOIC ERA				CENOZOIC ERA
SILURIAN	DEVONIAN	CARBONIFEROUS	PERMIAN	TRIASSIC	JURASSIC		CRETACEOUS	TERTIARY / QUATERNARY

Westlothiana ■ Near amniote
SYNAPSIDA — Mammalia
Archaeothyris
Protorothyris ■
CHELONIA
Proganochelys
Petrolacosaurus ■
ICHTHYOSAURIA
Grippia
PLESIOSAURIA
Pistosaurus
Archosaurus ■
CROCODYLOMORPHA
Hesperosuchus
PTEROSAURIA
Preondactylus
ORNITHISCHIA
Pisanosaurus
SAURISCHIA
Herrerasaurus
AVES
Archaeopteryx
RHYNCHOCEPHALIA
Brachyrhinodon
IGUANIA
(India – new genus)
SCINCOMORPHA
Bellairsia
GEKKOTA
Hoburogekko
ANGUIMORPHA
Parviraptor
SERPENTES
Coniophis
AMPHISBAENIA
Sineoamphisbaena

AMNIOTA
REPTILIA
DIAPSIDA
ARCHOSAURIA
DINOSAURIA
LEPIDOSAURIA
SQUAMATA

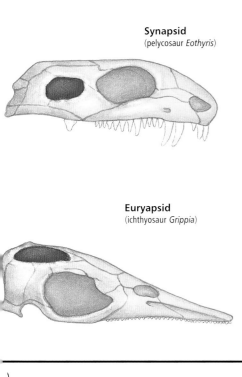

Synapsid
(pelycosaur *Eothyris*)

Euryapsid
(ichthyosaur *Grippia*)

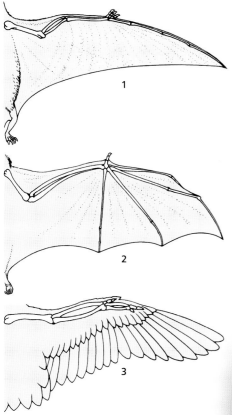

The earliest known crocodylomorphs were small terrestrial reptiles, but their descendants, the Crocodylia, are beautifully adapted for an amphibious, predatory mode of life. With birds, crocodiles are the only surviving archosaurs, but their continued existence is threatened by human activities such as the leather trade.

The great Jurassic marine reptiles, the fishlike ichthyosaurs and the long-necked plesiosaurs, have always proved difficult to classify. In the past, both were thought to possess a single temporal opening high up on the side of the skull and were grouped into a fourth skull category, euryapsid. However, both are now generally considered to be diapsid derivatives, although their precise relationships have yet to be resolved.

Finally there are the Chelonia, the tortoises and turtles, in some ways both the oddest and the most familiar of reptiles. They appear almost fully fashioned in the Triassic (*Proganochelys*, 215 million years ago) and this has given rise to much speculation about their origins. In a few forms, such as the sea turtles, the skull seems to resemble the primitive anapsid pattern, but in many others certain bones appear to have been "eaten away" or emarginated from behind or below to produce a form of temporal opening anatomically different from that of other amniotes. Some molecular and anatomical studies have placed chelonians as diapsid derivatives, but many palaeontologists consider them to have arisen from anapsid Permian reptiles (the parareptiles).

Beneath the Scales
SKIN

The skin is a fascinating organ with many functions. Besides acting as a barrier between the deeper tissues of the animal and the outside world, it can have a role in defense, concealment, mating, and locomotion. In reptiles, as in other vertebrates, it consists of two principal layers, the epidermis on the outside and the dermis beneath.

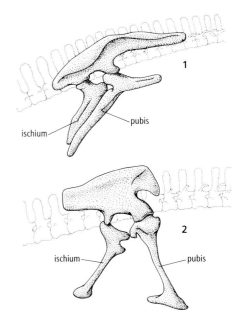

◐ Above Hip joints of the two lines of dinosaurs. **1** In ornithischian ("bird-hipped") dinosaurs, the pubis is parallel to the ischium. **2** In saurischian dinosaurs, the pubis is directed forward.

The epidermis is itself subdivided into various layers, the outer one composed of horny material called keratin. The scales of a reptile are localized thickenings of this keratin layer, connected by hinges of thinner material and often folded back so that they overlap each other. Unlike the scales of fish, they are not separate, detachable structures but parts of a continuous epidermal sheet. Their precise numbers and patterns on different parts of the head and body are of great value in reptilian classification, especially in distinguishing between different species.

Periodically the keratinous layer of the epidermis is shed and replaced through the activity of the deeper layers of cells. The keratin may come away piecemeal or in large flakes. In snakes, however, it is often shed as a single slough that is peeled

◐ Above Wing structure in **1** pterosaurs, **2** bats, and **3** birds. Like a bat's, the pterosaur wing was a membrane stretched over the bones of elongated fingers, but the longest finger was the fourth rather than the third.

◐ Right The head of a Green mamba (*Dendroaspis viridis*), displaying the intricate mosaic of scales that makes up a snake's skin. Pigment cells in the skin give each species its distinctive coloration.

off inside-out after the snake has rubbed it through at the snout. In these reptiles the old keratin layer on the surface is not shed until a new one has completely formed beneath it. Then a translucent zone of cleavage appears between the old and new layers so that they can be easily separated.

The brille – the spectaclelike eye-covering of a snake, formed from modified and fused eyelids – becomes blue and opaque some time before the animal is due to shed. Just before this happens, however, the brille becomes clear again, and this clearing seems to coincide with the establishment of the cleavage zone. Skin shedding in snakes may occur several times a year, being more frequent in young than in old animals. The process is influenced by the activity of the thyroid gland.

The dermis consists mainly of connective tissue and contains many blood vessels and nerves; it does not participate in the skin-shedding process. In some reptiles, including crocodilians and many lizards, the dermis contains plates of bone called osteoderms that lie beneath and reinforce the horny epidermal scales. In anguimorph lizards like the slowworm, and some scincomorphs such as cordylids, these osteoderms form a flexible bony covering for the body.

Both epidermis and dermis participate in that remarkable structure, the shell of a turtle. The horny scutes on the surface are keratinous epidermal formations with a layer of living cells beneath them, while the deeper, thicker shell layer is formed of bony dermal plates.

In reptiles the dermis also contains the majority of pigment cells. Many of these, the melanophores, contain black pigment, but there may also be white, yellow, and red pigment cells. The dispersal or concentration of the pigment within the melanophores, and the optical effects this creates when seen through the colored cells, are responsible for the changes in color for which chameleons and certain other lizards are noted. Color change in reptiles may be brought about by nervous activity, by the hormonal secretions of glands such as the pituitary, or by a combination of both; in chameleons, the action of the nerves seems to be the dominant factor.

Reptilian skin contains comparatively few glands, unlike that of many fish, amphibians, and mammals. Crocodilians possess a pair of glands beneath the throat that produce a musky secretion; this may play some part in sexual behavior. Some freshwater turtles have glands in the chin or the hindlimb pockets; in the musk turtles these produce a powerful smell. A few otherwise nonvenomous snakes have glands beneath the scales of the neck or back that secrete an irritant material that may act as a defense against predators and/or have some role in courtship behavior. In some geckos of the genus *Strophurus*, there are a series of large glands beneath the scales of the tail. When the lizard is threatened they squirt out filaments of sticky material that may deter such enemies as large spiders. Many lizards possess a series of curious, glandlike structures on the insides of

the thighs and sometimes in front of the cloaca. The function of these has been much debated; they seem to be related to mating since in male lacertid lizards they regress after castration. The most likely suggestion is that their secretions may assist in species or sexual recognition.

History in Bones
SKELETAL STRUCTURE

Compared to other living reptiles, the tuatara skeleton is probably closest to that of early reptiles, although there are important differences in detail. Other reptiles have departed in varying degrees from the basal pattern, and such departures are usually in the nature of adaptations to special modes of life.

One of the major trends in squamate evolution has been the development of additional joints and hinges within the skull, giving greater flexibility and efficiency to the upper and lower jaws during feeding (cranial kinesis). This skull mobility has been further elaborated in snakes, partly by the loss of any firm union between the two sides of the lower jaw. Overall, this is an adaptation to capturing and swallowing large prey. A second tendency within squamates has been toward the evolution of an elongated, serpentiform body with

⟳ **Below** *The shell of a tortoise (here, a Red-footed tortoise,* Chelonoides carbonaria) *is a complex structure made up both of bones and of scales (known as scutes) that develop from the epidermis.*

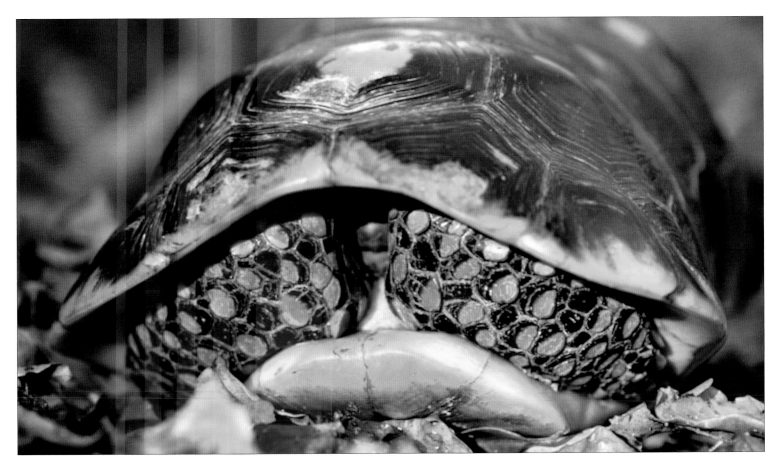

REPTILIAN BODY PLAN

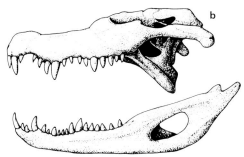

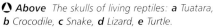

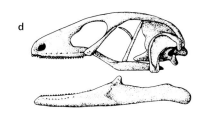

Above The skulls of living reptiles: **a** Tuatara, **b** Crocodile, **c** Snake, **d** Lizard, **e** Turtle.

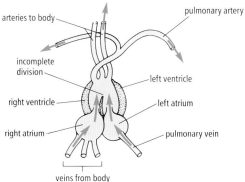

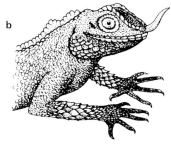

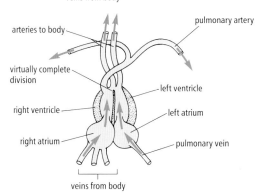

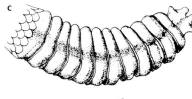

Above Reptilian hearts. **a** In most reptiles, the chambers of the ventricles are incompletely separated. **b** In crocodilians complete separation exists, although there is a small connection, the foramen of Panizza, between the outlet vessels. Even in the unseparated ventricle, a system of valves and blood pressure differences ensures that there is little mixing of arterial and venous blood under normal conditions. In all reptiles, however, the potential exists to shunt blood from one side of the heart to the other. This is adaptive, especially in aquatic animals, since blood can be recycled when breathing is interrupted.

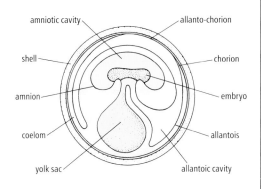

Above Developing egg, showing layers of membrane between shell and embryo. The partly fused chorion and allantois, on the inner surface of the shell, are richly supplied with blood vessels, enabling the embryo to breathe through pores in the shell. The allantois also acts as a repository for the embryo's waste products. The amnion is a fluid-filled sac around the embryo that prevents it drying out. The yolk-sac contains the embryonic food supply, rich in protein and fats. Eggs of this type, including those of birds, are called cleidoic ("closed-box") eggs, since apart from respiration and some absorption of water from the environment, they are self-sufficient. Water absorption by the eggs of many reptiles, especially the softer-shelled types, is higher than by birds' eggs.

Below Cross-sectional diagram of the skin of a slowworm. All anguimorphs such as this are heavily armored, having mostly non-overlapping scales with underlying osteoderms.

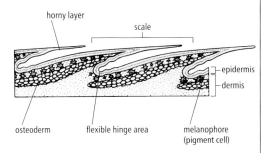

Left Modifications of the skin. The skin, particularly the epidermis, shows many modifications in reptiles. **a** It may be raised up into tubercles, as in the chisel-toothed lizard Ceratophora stoddartii, or into defensive spines, as on the tails of certain lizards. **b** It may form crests on the neck, back, or tail, often better developed in the male and perhaps assisting in sexual recognition, as in Lyriocephalus scutatus. **c** The rattlesnake's rattle, composed of interlocking horny segments, is a unique epidermal structure; a new segment is formed at each molt, though the end segments tend to break off when the rattle gets very long. In most snakes, the underbody scales are enlarged to form a series of wide, overlapping plates that assist in locomotion, especially in forms such as boas which can crawl stretched out almost straight. **d** The modified scales, or lamellae, on the toe pads of geckos have fine bristles (setae) that allow them to climb smooth surfaces.

reduced limbs. This development is estimated to have occurred independently more than 60 times, culminating in the evolution of snakes and amphisbaenians. As an adaptation, it has been extraordinarily successful, enabling the snake to "outswim the fish and outclimb the monkey," as the English zoologist Sir Richard Owen wrote with pardonable exaggeration. The remarkable position of the limb girdles of tortoises and turtles, which are situated within rather than outside the ribs, is another striking specialization associated with the incorporation of the ribs into the dermal plates of the shell.

The reptilian skeleton contains a number of bones that are not found in mammals, at least as separate elements. Some of these, such as the supratemporal, occur in the skull. Like birds, tuatara, lizards, and turtles have a series of small, bony plates, the scleral ossicles, around the eye to provide support and assist in focusing. In the shoulder girdle, most reptiles retain a midline bone, the interclavicle; among mammals, this is retained only in the egg-laying monotremes. In tuatara, crocodilians, and most extinct reptiles, the belly wall is reinforced by a series of abdominal ribs or gastralia. Many lizards have a system of cartilaginous bars called the parasternum in much the same position, but they have a different embryological origin.

Secondary centers of ossification (bony epiphyses) are found in most lepidosaurs (squamates and rhynchocephalians) but are lacking in the growing bones of turtles and crocodiles. In mammals, such as humans, the fusion of these epiphyses with the main shaft of the bone in early adult life sets a limit to the maximum size attained. The absence of bony epiphyses is perhaps one of the reasons why certain reptiles such as crocodiles and giant tortoises may be able to grow throughout much of their lives, albeit at a diminishing rate. If an individual is lucky enough to avoid life's hazards it may eventually become a giant, well above the usual size range of its species. However, many reptiles, including the smaller lizards and turtles, normally stop growing when they have reached a more or less definite size.

Moreover, reptiles do not lose their teeth when they get old in the way that mammals do; most reptiles have continuous, lifelong tooth replacement, while turtles, which did away with their teeth at an early stage in their history, have evolved a persistently growing horny beak. This feature removes a constraint on longevity, and therefore also on the ultimate potential size attainable by a species.

Egg-Layers and Live-Bearers
REPRODUCTION

In most reptiles, the two sexes differ to some extent in adult size, shape, or color. The males of many lizards and tortoises, and all crocodilians, are bigger than the females, whereas in most

snakes and some aquatic turtles it is the females that are bigger.

Differences between the sexes are particularly striking amongst iguanian lizards, which are predominantly visual animals. The males tend to be more brightly colored, especially in the breeding season, and in some species they possess erectile crests and throat fans that play a part in courtship and territorial display.

All reptiles practice internal fertilization, the sperm being introduced directly into the female's cloaca, the common opening that transmits both excretory products and eggs or sperm. In male turtles, tortoises, and crocodilians, there is a single penis, while male squamates possess paired organs called hemipenes – only one being used at a time (see Lizards). Tuatara possess rudimentary hemipenial structures, but fertilization is achieved by cloacal contact. The females of some snakes and some chelonians have the ability to store sperm in the reproductive tract, and instances of isolated females laying fertile eggs after a year or more in captivity have been observed.

Breeding in reptiles is greatly influenced by environmental factors such as temperature and the duration of daylight. Some tropical species may breed at intervals throughout the year, but in the majority breeding takes place only once, or perhaps twice, a year. Indeed, among reptiles that live in relatively severe climates, such as the adder in northern Europe, an individual female may breed only once every two years, or even less often.

Sexual reproduction is the general rule among vertebrates, but at least eight families of squamates include one or more all-female (parthenogenetic) species. This occurs among whiptail lizards in North America, and has been observed

◑ **Above** A python hatchling emerges from its egg, bursting through the amniotic membranes and shell with the tiny, pointed tooth on the tip of its snout. This projection is of no further use to reptiles once they have hatched, and falls off after a few days.

◐ **Right** The difference in color and size between male and female lizards is extreme in the Common agama (Agama agama). The vivid colors of the male and the iridescent green on the head of the female are stimulated by light and heat, and fade at night.

◓ **Below** Most reptiles exhibit no parental care, beyond building a nest for egg-laying. However, crocodilian species are a notable exception. Females (here, a Nile crocodile) carry their hatchlings and attend them closely, to guard against predation.

in some populations of rock lizards in the Caucasus. All members of a parthenogenetic population are genetically identical, as they are descended from a single founding female. The low level of genetic variability in such a species both confines it to a limited geographical area and probably limits its future in evolutionary time, because of its reduced ability (compared with a sexual species) to adapt to changing environmental conditions (see Unisexuality: The Redundant Male?).

Most modern reptiles lay eggs, which, unlike amphibian eggs, are resistant to drying out. Fossilized eggs of dinosaurs and more primitive extinct reptiles have been discovered. Reptilian eggs possess shells that may be pliable or parchmentlike in texture, as in many lizards, snakes, and aquatic chelonians, or hard and well calcified, as in tortoises, crocodilians, and many geckos. Hatching from the egg is facilitated by the presence in young lizards and snakes of a sharp, forward-pointing egg-tooth, which is later shed; in tuatara, tortoises and turtles, crocodilians, and birds, a horny outgrowth performs the same function.

The replacement of aquatic amphibian eggs by the shelled amniote egg played an important part in the successful colonization of the land. The eggs of reptiles are usually laid in holes, among rotting vegetation, or are buried in the soil. Seaturtles dig nests in sand on beaches and lay large clutches of up to 100 or more eggs. They resort to traditional nesting beaches, and Green seaturtles may migrate hundreds of kilometers to reach them. Among crocodilians, the Nile crocodile also digs nests in the sand, while others such as the American alligator construct more elaborate nests out of piled-up vegetation. Under the hot sun, these shelters act as highly effective incubators

for the eggs. Traces of vegetation within excavated dinosaur nests suggest they may have employed similar strategies.

In mammals and birds, sex is determined at fertilization by the possession of the requisite combination of sex chromosomes. Many reptiles, however, utilize temperature-dependent sex determination, whereby the sex of the hatchling depends on the prevailing temperature during a critical period of incubation. Thus, for example, in the American alligator temperatures of less than 30°C (86°F) between the seventh and twenty-first days of incubation produce all females, while those of more than 34°C (93°F) produce all males (see Temperature and Sex).

The majority of reptiles abandon their eggs after laying them, but in certain lizards, for example some skinks, and also in a few snakes such as cobras, the female remains with her eggs and may try to drive intruders from the nest. Attendance by the male has also been reported in cobras. Female pythons remain coiled round their eggs for up to several weeks, and in some species the egg temperature may be actively raised; this incubation may be assisted by muscular contractions of the mother's body. During recent years, remarkably elaborate forms of parental care have been observed in crocodilians (see Parental Care in Crocodilians).

A substantial number of lizards and snakes have forsaken egg-laying and gone over to live-bearing, the young escaping from their membranes at or shortly after birth. In such reptiles the egg-shell is lost or reduced, while a kind of placenta may be developed from the fused chorion and allantois and from the yolk sac. This development allows the exchange of some waste products and nutritional substances, as well as water and gases, between the embryo and the mother. Most viviparous species are "ovoviviparous" – the yolk remains large and is still the principal source of embryonic food – but in a few species with particularly well-developed placentae, the yolk is reduced.

Most seasnakes have become viviparous, and therefore do not have to come to land to lay their eggs, as the turtles do. The same was almost certainly true of extinct marine reptiles such as ichthyosaurs, as evidenced by preserved embryos found within the maternal body cavity. Today, this method of reproduction is also prevalent in reptiles that live under very severe climatic conditions, as in high latitudes or altitudes. It is found in three out of the six species of British reptiles. Under such conditions viviparity seems to be advantageous in enabling the mother to act as a mobile incubator, able to seek out sources of warmth that are optimal for the development of her embryos.   SEE/ADB

FEATURE

REPTILES

THE AGE OF REPTILES

When dinosaurs roamed the Earth

THE AGE OF REPTILES LASTED FOR ABOUT 215 million years, more than three times the length of time for which mammals have been in existence. Between 280 and 65 million years ago, a multitude of reptiles occupied ecological niches comparable to those filled by mammals and birds today. Although remains of reptiles have been found dating from the Carboniferous period, the "Age of Reptiles" really opened at the beginning of the Permian period (295–248 million years ago) and lasted throughout the Triassic and Jurassic periods (248–205 million years ago and 205–144 million years ago, respectively), ending with the close of the Cretaceous (144–65 million years ago). Pelycosaurs and therapsids, the so-called mammal-like reptiles, were the dominant amniotes for the Permian and much of the Triassic. When they declined, their places were taken by a group of diapsid reptiles called the rhynchosaurs, and when these suddenly died away towards the end of the Triassic their relatives the thecodonts (the ancestral order of the subclass Archosauria) expanded rapidly to fill most ecological niches. Thecodonts spawned the crocodilians, the pterosaurs (flying reptiles), and the dinosaurs that were to dominate

the greater part of the Mesozoic era. A few of the therapsids survived, and in the late Triassic they evolved into the first mammals.

Today's definition of dinosaurs includes many small creatures as well as the traditional giants. They evolved, along with the other thecodonts, as meat-eaters, but soon diversified into plant-eating as well. Changes in the limbs and girdles meant that the dinosaurs could walk like mammals, with limbs supporting the body from underneath. The early dinosaurs were bipeds, running and walking on their hind legs, but many later types, especially the heavier plant-eaters, reverted to a four-footed stance.

There are three main lines of dinosaur evolution: one, the theropods, were meat-eaters, while the other two were the plant-eating sauropodomorphs and ornithischians. These groups are sometimes regarded as three separate orders, but it is more customary to classify the theropods and sauropodomorphs together as a single order, the Saurischia, defined by the lizardlike arrangement of the bones of their hips. The order Ornithischia had a birdlike arrangement of hip bones, although this structure also evolved independently in some

members of the theropod group, particularly in the enigmatic segnosaurs, as well as in the birds themselves – the theropods' descendants. Flying pterosaurs comprise a separate order of the Archosauria, while the aquatic ichthyosaurs and plesiosaurs, prominent in the Jurassic, are even more distantly related, belonging to the subclass Euryapsida.

The smallest dinosaur known, the theropod *Compsognathus*, was no bigger than a chicken, although there are footprints of even smaller ones. For about a century, the largest meat-eater known was *Tyrannosaurus*, at 12m (39ft) long, but in the 1990s even longer theropods, such as *Carcharodontosaurus*, were discovered.

The largest complete skeleton known is of *Brachiosaurus*. It stands 12.6m (41ft) high, but the neck bones of closely related *Sauroposeidon* show that this animal was a good 20 percent larger, and the few bones known of the South American *Argentinosaurus* hint at an animal that weighed 100 tons or more. Such a staggering size would have presented formidable problems of support, nutrition, and growth. These giants were all sauropodomorphs, plant-eaters with long necks.

⊳ **Right** *A Late Jurassic landscape, showing a diversity of animal life:*
1 Rhamphorhyncus (pterosaur);
2 Kentrosaurus (stegosaur);
3 Elaphrosaurus (theropod dinosaur);
4 Ceratosaurus (theropod dinosaur);
5 Dicraeosaurus (sauropod dinosaur);
6 Brachiosaurus (sauropod dinosaur).

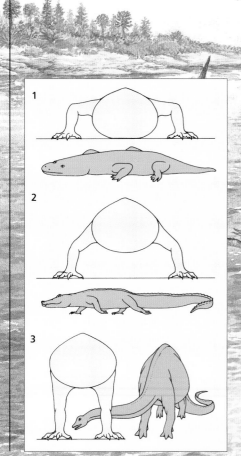

⊲ **Left** *Limb posture in extinct and living reptiles. 1 Sprawling stance of early reptiles and living lizards. 2 "Semi-improved" stance of early thecodonts and living crocodilians. 3 "Fully improved" stance of dinosaurs. which evolved independently in mammals.*

Among the plant-eating ornithischians, the ornithopods had a basically two-footed stance. The stegosaurs had plates and spines down their backs, while the ankylosaurs were armored. These two are often classed together as the thyreophorans. Ceratopsians, which had enormously thickened skulls and horns on the face, and pachycephalosaurs, with thick domes of bone on the tops of their heads, are jointly classified as the marginocephalians.

Most species of dinosaur flourished for only a short time, so that late Cretaceous dinosaurs were very different from their Triassic ancestors. At the end of the Cretaceous (about 65 million years ago) the dinosaurs became extinct, as did many other groups of organisms. The likely cause was major environmental change following a massive asteroid impact at Chicxulub on the

Yucatán Peninsula of Mexico. When the dinosaurs were gone, the tiny mammals that had scurried at their feet were able to evolve new forms to occupy the major land-animal niches. Ironically, then, the dinosaurs were superseded by descendants of the therapsids that had reached their peak long before they themselves evolved. Yet the dinosaurs' own descendants still flourish today, in the form of the birds and crocodilians.   AJC/DD

TEMPERATURE RELATIONS OF REPTILES

How reptiles depend on their environment to adjust body heat

TEMPERATURE AFFECTS VIRTUALLY ALL LIVING processes by changing physical properties and chemical reaction rates. Each species of reptile – indeed all organisms – can survive over a specific range of body temperature, usually from near freezing, or 0°C (32°F), to around 40°C (104°F). In general, species from colder climates tolerate lower temperatures better than do those from warmer climates, and vice versa. Some turtles can stand temporary freezing, whereas certain desert lizards can survive temperatures as high as 42°C (108°F). Within the range of tolerance, various life processes increase in rate or intensity with increasing temperature, with the maximal rates occurring near the upper end of the thermal range (see graph opposite). Metabolism increases in this manner, thereby affecting the energy use of an animal and its capacity for activities such as muscle movement and locomotion. Thus, within limits, a warm body speeds up living.

To stabilize chemical reaction rates and body processes, most vertebrates regulate their body temperature. Unlike birds and mammals, reptiles are not insulated with fur or feathers, and their metabolic rates do not produce large amounts of heat. Consequently, they depend on external sources in the environment for body heat – a condition called "ectothermy," in contrast to the "endothermy" that characterizes birds and mammals. One interesting exception is the python, which twitches its muscles to produce body heat. Brooding females employ this "facultative endothermy" by coiling around their clutch and generating muscle heat in order to elevate and stabilize the temperature of incubating eggs.

Several lines of evidence suggest that some of the dinosaurs were endothermic. This topic is controversial, but larger dinosaurs were almost certainly so by virtue of their large mass and low surface-to-volume ratio. Body temperatures in these "gigantotherms" were probably elevated above the environment by up to 10°C (18°F) or more because of the difficulty they experienced in losing heat. For similar reasons, some of today's sea turtles have body temperatures that become elevated several degrees above that of the surrounding water when they are swimming.

Most reptiles control their body temperature behaviorally by moving discriminately within the complex thermal structure of their habitat. Lizards are well-studied examples. In temperate habitats they often move between sun, which warms the body by radiation, and shade, where body heat slowly dissipates to cooler surroundings by conduction and radiation. If the shaded environment is at or near the lizard's body temperature, that temperature can remain stable for some time. Movements from shade to sun (or heat) are controlled by a neural system regulated by a "low body temperature set point," whereas the reverse movement is similarly regulated by a "high body temperature set point." Many terrestrial reptiles can regulate their body temperature very precisely within a narrow, preferred range by virtue of this control apparatus in their brains.

The body temperatures required for locomotion, foraging, and other behaviors are relatively high (commonly 30°–37°C/86°–99°F) in many species. But these so-called "activity temperatures" can be achieved only at certain times of day. Thus, in spring and fall, lizards in temperate climates are routinely active at midday, whereas during the hotter parts of summer they retreat from midday heat and are most active in the morning and late afternoon. Raising the body temperature by basking appears especially beneficial for juveniles, as it aids digestion and maximizes growth. It is perhaps hardly surprising that many reptiles choose somewhat higher body temperatures for brief periods after ingesting a meal. Higher temperatures are also important for the development of reproductive structures and the competent functioning of the immune system in various species.

In hotter regions such as deserts, many snake and some lizard species are nocturnal, remaining in seclusion during intense daytime heat. During winter, in hot or dry spells, and at other times of

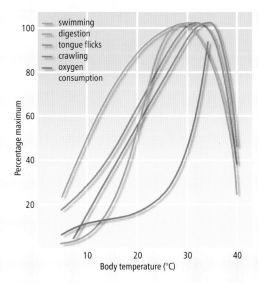

❶ **Above** This graph illustrates how various activities and functions in the Western terrestrial garter snake (Thamnophis elegans) are affected by changes in body temperature.

❷ **Right** In Western Australia, a Gould's monitor (Varanus gouldii) basks in the sunshine on a fallen tree trunk to raise its body temperature, thereby boosting its energy levels.

❷ **Right** Heat flows from warm to cold objects. The illustration shows the main avenues of heat exchange between reptiles and their surroundings; in regulating their body temperature, they take advantage of the fact that their environment is thermally diverse. As the graph indicates, the body temperature of lizards is relatively high and constant during the day, but it falls when the lizard takes shelter for the night. Body temperature records are gathered from small, temperature-sensitive radiotransmitters swallowed by the lizards. The pulse rate of the transmitters' signals increases as the temperature rises.

❶ **Left** Many snakes, such as this Timber rattlesnake (Crotalus horridus), assume flattened postures to increase heat absorption from sun or substrate during periods of warming.

❶ **Above** Incubating its eggs, a female python coils around them, flexing her muscles if necessary to generate the extra body heat required to keep the brood at the optimal temperature.

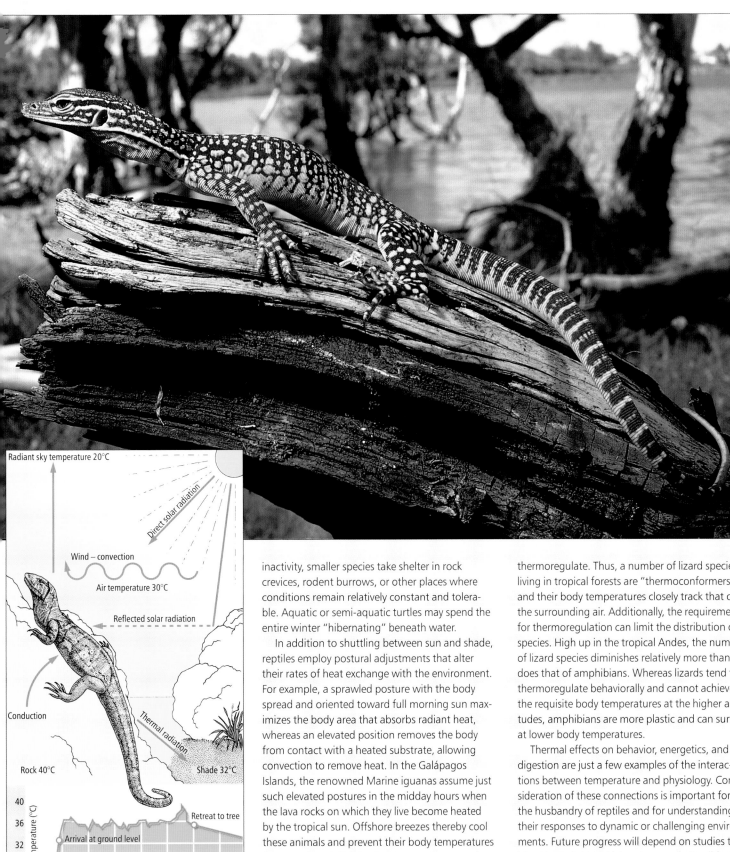

inactivity, smaller species take shelter in rock crevices, rodent burrows, or other places where conditions remain relatively constant and tolerable. Aquatic or semi-aquatic turtles may spend the entire winter "hibernating" beneath water.

In addition to shuttling between sun and shade, reptiles employ postural adjustments that alter their rates of heat exchange with the environment. For example, a sprawled posture with the body spread and oriented toward full morning sun maximizes the body area that absorbs radiant heat, whereas an elevated position removes the body from contact with a heated substrate, allowing convection to remove heat. In the Galápagos Islands, the renowned Marine iguanas assume just such elevated postures in the midday hours when the lava rocks on which they live become heated by the tropical sun. Offshore breezes thereby cool these animals and prevent their body temperatures from reaching lethal levels.

The level of body temperature selected, and the precision and timing of its regulation, vary in relation to the environment, the phylogenetic history (lineage), and the resource requirements of any given species. Animals living in relatively constant thermal environments, such as tropical forests or aquatic habitats, show little tendency to thermoregulate. Thus, a number of lizard species living in tropical forests are "thermoconformers," and their body temperatures closely track that of the surrounding air. Additionally, the requirements for thermoregulation can limit the distribution of species. High up in the tropical Andes, the number of lizard species diminishes relatively more than does that of amphibians. Whereas lizards tend to thermoregulate behaviorally and cannot achieve the requisite body temperatures at the higher altitudes, amphibians are more plastic and can survive at lower body temperatures.

Thermal effects on behavior, energetics, and digestion are just a few examples of the interactions between temperature and physiology. Consideration of these connections is important for the husbandry of reptiles and for understanding their responses to dynamic or challenging environments. Future progress will depend on studies that blend both field and laboratory work. Possibly the most important area of research integration will concern the way in which subjects such as thermal biology might relate to understanding and mitigating losses of biodiversity; in particular, herpetologists will want to predict the potential impact of localized drought and temperature changes related to global warming and climate change.   HBL

REPTILES AT RISK

The need to maintain sustainability in exploiting vulnerable populations

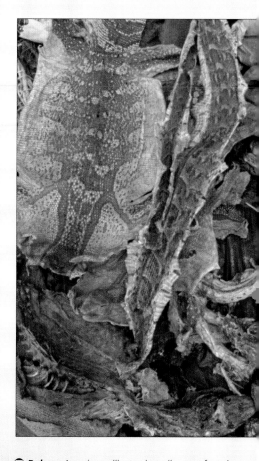

REPTILES HAVE BEEN EXPLOITED FOR FOOD EVER since humans began to share their living spaces, and they or their products have also been put to many other uses as well: in the cultural artifacts of indigenous peoples (such as the lizard-skin drumheads of New Guinea), in decorative objects (using tortoiseshell), in clothing and other articles in developed countries – footwear, jackets, watchstraps, handbags – as pets and curiosities, as rodent controllers, and for medicinal purposes or for medical research.

These various uses are not necessarily mutually exclusive. Snakes may be eaten for the supposed medicinal benefits of their flesh, for example, as well as for their nutritional value. In fact, belief in the curative or aphrodisiac properties of seaturtle eggs and snake organs drives much of the commercial exploitation of reptiles in Asia. Cultural factors may also play a part; for example, rattlesnake roundups retain an anachronistic ritual significance in the rural Southwest of the USA.

For millennia, while the numbers of people were low and technological capacity limited, local subsistence use of reptiles for the most part had little significant impact on the target species. However, as human populations have expanded and markets developed, such use has often been transformed into more systematic trading that in some cases has driven exploitation beyond the limits of sustainability. Sustainable levels are often difficult to assess, but in principle are those that can be maintained indefinitely without harming the target population. One reason why sustainability is difficult to measure is that populations are frequently affected at the same time by changes to their habitat.

Over the centuries, many reptile populations around the world have suffered from excessive use. Historic and recent population declines in seaturtles serve to illustrate the adverse effects of international trading. From the 17th century on, live Green turtles were shipped from the Caribbean

◐ Below *American alligator juveniles at a farm in Florida. Encouraged by the International Union for the Conservation of Nature (IUCN), the sustainable use of crocodilians as a renewable resource helps conserve wild populations. Crocodile and alligator ranching, in countries as diverse as the USA, Egypt, Papua New Guinea, and Australia, is now big business, generating over US$200 million annually. In Florida alone, over 30 farms produce 136,000kg (300,000lb) of meat and 15,000 skins per year.*

contributed to rural development in several places, including Papua New Guinea.

For many of the world's reptiles, food use may have affected local populations, but other factors such as habitat loss or collection for the pet or skin trade have been far more significant in hastening decline. Rural people in the tropics only occasionally keep tortoises or other reptiles, but this activity is widespread in developed countries, particularly in North America and Europe. While demand remains high, less wealthy countries have strong economic incentives for meeting it, even though some may lack the political will or the administrative means to keep the exploitation within sustainable limits. For instance, hundreds of thousands of tortoises were collected in Mediterranean countries and exported to northern and central Europe during the 1970s, and the impact on populations was so great that the trade was subsequently banned under CITES and local legislation.

Among specialist reptile-keepers, there is often a particular demand for species that are rare in the wild or have a restricted distribution, or those that are either attractively patterned or venomous.

Some species, including vipers in Turkey and Southwest Asia, combine all these properties, and many have been severely overcollected.

In recent years, major concern has developed for the enormous diversity of land tortoises and freshwater turtles in Asia. In many species there is evidence of widespread population decline, and even local extirpation, as a result of intensive collection for food and medicinal purposes. The trade is unregulated, so records are imperfect, but it appears that thousands of tons of live turtles have been entering China from neighboring South and Southeast Asia annually (see The Asian Turtle Crisis). There is similar concern for some turtle species in Africa and in the Americas, for example the Alligator snapping turtle in the USA.

At the same time, developments in medicine have given a new impetus to the exploitation of reptiles in the interests of medical research. Aspects of reptilian anatomy and physiology, including the circulatory and immune systems, have been investigated, and the properties of snake venom (as well as that of helodermatid lizards) have been examined by researchers interested in vitamin metabolism, neuromuscular transmission, blood pressure regulation, and blood clotting in humans. Isolation of critical venom components has in some cases led to the development of important drugs and new bioassay tools.   BG

to London, where they were made into soup, while in the early to mid-20th century nesting populations in many parts of the world supplied cities, mainly in Europe, with turtle products for the food trade. Besides meat, cartilage, and eggs, marine turtles also provided oil and leather, or were sold stuffed as curios.

International and local demand for these products, mainly derived from nesting females, in time brought about widespread decline in the turtle populations. Because females are late-maturing, potentially long-lived, and very rarely move to new beaches, depleted nesting populations do not readily recover; in parts of the Caribbean and in the eastern Mediterranean, they have still not been restored, even though most seaturtles have been listed by the Convention on International Trade in Endangered Species (CITES) since it came into force in 1975, and this has been effective in helping maintain some populations.

In contrast to the fate of seaturtles, traffic in crocodile, caiman, and alligator products has been greatly facilitated by CITES listings that permit trade from appropriately managed populations. These species mature relatively young, can be maintained in captivity, and can be monitored in the wild. Managed use of these species has

PLAYFUL REPTILES

Do turtles, lizards, and crocodilians indulge in play for its own sake?

FOR ALL THE FASCINATION PEOPLE HAVE FOR reptiles, they typically have not held their intelligence, cognitive abilities, or emotions in high regard. This view has started to change in recent years, however, as the animals' ability to learn many things including escape and migration routes and foraging behavior, and even to recognize individual conspecifics and keepers, has been increasingly acknowledged.

Even so, the psychological dimensions of reptiles are still largely ignored by science, although abundant anecdotal information suggests that there are several promising lines of investigation to be pursued. Among these is a key behavior that enthralls those with pet cats and dogs, as well as visitors to mammal exhibits at zoos: playfulness. "Playful" is not a term typically, if ever, used in describing reptiles, an apparent discontinuity that would seem to raise the question of what the differences are between ectothermic and endothermic vertebrates – specifically, reptiles on the one hand, and birds and mammals on the other – that might explain the difference. In fact, while most reptiles do not play in the sustained, vigorous manner of many mammals, it seems that some reptiles may actually do so.

But what is play, and how can we recognize it? One useful definition summarizes it as repeated, intrinsically rewarding but incompletely functional behavior, differing from more serious activities structurally, contextually, or ontogenetically, and initiated when the animal is in a relaxed or unstressed setting. Play behavior in animals is typically categorized under three headings: locomotor, object, and social. Locomotor play includes running, leaping, and rolling behavior. Object play is seen when animals push, hit, grasp, bite, or shake objects; these activities often relate to predatory behavior, as for example when cats repeatedly pounce on small, moving, inanimate objects or snatch and release live prey. Social play typically involves chasing or wrestling with conspecifics such as littermates or parents, but may also involve familiar humans, as in much playful interaction between humans and dogs.

Does anything comparable to the modes of play described above exist in turtles, lizards, snakes, or crocodilians? As it happens, Nile softshell turtles (*Trionyx triunguis*) will strike with their snouts at basketballs or plastic bottles floating on the surface of their tanks, and will also swim through hoops and play tug-of-war with their keepers. Such activities have been observed in these rarely-kept animals at the Toronto Zoo and the National Zoo in Washington, DC. Loggerhead and Green seaturtles have also been seen to manipulate objects, while Wood turtles (*Glyptemys insculpta*) have been observed repeatedly sliding down slopes into water, as river otters do. Sliding and swimming behavior might be causally similar to comparable behavior in mammals, but few close comparisons are available.

Komodo dragons (*Varanus komodoensis*), and perhaps also other species of monitor lizards, may play even more vigorously and repeatedly with objects. A dragon at the London Zoo reportedly pushed a shovel left behind by its keeper all around the cage, apparently enamored with the sound it made as it scraped over the rock substrate. A young female Komodo at the National Zoo would grab and shake various objects such as toy statues, beverage cans, plastic rings, and blankets. These objects were not confused with food, because she would only try to swallow them if they were coated with rat blood (in which case she would also become much more defensive, even to her keeper, to whom she was otherwise very attached). She would also repeatedly insert her head into boxes, shoes, blankets, and other items, seemingly for the stimulation offered by the experience. Social play was demonstrated by her interaction with objects worn or held by the keeper, playing gentle tug-of-war games, and otherwise acting in quite a doglike fashion. Viewing speeded-up videos of her behavior made the comparison with mammalian play even more compelling.

With their advanced parental care and close relationship to birds, crocodilians might be expected to play. There is, in fact, a field observation of an American alligator (*Alligator mississippiensis*) repeatedly circling around, and suddenly moving in and snapping at, water drops falling from a turned-off spigot in a pond. The behavior was not performed in order to gain food, but seemed rather to be a simulation of predatory behavior, just as in a cat.

Social play among reptiles has not been as well documented as object play. Rudimentary headbob displays in hatchling Eastern fence lizards (*Sceloporus undulatus*) have been compared with play, as has wrestling behavior in newborn Two-lined chameleons (*Chamaeleo bitaenatus*). However, the best-documented social play may be the precocious courtship seen in many North American emydid turtles, such as *Pseudemys nelsoni*. This behavior involves sexually immature animals of both sexes directing foreclaw vibration, typically seen only in males in courtship, toward each other and even at objects. Since it has been reported that much "play fighting" in mammals may really be more sexual than aggressive in nature, the existence of precocious sexual behavior in these turtles might be a form of play. To date, typical kinds of play have not been observed in any snake species, which are not as overtly social, nor even as active, as many lizards.

The play examples that have thus far been documented occur most often either in large, long-lived species or in those that exhibit relatively complex predatory or social behavior. Since adults

sometimes perform them, they cannot be regarded simply as practice for adulthood. Play may be particularly prevalent in well-fed, captive animals in rather spartan environments, as a response to boredom and stimulus deprivation. Even if that is the case, however, it in no way means that the activity is unimportant; in fact, just as enrichment is often provided for captive mammals and birds,

it may be that many reptiles have similar behavioral needs. Indeed, our lack of empathic understanding of reptiles may actually have prevented us from noticing commonalties with other vertebrates, as well as from grasping the importance of providing environments that develop and fulfill the animals' behavioral and psychological potential and needs.   GMB

○ Above *In their wild habitat of Indonesia, Komodo dragons feed on a deer they have killed. Research conducted with captive Komodos has provided convincing evidence that these reptiles will engage in play for its own sake, and do not treat all objects as surrogate prey, regardless of context.*

PRE-EJACULATORS, SNEAKERS, AND SHE-MALES

Alternative mating strategies in reptiles

ALTERNATIVE MATING STRATEGIES HAVE EVOLVED on a number of occasions in the animal kingdom and generally involve specialized morphologies, physiologies, or behaviors that permit successful reproduction by more than one male type. Among reptiles, the best-documented cases are among squamates (lizards and snakes). In contrast to systems where resources or mates are defended, mating systems with alternative strategies make it possible for "lesser" males to mate and produce offspring, despite their competitive disadvantage. Not surprisingly, alternative mating strategies have fascinated evolutionary biologists; a system in which all males are not equal in physical attributes, and use tricky means to compete for reproductive success, is certain to capture the attention of any natural historian. A result of this fascination is the colorful terminology that is often applied to different mating strategies: "sneakers," "dominant males," "satellites," "reproductive parasites," "pirates," and "she-males" are caught up in the most basic competition for reproductive success.

The genetic and environmental determinants of alternative mating systems vary among species. In some, alternative strategies exist because individuals with a low probability of success adopt new behavioral repertoires to increase their reproductive chances. In these cases, the strategies are condition-dependent, and increase the individual fitness of an otherwise unsuccessful male. In other cases, alternative strategies may be genetically determined. It is the relative fitness of individuals adopting each strategy that promotes the continued existence of morphological or behavioral polymorphism within a single population.

Three examples of alternative mating strategies underscore the varying ecological contexts in which strategies can evolve and be maintained. During the mating season, male Marine iguanas (*Amblyrhyncus cristatus*) form clusters of individual territories called leks. Large males have a competitive advantage and obtain the best territories in the lek, together with most of the copulations; thus, a small number of males father most of the offspring each year. Nonetheless, smaller, nonterritorial males also attempt to copulate with females. During copulation, males take approximately 3 minutes to ejaculate, and small males rarely achieve these durations because of harassment by larger males. The solution to this problem is an alternative behavioral strategy: small males pre-ejaculate and maintain sperm in their hemipenial sacs until the opportunity to copulate with a female arises. In this way, even though their copulation duration may be decreased, small males can immediately transfer sperm to any female they

encounter. This tactic compensates for their competitive disadvantage, and increases their reproductive success.

Alternative male mating strategies have also been observed in non-lekking species. In the Common Side-blotched lizard (*Uta stansburiana*) of North America, most males defend resource-based territories that overlap with the home ranges of one or more females with whom they copulate during the breeding season. Males in some populations of this species vary in physiology, coloration, and their degree of territorial investment. Orange-throated males are polygynous and maintain large territories with many females. Blue-throated males are mate-guarders. They also

maintain territories, but copulate with a smaller number of females and defend them after copulation. Finally, "sneaker" yellow-throated males are not territorial and mimic females, making forays into the territories of other males and copulating with their females. This system persists in nature in part because, in a manner similar to the "rock-paper-scissors" game, each male type has advantages that allow it to outcompete another morph, but weaknesses that leave it vulnerable to the tactics of the third. Sneaking yellow males are particularly successful when in competition with the territorial and polygynous orange males, but their antics are less successful against the blue-throated mate-guarders. Mate-guarding blue males are

● **Right** A mating ball of Red-sided garter snakes. Studies have shown that, in such aggregations, the confusion sown by estrogen-releasing "she-males" makes them twice as likely as normal males to achieve mating success. Counteracting this, however, is the fact that the estrogen reduces their sperm count.

● **Below** Marine iguanas on the Galapagos Islands are a lekking species, in which males congregate seasonally in large numbers at a sexual display area and compete to attract mates. Although the largest, most imposing males (which also take on a red and blue breeding coloration) have most success, smaller males improve their chances of mating through the strategy of pre-ejaculation.

successful in deterring sneakers, but are vulnerable to displacement by more aggressive orange males. Thus, there is no single winning strategy and all three male types share reproductive success.

Sneaking can be a high-cost strategy, since it often involves close proximity to a larger or more competitive male. A recurrent theme in alternative male mating strategies is female mimicry, to reduce the probability of detection and evade competition with dominant males. This strategy is taken to an extreme in the Red-sided garter snake (*Thamnophis sirtalis parietalis*). Early each spring, Manitoba garter snakes aggregate by the thousands to breed. Males outnumber females in these aggregations, and male–male competition is intense; anything from 10 to 100 males simultaneously court and form "mating balls" around receptive females. In this system, a small proportion of males, which scientists have termed "she-males," release a pheromone that attracts other males, as though they were females. Confused by such signals, normal males expend their reproductive effort fruitlessly on the bogus females; moreover, since they regard the she-male as a female, they do not try to interfere in its courting. She-males thus secure a clear advantage in the courting groups; in competitive mating trials, they mated with females significantly more often than did normal males, demonstrating both reproductive competence and a possible selective advantage to males with this female-like pheromone.

Sexual selection operates only in systems where individuals vary in phenotype, resulting in differential reproductive success. The three examples above illustrate that the seemingly "best" male is not always the victor. The success of alternative strategies promotes the maintenance of polymorphism within populations regardless of whether male strategy is genetically or environmentally determined.   KRZ

TEMPERATURE AND SEX

How slight variations in warmth determine the sex of some reptile embryos

TEMPERATURE AND SEX ARE ATTENTION-grabbing topics for biologists just as they are for the rest of the world. Because sexual reproduction is so fundamental an aspect of organisms, one might assume that the means by which sex is determined would have remained relatively constant in the course of evolutionary change. Surprisingly, however, that assumption turns out to be incorrect. In fact, sex is determined in various different ways, and reptiles are classic examples.

Sex is fixed in reptiles in two main ways. First and most familiar is genotypic sex determination (GSD), in which sex is determined at conception, for example by sex chromosomes, as in humans. The other system is more remarkable altogether: in temperature-dependent sex determination (TSD), whether an offspring is born male or female depends on temperatures experienced during roughly the middle third of embryonic development. Biologists interested in understanding such mechanisms have placed considerable emphasis on studying reptiles, partly because they offer a diverse range of sex-determining mechanisms but also in view of the fact that we increasingly know how the major reptile groups are related ancestrally. These characteristics permit rich exploration of the biology of unusual mechanisms like TSD.

Sex-determining mechanisms are not distributed uniformly among vertebrates. Amphibians, snakes, birds, and mammals, as well as nearly all fish, follow the genotypic sex route. In contrast, all tuatara and crocodilians are temperature-dependent, while lizards and turtles exhibit both mechanisms in different species. Genotypic sex determination occurs much more frequently in lizards than in turtles, while the opposite is true

for TSD. Phylogenetic analyses suggest that genotypic determination is probably the ancestral mechanism for vertebrates – dinosaurs probably had GSD – but that temperature-dependent mechanisms have evolved independently in reptiles on several occasions.

Neither form of sex determination conforms to a single pattern. In many turtles, low temperatures produce male offspring and high temperatures produce females, a situation categorized as Pattern Ia TSD; but the converse (Pattern Ib TSD) is thought to be true of some lizards, and perhaps also of tuatara. Finally, females are produced at low and high temperatures (and males at intermediate temperatures) in crocodilians, in many TSD-exhibiting lizard species, and in a few turtles (Pattern II TSD). Recent results suggest that this last pattern is the ancestral state in reptiles, with perhaps a single subsequent origin of Pattern Ia in turtles.

Of greater interest is the adaptive significance of TSD. It is easy enough to grasp the link between GSD and 1:1 sex ratios, but the potentially skewed sex ratios implicit in TSD are harder to explain. Why should a trait as fundamental to an organism as its sex be left to the vagaries of the environment? Understanding this conundrum has posed a considerable challenge.

One promising line of research has set out to investigate whether some developmental temperatures enhance the fitness of males more than females, and vice versa. For example, young snapping turtles bred from incubation temperatures that produce both sexes are more active than their consexuals from temperatures producing only one sex – and as a result they seem to suffer greater mortality from visually-oriented predators. At

○ **Above** Research on Leopard geckos, which have Pattern II TSD, shows that topical application of estrogen counteracts the bias toward male offspring at intermediate temperatures. This technique can be used to halt the decline of some endangered species, by skewing the sex ratio toward more breeding females.

◖ **Left** Graphs showing the gender ratios produced by the three different patterns of temperature-dependent sex determination and genotypic sex determination. The TSD curves and GSD line are plotted against a horizontal axis of egg incubation temperature, while the vertical axis shows the incidence of male hatchlings, expressed as a percentage of all offspring.

◗ **Right** A crocodile embryo shortly before hatching. Pattern II TSD is ubiquitous in crocodilians; for example, a study of Nile crocodiles undertaken in the Lake St. Lucia game reserve in KwaZulu–Natal, South Africa, demonstrated that female hatchlings were in the majority at temperatures below 31.7 °C (89.1 °F) and above 34.5 °C (94.1 °F). Between these temperatures, male offspring predominated.

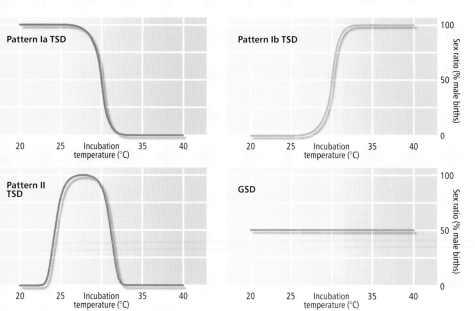

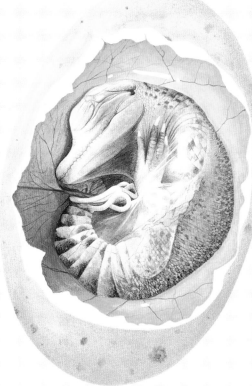

present, however, there is insufficient evidence to indicate how widely such findings might apply. In certain lizard species, for example, some incubation temperatures have been shown to favor male and others female offspring – but the species concerned have GSD! Despite extensive effort, we do not yet have a comprehensive answer as to why TSD exists in reptiles.

Even so, emerging results are providing remarkable insights into certain aspects of TSD, including some intriguing maternal effects. In the laboratory, individuals from some gecko species with TSD have been shown to seek out nest sites with particular thermal regimes. Turtles not only display homing behavior when preparing to lay eggs; some species such as Painted turtles also regularly choose nest locations with characteristics that correlate with a particular offspring sex ratio, so their broods may exhibit a consistent bias throughout their lives. Turtles, and possibly other reptiles that have TSD, may be able to manipulate offspring sex ratio further by altering the levels of hormones that they allocate to their egg yolks!

Of course the peculiar sensitivity of this form of sex determination makes animals exhibiting TSD very vulnerable to environmental disturbances, especially if they are already imperiled, as is the case with many seaturtles, crocodilians, and tuatara. If offspring sex ratio in the laboratory is profoundly affected by even minor changes in incubation temperature, what happens in the outside world when the climate warms a few degrees or a wooded area that once provided shade for a nesting area is cleared? Conversely, what is the impact when shade trees are planted, or a multi-story condominium is built next to such a spot? All research to date on these questions reveals skewed offspring sex ratios in reptiles with TSD. At the same time, the embryonic sexual development of American alligators and Red-eared slider turtles is also known to be highly susceptible to hormone-mimicking chemicals that are breakdown products of some commonly-used herbicides and pesticides. Although reptiles with TSD have survived substantial environmental upheaval over hundreds of millions of years, the sheer speed of current changes and the diversity of threats affecting the sex ratios of their young may be unprecedented.　　FJJ

Turtles and Tortoises

URTLES ARE PERHAPS THE MOST RECOGNIZABLE *backboned animals on the planet. They are the only vertebrates with a shell of ribs and dermal bone, inside of which lie the shoulder bones. This critical innovation appeared during the Triassic Period, over 220 million years ago, before there were mammals, birds, lizards, snakes, crocodiles, or flowering plants, but at the same time as the earliest dinosaurs. This novel adaptation probably accounts (at least in part) for the chelonians' subsequent success during and beyond the Age of Dinosaurs.*

All living turtles have many features in common. They lack teeth, have internal fertilization, lay shelled (amniotic) eggs in a nest constructed by the female, and share a number of life-history characteristics (for example, late maturity, extreme longevity, and low adult mortality) that make them especially vulnerable to the activities of humans, the only organism to challenge their existence in over 200 million years. In fact, human impact on these animals has now increased to a point at which over 44 percent of known turtle species are officially considered critically endangered, endangered, or vulnerable to extinction.

This scenario is all the more unfortunate because of the turtles' extraordinary diversity.

◐ **Below** *The plastron and carapace of softshell turtles are covered with a leathery skin rather than horny scutes. As highly aquatic forms, they are equipped with long necks and snorkel-like snouts to enable them to breathe without surfacing. Shown here is a Spiny softshell turtle (Apalone spinifera).*

◑ **Right** *Named for the blotched patterning on its dorsal scutes, the Leopard tortoise (Psammobates pardalis) is a large species from south and east Africa.*

There are species that migrate as far as 4,500km (2,800 miles) to nest, using ocean currents and the earth's magnetic field as navigation aids; species that have mass nesting frenzies, with up to 200,000 females nesting on the same small beach in a period of less than 48 hours; and species that can lay as many as 11 clutches of over 100 eggs each in a single season (others may lay 258 eggs in a single clutch).

The chelonians' uniqueness also stretches well beyond their reproductive behavior. Some can survive their first winter as hatchlings even though temperatures in the nest may drop as low as –12°C (10°F); others can hybridize and produce viable offspring despite the fact that they belong to distinctly different genera. Some can survive indefinitely under water without access to air or can dive to over 1,000m (3,300ft) below sea level, while others live at altitudes up to 3,000m (10,000ft). In size they can vary from a shell length of only 8.8cm (3.5in) to a massive 244cm (96in). How could we possibly tolerate the loss of such rich variety?

Armor on Legs
THE TURTLE'S SHELL

No other vertebrate has evolved an armor quite like the turtle's shell. Generally comprising a total of some 50–60 bones, the shell consists of two parts: a carapace covering the animal's back and a plastron (made up of 7–11 bones) covering its underside. The two are connected on each side by a bony bridge formed by extensions at the sides of the plastron.

The carapace is formed from bones, originating in the dermal layer of the skin, that are fused with one another and with the animal's ribs and vertebrae. Large scales, or scutes, derived from the epidermal layer of the skin, cover and strengthen the bony constituents of the shell. The plastron is derived from certain bones of the shoulder girdle (the clavicle and interclavicle), the sternum, and the gastralia (abdominal ribs, like those found in crocodilians and tuatara today). The remaining portion of the shoulder girdle has shifted inside the turtle's ribs, a unique feature found in no other vertebrate, present or past.

So successful has this protective armor been that it has become the cornerstone of turtle architecture. Other adaptations have been built around it, and it accounts both for the longevity of the

line and its limited variation in locomotion. Because of the shell, running, jumping, or flying turtles were not viable evolutionary options, but moderate adaptive radiation has occurred within the group. Beginning as semiaquatic marsh dwellers, some turtles evolved to become totally terrestrial, inhabiting forests, grasslands, and deserts. Others became more thoroughly aquatic, invading lakes, rivers, estuaries, and the sea.

Ironically, the massive shell that led to their initial success has become much reduced in most modern lines. The larger tortoises have retained an extensive shell, while greatly reducing its weight through marked thinning of the shell bones; strength is provided by the shinglelike covering of

◑ **Right** *The structure of the chelonian shell:*
1 Transverse section; 2 Longitudinal section, showing the arrangement of pelvic and pectoral girdles.

TURTLES AND TORTOISES

Order: Testudines (Chelonia, Testudinata)

310 species in 96 genera and 14 families

DISTRIBUTION Temperate and tropical regions, on all continents except Antarctica and in all oceans.

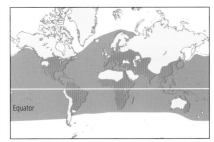

Equator

HABITAT Marine, freshwater aquatic and semiaquatic, terrestrial.

SIZE Length (maximum straight-line shell measurement) 11–244cm (4–96in).

COLOR Highly variable, ranging from dull, dark, and cryptic in bottomd-wellers to brightly patterned conspicuous forms. Browns, olives, and shades of gray or black predominate on upper surfaces, with yellows, reds, and oranges common for pattern markings. Yellows, with brown, black, and white, predominate on undersurfaces.

REPRODUCTION Fertilization internal; all lay eggs on land; parental care in 1 species.

LONGEVITY Occasionally more than 150 years in captivity (best documented record c.200 years); some American box turtles thought to survive up to 120 years in wild; aquatic species probably shorter-lived.

CONSERVATION STATUS Almost half of all turtle and tortoise species are considered to be at risk; the IUCN currently lists 24 species as Critically Endangered, 48 as Endangered, and 60 as Vulnerable. In addition, 7 Recent species including the Yunnan box turtle are now identified as Extinct.

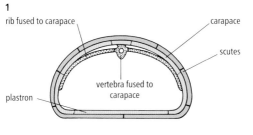

1
rib fused to carapace / carapace / scutes / vertebra fused to carapace / plastron

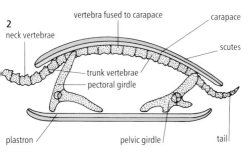

2
neck vertebrae / vertebra fused to carapace / carapace / scutes / trunk vertebrae / pectoral girdle / plastron / pelvic girdle / tail

durable, lightweight scutes and by the arching shape of the shell rather than by heavy bone. In some turtles, especially highly aquatic forms such as softshells and seaturtles, the bones of the shell have decreased in size, leaving large spaces (fontanelles) between them. The most extreme case is the Leatherback seaturtle, which as an adult lacks epidermal scutes and has only small, bony platelets embedded in its leathery skin. The chief advantages of shell reduction seem to be the reduced physiological costs associated with building and maintaining the heavy shell, as well as a lower energy cost for locomotion in terrestrial species and greater buoyancy in the aquatic forms.

Different lifestyles and ecologies have led to other alterations in shell structure. Land turtles typically have high-vaulted shells as a defense against their predators' crushing jaws. Water turtles have lower, more streamlined shells that offer less water resistance while swimming. Extreme flattening is found among the softshells, helping them to hide beneath sand and mud on the bottom of their watery environment. Exceptions exist in both categories, however. The shell of the Pancake tortoise of East Africa is not domed but extremely flattened, allowing it to squeeze into narrow crevices within its rocky habitat; with the force generated by its legs and the natural elasticity of the shell bones, this tortoise becomes extremely difficult to extract once it has become wedged between rocks. On the other hand, aquatic species such as the River terrapin of Asia, the Annam turtle (*Mauremys annamensis*), and the Florida red-bellied cooter, which cohabit with large, voracious crocodilians, have forfeited some streamlining and have evolved high-vaulted, strongly buttressed shells for protection.

The shell is least developed in young turtles, in

TELLING A TURTLE'S AGE

All turtles but the softshells, the Pig-nosed turtle, and Leatherback seaturtles have a bony carapace and plastron covered by epidermal scales called scutes, which are similar in structure to human fingernails. As the turtle grows during each active season, the scutes grow outward, typically from one corner. However, when growth ceases for the winter (or for some tropical species, during the dry season), a ring, or annulus, develops, which marks the period of no growth. These annuli are most obvious in juvenile turtles, because they are typically separated by a wide band of lighter-colored scute growth. In most turtles the annuli are most easily seen on the central (abdominal) scutes on the plastron.

As the turtle reaches maturity, the rate of body and scute growth slows. Annuli are no longer separated by wide bands of growth. Rather, they are laid down immediately adjacent to one another, making them very difficult to count. In addition, as the turtle ages, increasing abrasion of the plastron against the substrate can wear the scutes smooth, obscuring the annuli. Finally, an unusually cold, or hot and dry, period during the growing season can cause scute growth nearly to cease, resulting in the formation of a weak, "false" annulus. Thus, aging turtles by counting scute annuli is most accurate for juveniles, especially those with distinct alternate growth and non-growth seasons. Yet the only sure way to age a turtle precisely is to recapture it regularly through its life.

The figure illustrates the plastron of a Yellow mud turtle from a field site in Nebraska, USA, that was photographed on 4 May 2000 when it was 9 winters old (i.e., after 8 growing seasons) and just under 7.1cm (3in) in shell length. This turtle was first caught on 15 May 1993, when it was just over 3cm (1.2in) long and was emerging from torpor after its second winter of life (i.e., after one growing season). It was uniquely and permanently marked with tiny notches in the shell (one is visible at the margin of the plastron at the bottom of the picture).

The scute areas marked "H" show the scute sizes when this turtle hatched during the fall of 1991.

Arrows indicate the summer growth periods by year. Growth in 1993 was minimal, as was that in 1994. Yet 1995 and 1996 were years of very good growth. However, there was almost no growth in 1997 or 1998, resulting in what appears to be a single annulus, but is in fact three annuli. During the cool spring of 1998 the turtle (then seven winters old) did not emerge from hibernation until 28 May. The water level in its wetland was unusually low, and as it fell further in the summer heat, the turtle left the marsh on 23 June (after a total active season of only 26 days), buried itself in the adjacent sandhill, did not emerge again until 19 May 1999, and grew hardly at all. As a result, the 7th and 8th annuli cannot be individually distinguished from the 6th. So, although this turtle appears to have 7 true annuli, it is actually 9 winters old.

The evidence from annuli counts, plus over 20 years of recaptures, suggests that many Yellow mud turtles exceed 35 years in nature. However, some turtles have been known to live for well over a century. Reliable records about pet tortoises that outlive their original owners are difficult to obtain. One dubious record concerns a Radiated tortoise supposedly given to the King of Tonga by Captain James Cook in 1773 or 1777. The tortoise was highly prized by successive Tongan rulers. After surviving two forest fires and other mishaps, it died in 1966, having spent at least 189 years in captivity. Unfortunately, there is no record of the gift in James Cook's journals.

The so-called "Marion's" tortoise is better documented. Five tortoises collected by the French explorer Marion de Fresne in 1776 from the Seychelles were taken to Mauritius. The last of these finally perished in 1918, after falling into a gun emplacement at the barracks where it lived. The tortoise was presumably an adult when caught and so could have been nearing 200 years old when it died.

◗ **Right** *The aptly-named Northern Australian snake-necked turtle (Chelodina rugosa) protects its head and long neck by withdrawing them horizontally into the outer margin of its shell. This capacity, common to all members of the suborder Pleurodira, has earned them the designation "side-necked turtles."*

which the bones are not yet fully formed. However, the protective importance of the bony shell improves with age. The small, soft-boned young of certain species have evolved prominent epidermal spines, most commonly around the edge of the shell, to discourage predators. An extreme example of this adaptation is found in the Spiny turtle of Southeast Asia. Nearly circular in outline, with each marginal scute attenuated into a spine, juveniles of this species resemble sharply-toothed cogwheels. The shells of the Matamata and the Alligator snapping turtle are disguised with bumps and ridges, giving them an inanimate appearance.

A number of turtles have evolved non-rigid shells with varying degrees of movement (kinesis) between the bones. A common modification is the hinged plastron. Various turtles, including Indian flapshells, American and African mud turtles, Asian and American box turtles, and Madagascan and Egyptian tortoises, have evolved one or two plastral hinges. African hinge-backed tortoises have a hinge on the carapace rather than on the plastron. Hinges give these turtles the capability of closing the shell with the vulnerable parts safely within. While this no doubt provides protection from predators, protection against loss of moisture is probably an equally important function: few species with hinges are fully aquatic.

A lesser degree of movement is found among certain pond and river turtles. Some, such as the Asian leaf turtles, the Flat-shelled turtle of Southeast Asia, and the Neotropical wood turtles, have partial hinging of the plastron, and connections of ligament rather than bone between the plastron and carapace. Their plastrons can move a little but cannot close the carapacial opening. This flexibility appears necessary to allow the turtles to lay their huge, brittle-shelled eggs, which could not otherwise fit through the shell opening. In several Asian species, including the Cochin Forest cane turtle, Spiny turtle, and Tricarinate hill turtle, only mature females develop the kinetic plastron.

Certain large-headed, aggressive species, such as Snapping turtles and Neotropical musk turtles, also have movable plastrons due to a reduction in the bones and a ligamentous connection to the carapace. In these species the flexibility of the plastron allows the turtle to retract its large head within the shell with jaws agape, thus providing a formidable and nearly impregnable defense.

Early on, turtles evolved two separate mechanisms for retracting their necks within the anterior opening between the carapace and the plastron. All turtles have eight cervical (neck) vertebrae, but one major group (the Pleurodira, or

side-necked turtles) retracts the head horizontally to the side, leaving the neck and head somewhat exposed at the front of the shell. This group includes the snake-necked turtles, some of which have necks that exceed the shell in length. The other group (the Cryptodira, or hidden-necked turtles) retracts the head by folding the neck back on itself in a tight vertical "S" shape. Most of these turtles can withdraw their heads completely inside the shell, and even bring their elbows together in front of the nose to further protect the head. Three-fourths of all living turtles belong to this latter group, including all of those in North America, Europe, and mainland Asia.

Showing their Age
GROWTH

Most fully-grown turtles and tortoises attain a size of at least 13cm (5in) in shell length. Exceptions include the Speckled cape tortoise, Flattened musk turtle, and Bog turtle, among the world's smallest species with maximum shell lengths of less than 12cm (4.7in). The giant of living turtles is the Leatherback seaturtle, which attains a shell length of 244cm (96in) and may weigh up to 867kg (1,907lb). Other impressively large species are the Alligator snapping turtle at 80cm (31in) and 113kg (249lb); the South American river turtle at 107cm (42in) and 90kg (198lb); the Asian Narrow-headed softshell at 120cm (47in) and over 150kg (330lb); and the Aldabra giant tortoise at 140cm (55in) and 255kg (561lb).

Growth rates of turtles vary considerably even within members of the same clutch. Habitat, temperature, rainfall, sunshine, food type and availability, and sex have all been associated with the rate of growth. Generally, a species grows rapidly to sexual maturity; then slows down markedly. In later years small species may stop growing altogether, but larger turtles can grow throughout life. Not surprisingly, turtles that live long lives also take longer to reach sexual maturity.

Growth can be more conveniently studied in turtles than in most reptiles because many species carry growth records on their horny scutes (see Telling a Turtle's Age). The rate of growth of a turtle is also reflected in the growth layers deposited in its long bones (especially the femur and

humerus). As with trees and scutes, distinct rings are added to these bones as growth slows during the winter or dry season; because of natural bone remodeling, however, the inner (younger) growth layers are constantly being eliminated. As a result of this natural source of potential error and because the technique requires destructive sampling, it has not been applied extensively.

Not So Slow in the Water
LOCOMOTION

Turtles are reputedly slow-moving animals. Certainly most tortoises, with their mobility greatly restricted by a large, cumbersome shell, are slow. The Desert tortoise moves at speeds of 0.22–0.48km/hour (0.13–0.3 mph). Charles Darwin clocked a Galapagos giant tortoise at 6.4km (4mi) per day. However, seaturtles can move as swiftly through water as humans can run on land, at speeds in excess of 30km/hour (18.6mph).

The limbs of a turtle are good indicators of its habitat and means of locomotion. Tortoises, the most terrestrially adapted group, have elephantine feet in which the toes are very short and lack all traces of webbing. In the Gopher tortoise, an accomplished burrower, the front limbs are much flattened and serve as digging scoops.

The feet of aquatic turtles differ by having longer toes joined together by a fleshy or membranous web, providing the feet with additional thrust through the water. Aquatic turtles move either by bottom-walking or by swimming. Bottom-walkers move over the water's bed in much the

▷ **Right** *Acting like paddles, the powerful forelimbs of the Hawksbill seaturtle* (Eretmochelys imbricata) *allow it to "fly" gracefully through the water. Sponges and soft corals form a major part of this species' diet.*

same manner that they walk on land. The Mata-mata, the Southeast Asian box turtle, snapping turtles, and mud turtles all use bottom-walking as their primary mode of locomotion. Swimming species such as the Smooth softshell, the River terrapin, and the Central American river turtle may bottom-walk but usually paddle through the water, using all four limbs in an alternating sequence. Much of the power stroke is accomplished by simultaneous retraction of the opposing front and hind foot, which provides thrust while maintaining the animal's direction of travel.

Seaturtles and the Pig-nose turtle are the most specialized swimmers. Their forelimbs, modified into flipper-shaped blades, move synchronously and gracefully in what might best be termed "aquatic flying." The hind feet provide little propulsion and serve chiefly as rudders.

Deep Breathing
RESPIRATION

As descendants of terrestrial ancestors, turtles not surprisingly respire with lungs. Due to their rigid shell, however, their breathing differs from that of other vertebrates. Pressure changes within the lungs of most turtles are created by muscles that expand and then retract within the front- and hind-limb pockets. Expiration is aided by abdominal muscles that compress the lungs by forcing the internal organs against them. Movements of the limbs and girdles augment these actions.

Lungs are not the only organs of respiration in turtles. Aquatic species can also respire through their skin, the lining of the throat, and through thin-walled sacs, or bursae, in the cloaca. The degree to which these accessory devices are used varies with the species. In Nile softshells, which lack cloacal bursae, 70 percent of the submerged oxygen intake is through the skin and 30 percent is through the lining of the throat. Because of their dependency upon underwater respiration, softshells tend to be more sensitive to rotenone (a selective poison for gill-breathing vertebrates) than are cohabiting species of pond turtles. Cloacal bursae are common to most other aquatic species, including pond, snapping, and side-necked

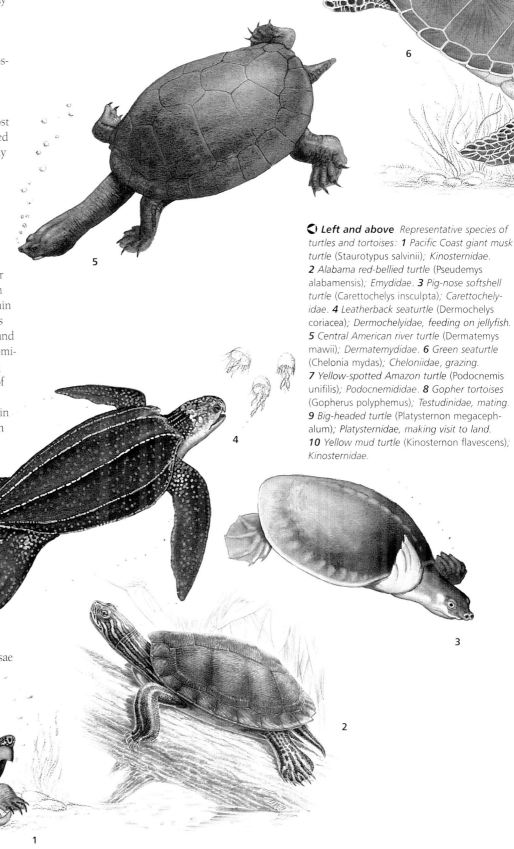

◖ **Left and above** *Representative species of turtles and tortoises:* **1** *Pacific Coast giant musk turtle* (Staurotypus salvinii); *Kinosternidae.* **2** *Alabama red-bellied turtle* (Pseudemys alabamensis); *Emydidae.* **3** *Pig-nose softshell turtle* (Carettochelys insculpta); *Carettochelyidae.* **4** *Leatherback seaturtle* (Dermochelys coriacea); *Dermochelyidae, feeding on jellyfish.* **5** *Central American river turtle* (Dermatemys mawii); *Dermatemydidae.* **6** *Green seaturtle* (Chelonia mydas); *Cheloniidae, grazing.* **7** *Yellow-spotted Amazon turtle* (Podocnemis unifilis); *Podocnemididae.* **8** *Gopher tortoises* (Gopherus polyphemus); *Testudinidae, mating.* **9** *Big-headed turtle* (Platysternon megacephalum); *Platysternidae, making visit to land.* **10** *Yellow mud turtle* (Kinosternon flavescens); *Kinosternidae.*

turtles. These structures are particularly well developed in the Fitzroy River turtle. This Australian side-necked turtle, which lives in well-oxygenated streams, continually maintains a widely gaping cloacal orifice and rarely surfaces.

Turtles are exceptionally tolerant of low oxygen levels, and individuals have survived for up to 20 hours in an atmosphere of pure nitrogen. How long a turtle can stay submerged depends upon the species, the temperature, and the amount of oxygen dissolved in the water. Pond sliders can survive submerged in water saturated with oxygen for a maximum of 28 hours, whereas Loggerhead musk turtles seem capable of surviving indefinitely under similar conditions. Species that overwinter underwater in a torpid state can survive for weeks or months without surfacing.

Unhurried Hunters
DIET

Many turtles (for example, seaturtles and tropical species) are active, and hence feeding, year-round, while species at higher latitudes may spend more than half the year inactive in a state of winter torpor underwater or underground. A few species in extremely arid environments (for example, some American mud turtles) may only be active for three months or less each year, during which time they must feed, grow, mate, and lay eggs, all

before the short "rainy" season ends.

Relatively few turtles have the speed and agility to catch fast-moving prey. Hence, most feed either on vegetation or on more sedentary animals (for example, mollusks, worms, and insect larvae). However, opportunistic events such as a dead animal or ripened fruit under a riverside tree are rapidly exploited and often attract large numbers of turtles.

Diets of omnivorous species frequently change with age. Typically, juveniles tend to be highly insectivorous, while adults are either more herbivorous (for example, the Pond slider and the Painted turtle) or exploit a more specialized diet such as mollusks (the Loggerhead musk turtle or the Snail-eating turtle). In species where the sexes differ greatly in size, diets of males and females may also differ. Female Barbour's map turtles (*Graptemys barbouri*) are principally mollusk-eaters, whereas the much smaller males mostly consume arthropods.

For the most part, turtles feed in a simple and straightforward fashion. However, a few have special techniques and strategies to obtain their food, such as ambush, gape and suck, and luring (fishing) methods. Turtles that use ambush generally lie in wait rather than pursuing their prey. However, many species employ several strategies. Commonly, ambush feeders have cryptic coloration

and/or cryptic shapes, along with long, muscular necks that can strike out for prey at some distance. The Snapping turtle, with its long, tubercle-covered neck, mud-colored skin, and algae-festooned shell, illustrates these characteristics well. The Narrow-headed softshell of southeastern Asia is an ambush feeder with a smooth, brightly patterned carapace; however, when it is lying on the bottom of its riverine habitat, partially covered with sand or mud, the turtle's dark stripes and patches blend well with shadows cast on the uneven river bed.

Most aquatic species use the gape and suck technique to some degree. By quickly opening the mouth and simultaneously expanding the throat, they create an area of low pressure that can pull small food items into the gullet, along with a rush of water that is later expelled.

The most adept practitioner of this technique, the bizarre Matamata, is extremely well camouflaged. The shell is flattened, with bumpy ridges, and is usually covered with algae. The unusually broad head and long, muscular neck are adorned with an array of irregular flaps and projections. The tiny, beady eyes are set far forward and flank an attenuated, snorkel-like snout. The mouth is preposterously wide, and the jaws lack the horny covering of other turtles.

The Matamata is also an ambush feeder. It waits passively on the bottom of its aquatic haunts for fish to approach. Experiments have shown that certain of the turtle's skin flaps on the chin and neck serve for more than just camouflage. Rich in nerve endings, they respond to slight disturbances in the water, alerting the turtle even in murky waters to the approach of prey. It has also been suggested that passive movements of the skin flaps by water currents might serve as a lure to attract fish. Once a fish comes in range, the turtle rapidly strikes while expanding its widely distensible mouth and throat, thus pulling in the day's dinner.

Matamatas have also been observed actively herding fish into shallow water where they could be more easily captured.

Only slightly less cryptic in appearance and occupying a similar niche, the Alligator snapping turtle of the United States uses a lure to attract fish. The Alligator snapper has a small, wormlike projection on its tongue that becomes bright pink when filled with blood. The rest of the oral cavity is darkly pigmented so as to exaggerate the lure even more. By moving underlying muscles, the turtle can make the lure wriggle. When fishing, the snapper sits quietly on the bottom with jaws agape, moving the lure. A fish swimming between the sharp, horny jaws to investigate rarely escapes the swift, snapping response. If the prey is small enough, it is swallowed whole; if it is too large, the turtle will use its jaws to hold the fish while the forefeet are used alternately to tear it apart. The lure darkens with age and may be of less importance to adults.

An additional feeding adaptation worthy of mention is the broad alveolar shelf, or secondary

⚫ *Below* The bizarrely shaped Matamata (Chelus fimbriatus) *inhabits lakes and sluggish creeks in the Amazon basin. It employs a highly effective "gape and suck" method of feeding, opening its mouth to create a vacuum that draws the prey in.*

palate, in the upper jaw of certain turtles. Species that eat snails and clams (for example, map turtles, the snail-eating turtles, and some American musk turtles) use this shelf for crushing the calcareous shells of their prey. A similar shelf is also present in certain herbivorous species, among them the cooters of America and the River terrapin and roofed turtles of Asia. These turtles have one or two serrated ridges on this shelf, as well as serrated edges on their jaws, which are used together for cutting and crushing stems and fruits.

Preserving the Eggs
MATING AND REPRODUCTION

Sexing turtles is often difficult. Males usually have longer, thicker tails, with the vent somewhat farther back than in females. In many aquatic swimming species, the males are considerably the smaller sex, but males in terrestrial and semiaquatic forms tend to be as large as or larger than females. To accommodate the female's high-vaulted shell during copulation, the male's plastron is often concave. Elongated foreclaws, attenuated snouts, patches of scales behind the knees, or a spine-tipped tail distinguish males of certain species.

Coloration can be used for sexing some species but, in the majority, the sexes are similarly colored. Even in sexually dichromatic species, the

differences are often subtle. The male Eastern box turtle typically has red eyes, the female brown eyes. The female Spotted turtle has a yellow chin and orange eyes; the male has a tan chin and brown eyes. Certain tropical Asian river turtles are exceptional, however, in that the males exhibit spectacular seasonal breeding coloration in contrast to the drab females.

All species lay shelled eggs on land. Nesting in most is annual and seasonal, although individual females may not reproduce every year. Seaturtles typically nest every second or third year. In turtles from temperate zones, courtship may occur in the fall or the spring, but nesting usually takes place in the spring only. In tropical species, courtship and nesting may occur in either the wet or the dry season. A few tropical and subtropical species have been found to nest nearly year-round, although it is unlikely that any turtle has truly continuous reproduction. Females of many species (for example, the Diamond-backed terrapin and Eastern box turtle) can store sperm for years and do not need to mate annually. In addition, paternity analysis using DNA fingerprinting techniques has shown that different eggs in the same clutch may have had different fathers.

The majority of turtles nest in the vicinity of their foraging areas. However, some sea and river turtles make extensive migrations in order to nest.

Green seaturtles inhabiting the Brazilian coast of South America may migrate over 4,500km (2,800mi) to nest on Ascension Island. Certain migrating species nest in large numbers over a short period of time, the most spectacular example being the mass nesting ("arribada") of the Olive ridley seaturtles. One of the largest arribadas occurs in Orissa, India, where up to 200,000 Olive ridleys nest on 5km (3mi) of beach over a period of one or two days. A few freshwater turtles such as the South American river turtle and the River terrapin of Southeast Asia similarly nest en masse. An advantage of mass nesting is that predators become overwhelmed by the sheer volume of the reproductive output, with many nests escaping detection.

In many aquatic swimming species, including some pond and river turtles, side-necked turtles, and seaturtles, males are usually smaller than females and have elaborate courtship behavior. In semiaquatic bottom-walking species (for example, mud turtles and snapping turtles) where males are as large as or larger than females, courtship displays are generally minimal. Among male tortoises, combat for territories and mates is common.

In temperate-zone pond and river turtles (the Painted turtle, the Pond slider, and the map turtles), males employ elongated claws on the

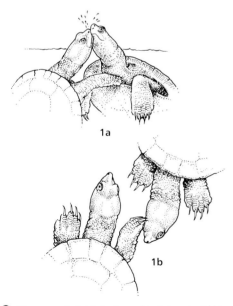

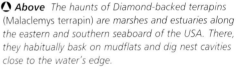

▶ **Above and right** Turtle and tortoise courtship: **1** Head-bobbing at surface **a** and head-stroking beneath surface **b**, during courtship of the side-necked turtle Emydura macquarii; **2** Male Gopher tortoise head bobs and circles female, then bites her on the shell and limbs; **3** Mating positions in Eastern box turtles: **a** male biting female's head and **b** gripping her shell with his hindfeet.

forefeet in their courtship displays. While swimming backward in front of (or in some species just above) the female, the male fans his claws across the female's snout and chin in highly stereotyped patterns. When receptive, the female sinks to the bottom. The male then mounts from the back and, using his foreclaws to grip the anterior edge of the shell, forces his tail beneath that of the female, bringing their vents together. Intromission is achieved by the male's single penis. Copulation may require an hour or more, during which time the participants occasionally surface for air.

Courtship in tortoises typically consists of some head-bobbing by the male, the male butting and biting the female to immobilize her, and finally mounting her shell from the rear. The courtship and mounting of some giant tortoises is accompanied by bellowing that would make elephants envious.

In order to achieve intromission, males may have to incline their body toward the vertical. The most extreme examples are American box turtles, which lean back somewhat beyond the vertical during mating.

The eggs may be laid beneath decaying vegetation and litter (for example, by the Spot-legged turtle Rhinoclemmys punctularia and the Stinkpot);

◑ **Above** The haunts of Diamond-backed terrapins (Malaclemys terrapin) are marshes and estuaries along the eastern and southern seaboard of the USA. There, they habitually bask on mudflats and dig nest cavities close to the water's edge.

in nests of other animals (as is the case with the Florida red-bellied cooter in alligator nests); in specially dug burrows (the Yellow mud turtle); or in a nest constructed while the female is completely underwater (the Northern Australian snake-necked turtle), but usually in carefully constructed, flask-shaped nest chambers excavated into the surface of the ground with the turtle's hind feet. Some species (for example, most softshells and the Painted terrapin) quickly cover the eggs and leave the area, but others spend considerable time concealing the nest; Leatherback seaturtles may spend an hour or more traveling over the site kicking sand in all directions before returning to the sea. The River terrapin often digs a false nest some distance from the first, and in

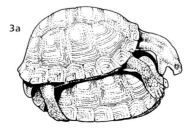

certain areas divides the clutch between two or three nests, serving to confound predators. Parental care is rare in reptiles and largely lacking among turtles, but the Asian giant tortoise, which fashions a mound nest of leaf litter, will defend eggs from potential predators for several days following laying.

Fecundity generally correlates with body size both within and across species. Smaller species lay 1–4 eggs per clutch, whereas large seaturtles regularly lay over a hundred eggs at a time, the record being held by a Hawksbill turtle that laid 258 eggs in one clutch! The majority of species lay twice or even more often each nesting season. The Green seaturtle is the most prolific reptile known. In Sarawak it lays up to 11 clutches of over 100 eggs each at 10.5 day intervals during a single nesting season. Both within and across species there is also a general trend toward the production of more (and smaller) eggs at higher latitudes.

Turtle eggs are of two shapes. Pond turtles, river turtles, wood turtles, and American musk and mud turtles lay elongate eggs, whereas softshells, snapping turtles, and seaturtles lay more spherical ones. Members of other diverse groups such as the tortoises and Australo-American side-necked turtles have members laying eggs of both shapes. Turtles laying the largest clutches (50 eggs or more) have spherical eggs, suggesting that this shape is more efficient for packing large numbers into a confined space. The spherical shape has the lowest possible surface-to-volume ratio, and hence is also less subject to dehydration. Possibly this explains why many tortoises lay such eggs.

The egg shell also varies. Although some exceptions and intermediate types exist, temperate pond and river turtles, snapping turtles, seaturtles, and African and American side-necked turtles tend to have eggs with flexible, leathery shells (indentable by thumb pressure), while the eggshells of tropical pond and river turtles, American musk and mud turtles, softshells, and Austro-American side-necked turtles are more inflexible and are often brittle.

Eggs with brittle shells tend to be more independent of the environment, losing and absorbing less water than eggs with flexible shells. However, those with flexible shells often develop faster. Species that do not dig sophisticated nests or that nest in particularly dry or very moist soils tend to lay eggs with brittle shells. Conversely, those nesting on beaches prone to flooding, or in areas with limited growing seasons where rapid egg development is important, are more likely to lay eggs with flexible shells.

Egg size tends to increase with body size, both within and across species, but there are exceptions. The enormous Leatherback seaturtle and the Galapagos tortoise lay the largest spherical eggs, with diameters of 5–6cm (2–2.4in) and weights of up to 107g (3.8oz). The smallest spherical eggs are those of the Chinese softshell, at

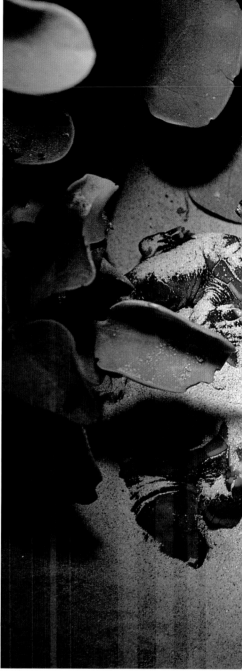

⬗ **Above** *Under cover of darkness, a female Green seaturtle* (Chelonia mydas) *lays her eggs on a beach in Sabah, northern Borneo. This species has been known to migrate for some 4,500km (2,800mi) in order to nest. After burying the eggs, she returns to the sea and plays no further part in their development. The eggs, up to 100 in a clutch, hatch in about 2 months.*

⬖ **Right** *Green seaturtle hatchlings on the Galapagos islands. Once they have dug themselves out of the nest, newborn turtles face a perilous dash to the sea. Attrition rates are high. Those that survive return to their "home" beach after many years at sea.*

⬗ **Below** *A female Loggerhead seaturtle* (Caretta caretta) *searching for a nesting site on an island off the coast of Australia.*

2.1cm (0.8in) and 5.1g (0.18oz). The largest elongate eggs, however, occur in tropical pond turtles, many only small to moderate-sized as adults. The Malaysian giant turtle, largest of the pond and river turtles, with a shell length of 80cm (31.5in), lays eggs up to 4.4 by 8.1cm (1.7 x 3.2in). However, the Black wood turtle (*Rhinoclemmys funerea*), with far less than half the maximum shell length (32.5cm/12.8in), lays only slightly smaller eggs (up to 3.9 x 7.6cm/1.5 x 3in). The Stinkpot lays the smallest known eggs, some only 1.4 x 2.4cm (0.5 x 0.9in) and weighing just 2.6g (0.09oz).

Speed of incubation is positively correlated with temperature within species; however, incubation typically requires two to three months in temperate species, but from four months to over a year among tropical species. The shortest known incubation period is for the Chinese softshell (23 days) and the longest is for the Leopard tortoise (420 days). It seems that longer-incubating eggs often undergo periods of arrested development, and even within the same clutch incubation times can vary widely. Arrested development may be due to cold torpor (when temperatures are too low for development to proceed – a common phenomenon in temperate species); to diapause, when early embryonic development ceases despite normal incubation conditions; this condition is found in Chicken turtles, several subtropical American musk and mud turtles, some softshells, some Asian Pond turtles, some Neotropical turtles, some Austro-American side-necked turtles, and perhaps some tortoises; and/or delayed hatching or estivation, when fully developed embryos remain in the egg without hatching until conditions are favorable, as in some tropical American mud turtles, some Austro-American side-necked turtles, the Pig-nose turtle, and the Wood turtle). These intriguing patterns of development remain poorly studied for most turtles.

In most turtles, temperature during incubation also determines the sex of the hatchling. In species with so-called "temperature-dependent sex determination" (TSD), the temperature during approximately the middle third of incubation drives gonadal development that leads eventually to the sex of the hatchling. Two patterns of TSD have been described for turtles: Type I (found in New World pond turtles, neotropical wood turtles, sea-turtles, tortoises, and the Pig-nosed turtle) includes species that exhibit a narrow pivotal temperature range (usually between 27° and 32°C/81°–90°F) above which only females are produced and below which only males result. Type II (American musk and mud turtles, African side-necked turtles, American river turtles, snapping turtles, and Eurasian pond and river turtles) includes species that exhibit two pivotal temperature ranges, with males predominating at intermediate temperatures and females at both extremes. However, sex determination appears to be genetically determined in the softshells, Austro-American side-necked turtles, the Wood turtle, the two Giant musk turtles, the Brown roof turtle (*Kachuga smithii*), and the Black marsh turtle. Furthermore, only the latter four species have dimorphic sex chromosomes; all others have identical chromosome sizes in males and females. The evolutionary advantage of this puzzling diversity of sex determination remains an enigma (see Temperature and Sex).

When they are fully developed, baby turtles use a small, horny tubercle of epidermal origin (the caruncle) on the upper beak to tear or break their way through the embryonic membranes and eggshell. Following hatching, most neonates dig out of the nest and head directly for cover of water or vegetation; however, some hatchlings delay emergence from the nest. After hatching, a few temperate species including the Ornate box turtle and the Yellow mud turtle immediately dig downward 1m (3.3ft) or more below the nest, presumably an adaptation to avoid the impending lethal winter temperatures in shallow water or soil. Hatchlings of a few other temperate species (for instance, the Painted turtle) remain in the nest over the winter, where they are exposed to temperatures of –12°C (10°F) or lower. Although they are capable of surviving a freezing event at high subzero temperatures (for example, to –4°C/25°F), they must supercool (without the tissues freezing) in order to survive the colder temperatures. The precise mechanism by which they accomplish this is not yet known. Still other turtles (particularly in highly seasonal tropical environments) must remain in their nests until rain softens the soil, allowing them to dig out. One anecdotal report claimed that the young of Broad-shelled snake-necked turtles were entrapped in their nest for 664 days during an Australian drought – and nevertheless survived!

A Gradual Build-up of Biomass
POPULATION BIOLOGY

In terms of body mass per area of habitat (i.e. standing crop biomass), turtles are often the dominant vertebrate in undisturbed ecosystems. The maximum known biomass for turtles is a huge 584kg/hectare (520lb/acre) for the Aldabra tortoise, although an average value for turtle populations is more like 100 kg/ha (90lb/acre). These values equate to those of other cold-blooded, aquatic vertebrates, but are an order of magnitude higher than is typical for mammals, and two orders of magnitude higher than for birds! Unfortunately, since turtle growth is so slow, this biomass accumulates literally at a turtle's pace. Thus, it may take decades or even centuries for a turtle population to recover from even a single incident of overharvesting, as it is usually large, reproductive adults that are taken.

Turtles are able to live so long in part because mortality rates in natural populations are generally quite low, especially in adults. Adult survivorship often exceeds 80 percent per year, and sometimes 99 percent. However, most of the mortality in

NATURAL SURVIVORS

In captivity, African spurred tortoises have lived for more than 54 years, Desert tortoises for nearly 57, Galapagos tortoises for over 62, and Aldabra tortoises for almost 64 years, while Spur-thighed tortoises occasionally survive for over a century. Freshwater species are nearly as long-lived, with Nile softshells having survived for 50 years, Stinkpots for 54, Wood turtles for 58, and European pond turtles and Alligator snapping turtles for 70 years. At least six additional freshwater species (including emydids, kinosternids, geoemydids, and pelomedusids) have lived 30–40 years in captivity. Maximum captive longevity among marine turtles is 33 years for both Green and Loggerhead seaturtles.

Estimates of longevity in the wild are consistently lower than for animals in captivity, in part because there are so few longterm field studies of turtles. In North America, Eastern box turtles are often found with initials and dates carved in their shells. The better substantiated of these records suggest that some box turtles might survive in the wild for over 100 years. However, other field studies of the same species estimate natural longevity at only 60 years, and no-one has actually recaptured an individual over an interval longer than about 32 years. A Snapping turtle (41 years old at time of writing) captured as a hatchling in a Wisconsin lake was released at the capture site after seven years and has been recaptured every year since.

Based on counts of scute annuli and long recapture histories, some Blanding's turtles are estimated to live for as long as 77 years in the wild. However, a number of longterm field studies of other turtles have recaptured individuals after 20 to 30 years, but none has provided strong evidence that they attain ages over 50 years. Another 20 to 30 years of field work will be necessary to establish actual longevities of turtles in nature.

natural populations occurs during the first year of life (in other words, as an egg or hatchling), with terrestrial and marine species experiencing less mortality (average 50–55 percent) than freshwater species (average 80 percent). Because mortality is higher during a turtle's early life, many wildlife managers have set up programs to protect eggs and young turtles. So-called "headstart" programs protect and hatch eggs and raise the young until they are big enough to avoid many of the hazards of today's disturbed environments. Although this is often a practical (and viable) conservation step, population biologists have shown that, if it is the only protection given, most turtle populations will continue to decline. Protection of subadult and adult females must occur if the decline in many turtle populations is to be reversed.

Our understanding of turtle population biology (and taxonomy) has recently been complicated by the discovery that a number of distantly related species of Asian pond turtles are capable of hybridizing and producing unique hybrid offspring and, in addition, that those hybrids can themselves produce viable young identical to their hybrid parents. It is now known that some of these hybrids were produced accidentally and/or intentionally on turtle breeding farms in Asia, but it has been claimed that others were collected in the wild and may represent natural hybrid events. Several of the resulting hybrid forms have even been given scientific names. Because we know so little about the population biology of most Asian turtles, this hybridization phenomenon complicates turtle conservation efforts. There is considerable uncertainty about which forms represent natural species (and hence deserve protection), and which were artificially produced to feed the lucrative demands of the pet trade.

Turtles occupy nearly every conceivable habitat except those that are permanently or nearly permanently frozen. They have successfully exploited nearly every aquatic environment as well as most terrestrial ones, including some of the driest deserts on earth. The Loggerhead seaturtle can even dive to over 1,000m (3,300ft) below sea level, and Bell's hinge-back tortoise (*Kinixys belliana*) lives in Africa up to altitudes of 3,000m (9,800ft) above sea level.

Turtles reach their greatest species diversity in the lower Ganges–Brahmaputra River basin, where the ranges of at least 19 species overlap. The lower Mobile River basin in Alabama in the United States ranks a close second, with 18 species. Worldwide, turtle species diversity is related most strongly to precipitation patterns; in general, the more rain in an area, the greater the turtle diversity.

When Turtles Interact
SOCIAL BEHAVIOR

There is little evidence to indicate that turtles defend territories. Many appear to have definite home ranges and some species, particularly terrestrial and semiaquatic ones with overlapping ranges, may develop dominance relationships. The Galapagos tortoise establishes dominance hierarchies based on the height to which each animal can extend its head. Geographic races on certain islands have evolved a narrow, saddle-backed shell with a high anterior notch that allows for greater vertical extension of the head and neck. In captive groups of Galapagos tortoises, saddlebacks typically establish dominance over races from other islands that have more restrictive, dome-shaped shells.

Although not highly social, aquatic turtles also gather and interact for purposes other than mating and egg-laying. Some species congregate to overwinter (e.g., Green and Loggerhead seaturtles); others, including the Painted turtle, pond sliders, map turtles, and the Indian tent turtle, may congregate on exposed sites for basking. Whether they are being gregarious or simply selecting the most favorable site is debatable.

Juvenile sliders in the New World tropics sometimes seek out conspecifics for mutual cleaning behavior. One turtle uses its jaws to pull algae (and pieces of scute) from another's shell, positioned at right angles to it; then the positions are reversed and the other reciprocates. Similar behavior has been observed in captive roofed turtles and South American river turtles. Helmeted turtles in Africa have even been reported to clean ectoparasites from rhinoceroses that enter their water holes.

The Human Impact
CONSERVATION AND ENVIRONMENT

Almost every aspect of turtles' biology makes them vulnerable to human impact, and very few species are now free from its effects; many local populations have been extirpated. Moreover, their slow individual and population growth rates and long lives make recovery from decline very difficult.

Most turtles nest in restricted areas and at regular times, and hence both eggs and breeding females are easily overexploited by humans and feral animals. Burgeoning human populations in developing countries exacerbate this problem as people struggle to survive by destroying turtle habitat, by cultivating turtle competitors (e.g., goats, cows, and burros) or predators (e.g., cats, dogs, and pigs) for their benefit, by contributing indirectly to growth in the populations of natural predators such as raccoons, or by directly consuming eggs, young, or adults for food or for traditional medicines. Developed nations also play a part, by providing markets for luxury items such as tortoiseshell jewelry from the Hawksbill seaturtle or leather goods from the Olive ridley, as well as for pets (some fetching over US$1,000 each).

All these factors have conspired to create a particular crisis for Asian turtles (see The Asian Turtle Crisis). Indeed, the situation is so dire that an international Turtle Survival Alliance of conservationists, field biologists, population demographers, geneticists, zoological park curators, private captive breeders, law enforcement authorities, public educators, and wildlife managers was established in 2001 to try to slow or even reverse what would otherwise be the systematic elimination of turtles from all of Southeast Asia. This effort has been made all the more difficult because of our lack of knowledge about even the basic natural history of many of the involved species, as well as the hybrid issue mentioned earlier. Without this kind of intervention, the 220-million year legacy of turtles may well end during this century.   JI/EOM

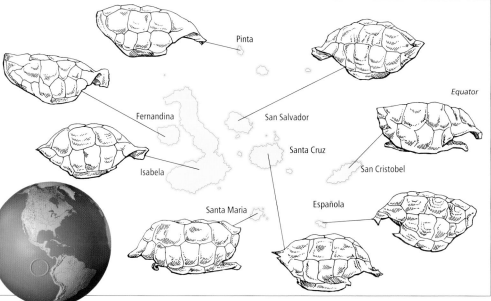

◐ **Above** *Several geographic races of the Galapagos giant turtle* (Chelonoides nigra) *live in geographic isolation from one another on different islands in the Galapagos group off South America. The correlation between the different conditions prevailing on each island and the marked divergence in the shape of the turtles' shells helped Charles Darwin formulate his ideas on evolution. All subspecies are now endangered.*

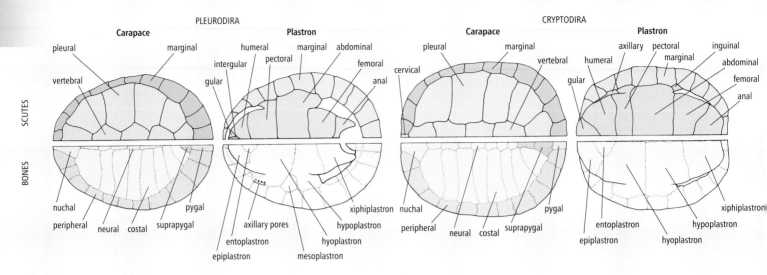

Turtle and Tortoise Families

PLEURODIRA

Carapace — pleural, marginal, vertebral

SCUTES

BONES — nuchal, peripheral, neural, costal, suprapygal, pygal

Plastron — humeral, intergular, pectoral, marginal, abdominal, femoral, anal, gular; axillary pores, entoplastron, epiplastron, mesoplastron, hyoplastron, hypoplastron, xiphiplastron

CRYPTODIRA

Carapace — pleural, marginal, vertebral, cervical, nuchal, peripheral, neural, costal, suprapygal, pygal

Plastron — axillary, pectoral, inguinal, marginal, abdominal, humeral, femoral, anal, gular; entoplastron, epiplastron, hyoplastron, hypoplastron, xiphiplastron

THE ORDER TESTUDINES IS DIVIDED INTO 14 families, grouped into two suborders: hidden-necked turtles (suborder Cryptodira) and side-necked turtles (suborder Pleurodira). Hidden-necked turtles retract their heads into the shell by bending the neck into a vertical S-shaped curve. They have 11–12 plastral scutes and 8–9 plastral bones. Side-necked turtles retract their heads under the lip of the shell by bending the neck to the side in a horizontal plane. They have 13 plastral scutes and 9–11 plastral bones, and the pelvis is fused to the shell.

SUBORDER CRYPTODIRA

Pig-nosed turtle
Family Carettochelyidae
Carettochelys insculpta
Pig-nosed turtle or New Guinea plateless turtle

1 species. S New Guinea, N Australia. Rivers, lakes, lagoons.
SIZE: Length to 56cm
COLOR: Unpatterned gray, gray-olive, or gray-brown above; white, yellowish, or cream below; pale streak or white blotch between eyes.
BODY FORM: Shell covered with soft, pitted skin; limbs paddle-shaped; snout a fleshy, piglike proboscis.

EGGS: Spherical, brittle; about 4cm in diameter; clutches of 7–20.
DIET: Crustaceans, insects, mollusks, fish, aquatic plants, fruits.
CONSERVATION STATUS: Vulnerable

Snapping turtles
Family Chelydridae

2 species in 2 genera. From S Canada across the E USA to NW S America. Freshwater bottoms. Species: **Alligator snapping turtle** (*Macrochelys temminckii*), **Snapping turtle** (*Chelydra serpentina*).
SIZE: Length from 49cm in Snapping turtle to 80cm in Alligator snapping turtle; weight: 10–113kg.
COLOR: Drab, with gray, black ,or brown top.
BODY FORM: Head large, with strong, hooked jaw; shell with three knobby keels; tail long; plastron reduced, cross-shaped; abdominal scutes widely separated, chiefly confined to bridge; 24 marginal scutes.
EGGS: Spherical, thick, and leathery; 2.3–3.6cm; clutches of 6–80 (maximum 109) in Snapping turtle; 3–5cm, 30–40 (maximum 52) in Alligator snapper.
DIET: Carrion, insects, fish, turtles, mollusks, plant food.
CONSERVATION STATUS: The Alligator snapping turtle and 1 other are Vulnerable.

◑ Above *Turtle shell structure.*

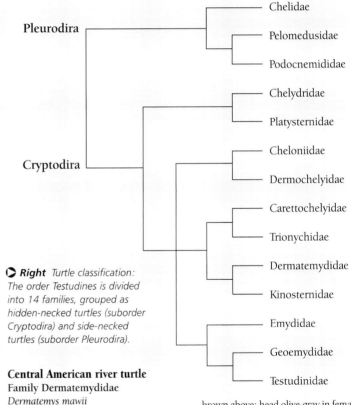

◑ Right *Turtle classification: The order Testudines is divided into 14 families, grouped as hidden-necked turtles (suborder Cryptodira) and side-necked turtles (suborder Pleurodira).*

Pleurodira — Chelidae, Pelomedusidae, Podocnemididae

Cryptodira — Chelydridae, Platysternidae, Cheloniidae, Dermochelyidae, Carettochelyidae, Trionychidae, Dermatemydidae, Kinosternidae, Emydidae, Geoemydidae, Testudinidae

Central American river turtle
Family Dermatemydidae
Dermatemys mawii

1 species. From Veracruz, Mexico, S to NW Honduras. Rivers and lakes.
SIZE: Length up to 65cm
COLOR: Unpatterned dark gray to gray-brown above; head olive-gray in females, with yellowish to reddish-brown back of head in males.
BODY FORM: Shell well developed, streamlined, covered with thin scutes fusing with age to obliterate seams; 24 marginal, 4–5 inframarginal scutes. Head relatively small with moderately upturned, tubular snout.
EGGS: Elongate, brittle, averaging 6.4cm x 3.2cm; clutches of up to 20.
DIET: Juveniles omnivorous; adults herbivorous, feeding on fruits and aquatic plants
CONSERVATION STATUS: Critically Endangered

NOTES	Length = maximum straight-line shell length.
	Approximate nonmetric equivalents: 10cm = 4in. 1kg = 2.2lb.

Seaturtles
Family Cheloniidae

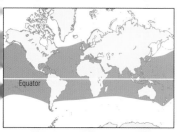

6 species in 5 genera. Circumtropical, extending into subtropical and temperate oceans. Species include: **Kemp's ridley seaturtle** (*Lepidochelys kempii*), **Olive ridley seaturtle** (*L. olivacea*), **Flatback seaturtle** (*Natator depressus*), **Green seaturtle** (*Chelonia mydas*), **Hawksbill seaturtle** (*Eretmochelys imbricata*), **Loggerhead seaturtle** (*Caretta caretta*).

SIZE: Length from 75cm in Kemp's ridley to 213cm in the Loggerhead seaturtle; weight up to 450kg in the Loggerhead.

BODY FORM: Shell low, streamlined, scute-covered; limbs paddle- or flipperlike; skull completely roofed, cannot be retracted within shell.

EGGS: Spherical, leathery; usually 50–150 (258 maximum), laid in multiple clutches (up to 7 or more) at intervals of 10–30 days. Both sexes migrate between feeding and nesting areas, female reproducing on 1–3 year cycle.

DIET: Chiefly carnivorous (except adult Green seaturtle, which grazes on sea grasses), including sponges, jellyfish, mollusks, crabs, barnacles, sea urchins, and fish.

CONSERVATION STATUS: The Hawksbill and Kemp's ridley are Critically Endangered. 3 other species are Endangered, and the sixth – the Flatback – is Vulnerable.

Leatherback seaturtle
Family Dermochelyidae
Dermochelys coriacea

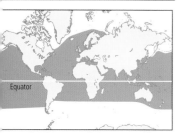

1 species. Tropical waters, with periodic occurrences in temperate and subarctic seas.

SIZE: Largest living turtle; length to 244cm; weight to 867kg.

COLOR: Black.

BODY FORM: Adult shell without scutes, covered with oily skin, reduced to mosaic of small bony platelets embedded in the skin – carapace with 7 prominent ridges,

plastron with 5; forelimbs modified into broad flippers. Can maintain body temperature to several degrees above surrounding water even in subarctic.

EGGS: Spherical, 5–6cm; several clutches of 50–170 laid at intervals of about 10 days on tropical beaches; females reproduce on 2–3 year cycle.

DIET: Jellyfish and other soft invertebrates.

CONSERVATION STATUS: Critically Endangered.

Pond turtles
Family Emydidae

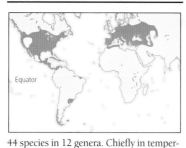

44 species in 12 genera. Chiefly in temperate N America, with European pond turtle over most of Europe, W Asia, and NW Africa, and sliders S to N and SE South America. Fully aquatic species in estuaries and coastal waters or freshwater rivers and lakes; semiaquatic species in forested ponds and streams, marshes and bogs; also fully terrestrial species. Species and genera include: **Blanding's turtle** (*Emydoidea blandingii*), **Bog turtle** (*Glypterys muhlenbergii*), **Spotted turtle** (*Clemmys guttata*), **box turtles** (genus *Terrapene*), **Eastern box turtle** (*T. carolina*), **Chicken turtle** (*Deirochelys reticularia*), **Pondslider** (*Trachemys scripta*), **Diamond-backed terrapin** (*Malaclemys terrapin*), **European pond turtle** (*Emys orbicularis*), **Florida red-bellied cooter** (*Pseudemys nelsoni*), **Northern Red-bellied cooter** (*P. rubriventris*), **River cooter** (*P. concinna*), **map turtles** (genus *Graptemys*), **Painted turtle** (*Chrysemys picta*).

SIZE: Length from 10cm in the Bog turtle to 43cm in the River cooter.

COLOR: Variable; browns, olives, and shades of gray and black above, often marked with yellow, orange, red, or white; yellow, white, brown, and black most common below; striped and spotted patterns common on soft parts.

BODY FORM: Shell well developed, with 24 marginal and 12 plastral scutes, pectorals and abdominals meeting marginals; feet usually have some webbing between toes; some with hinged plastron; a double articulation between the fifth and sixth cervical vertebrae; posterior marginal scutes not extending up onto suprapygal bone.

EGGS: Elongate, 3–5cm; all but 1 leathery; usually 2 or more clutches per year of 1–2 eggs, in small species, up to 27 in large

riverine forms.

DIET: Large species mostly herbivorous, others chiefly omnivorous, including insects, mollusks, aquatic vertebrates, and plants.

CONSERVATION STATUS: 7 species are Endangered and 9 are Vulnerable.

Eurasian pond and river turtles, Neotropical wood turtles
Family Geoemydidae

69 species (excluding possible hybrid forms) in 22 genera. Tropical to subtropical Asia, N Africa, S Europe, and tropical America. Fully aquatic species in freshwater rivers, streams, lakes, and ponds; some found in estuaries and coastal waters; other species semiaquatic to totally terrestrial. Species include: **Asian leaf turtle** (*Cyclemys dentata*), **Black marsh turtle** (*Siebenrockiella crassicollis*), **Black-breasted leaf turtle** (*Geoemyda spengleri*), **Giant Asian pond turtle** (*Heosemys grandis*), **Spiny turtle** (*H. spinosa*), **Southeast Asian box turtle** (*Cuora amboinensis*), **Malayan flat-shelled turtle** (*Notochelys platynota*), **Malaysian giant turtle** (*Orlitia borneensis*), **Malayan snail-eating turtle** (*Malayemys subtrijuga*), **stripe-necked turtles** (genus *Mauremys*), **Mediterranean turtle** (*M. leprosa*), **neotropical wood turtles** (genus *Rhinoclemmys*), **roofed turtles** (genus *Kachuga*), **Painted roof turtle** (*K. kachuga*), **Painted terrapin** (*Callagur borneoensis*), **Reeves' turtle** (*Chinemys reevesii*), **River terrapin** (*Batagur baska*), **Three-striped box turtle** (*Cuora trifasciata*), **Tricarinate hill turtle** (*Melanochelys tricarinata*).

SIZE: Length from 11cm in Black-breasted leaf turtle to 80cm in Malaysian giant turtle; weight up to 50kg in Malaysian giant turtle.

COLOR: Highly variable; browns, olives, grays, or black above, some marked with yellow, orange, red, or black; generally yellow, white, brown, or black below, often with dark markings; soft parts uniform to striped to spotted, often with striking colors.

BODY FORM: Shell well developed, with 24 marginal and 12 plastral scutes, pectorals and abdominals meeting marginals; feet usually with some webbing between toes; some species with hinged plastron; a simple articulation between the fifth and sixth cervical vertebrae; posterior marginal

scutes extend up onto suprapygal bone. Hybridization between species common, making identification difficult.

EGGS: Elongate, 3–8cm; leathery to brittle shells; 1–2 to 35 eggs per clutch, generally correlated with body size; multiple clutches (up to 9) per year typical.

DIET: From strictly herbivorous to strictly carnivorous, including insects, mollusks, aquatic vertebrates, seeds, and leaves; from specialists to generalists.

CONSERVATION STATUS: 13 species are Critically Endangered, 18 are Endangered, and 10 are Vulnerable.

American mud and musk turtles
Family Kinosternidae

25 species in 4 genera and 2 subfamilies. E Canada to Argentina. Permanent to temporary fresh water. Species and genera include: **Pacific Coast giant musk turtle** (*Staurotypus salvinii*), **Mexican giant musk turtle** (*S. triporcatus*), **Eastern mud turtle** (*Kinosternon subrubrum*), **Scorpion mud turtle** (*K. scorpioides*), **Yellow mud turtle** (*K. flavescens*), **Flattened musk turtle** (*Sternotherus depressus*), **Razor-backed musk turtle** (*S. carinatus*), **Loggerhead musk turtle** (*S. minor*), **Stinkpot** (*S. odoratus*), **Narrow-bridged musk turtle** (*Claudius angustatus*).

SIZE: Length from 11cm (Flattened musk turtle) to 38cm (Mexican giant musk turtle).

COLOR: Predominantly drab brown, yellow, olive to black above, with white, yellow, red, or orange markings on the head of some species; plastron white to yellow to brown or black.

BODY FORM: Carapace with 22 marginal scutes and cervical scute; small, fleshy barbels on chin; scent glands produce malodorous secretions on skin near bridge; 11 or fewer plastral scutes; subfamily Staurotypinae has reduced plastron, cross-shaped, with entoplastron present; subfamily Kinosterninae with hinged plastron lacking entoplastron.

EGGS: Elongate, with brittle shells; multiple clutches of 1–12 eggs, from 2.3 x 1.4 to 4.4 x 2.6 cm.

DIET: Primarily carnivorous, including mollusks, arthropods, annelids, fish, and also aquatic plants (especially seeds).

CONSERVATION STATUS: 4 species are Vulnerable.

Tortoises
Family Testudinidae

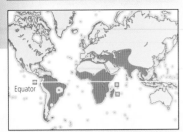

51 species in 15 genera. Chiefly tropical and subtropical, on all major land masses except Australia and Antarctica. Terrestrial. Species include: **Aldabra tortoise** (*Dipsochelys hololissa*), **Angonoka** or **Ploughshare tortoise** (*Angonoka yniphora*), **Radiated tortoise** (*Astrochelys radiata*), **Galápagos tortoise** (*Chelonoides nigra*), **Asian brown tortoise** (*Manouria emys*), **Gopher tortoise** (*Gopherus polyphemus*), **Bolson tortoise** (*G. flavomarginatus*), **Desert tortoise** (*G. agassizii*), **Egyptian tortoise** (*Testudo kleinmanni*), **Elongated tortoise** (*Indotestudo elongata*), **hinge-backed tortoises** (genus *Kinixys*), **Madagascar spider tortoise** (*Pyxis arachnoides*), **Pancake tortoise** (*Malacochersus tornieri*), **Speckled Cape tortoise** or **Padloper** (*Homopus signatus*).
SIZE: Length from 9cm in Speckled Cape tortoise to 140cm in Aldabra tortoise; weight up to 255kg in Aldabra tortoise.
COLOR: Predominantly shades of brown, olive, yellow, and black above; predominantly yellow, brown, and black below; scutes plain or commonly patterned with bright rays or concentric rings; head and neck rarely striped or spotted.
BODY FORM: Hind legs columnar, elephantine; forelimbs somewhat flattened, armored with large, bony-cored scales; short, unwebbed toes, each with no more than 2 phalanges; most with high-arched or domed shells.
EGGS: Elongate to spherical, greatest diameter 3–6cm; leathery to brittle shells; 1–51 per clutch, multiple clutches known in some species.
DIET: Chiefly herbivorous; a few forms omnivorous.
CONSERVATION STATUS: The Burmese starred tortoise (*Geochelone platynota*), the Egyptian tortoise (*Testudo kleinmanni*), and the Negev tortoise (*T. werneri*) are Critically Endangered; 6 other species are Endangered and 17 Vulnerable.

Softshell turtles
Family Trionychidae

30 species in 14 genera and 2 subgroups: flapshells and typical softshells. Widespread in temperate to tropical N America, Africa, Asia, and Indo-Australian Archipelago. Fresh water, but Asian giant softshell commonly enters estuaries and has occasionally been found at sea. Species include: **Asian giant softshell** (*Pelochelys cantorii*), **Asian narrow-headed softshell** (*Chitra chitra*), **Black softshell** (*Aspideretes nigricans*), **Central African flapshells** (genus *Cycloderma*), **Chinese softshell** (*Pelodiscus sinensis*), **Indian flapshell** (*Lissemys punctata*), **Nile softshell** (*Trionyx triunguis*), **Smooth softshell** (*Apalone mutica*), **Spiny softshell** (*A. spinifera*), **Sub-Saharan flapshells** (genus *Cyclanorbis*).
SIZE: Length from 25cm in Chinese softshell to 120cm (and 150kg or more) in Asian narrow-headed softshell.
COLOR: Predominantly shades of brown, olive, or gray above; some with white, black, yellow, red, or orange markings; white, yellow, or gray most common below.
BODY FORM: Characterized by flattened, reduced shell lacking peripheral bones (except in Indian flapshells) and covered by a leathery skin instead of scutes; elongated, retractable neck; limbs somewhat paddle-like with 3 claws per foot; snout an elongated proboscis; plastron reduced, typically with large spaces (fontanelles) between bones, and with ligamentous or cartilaginous rather than bony connection to carapace. Flapshell subgroup named for a pair of fleshy cutaneous flaps on rear of plastron covering hind limbs when withdrawn.
EGGS: Spherical, hard- to brittle-shelled; 2.5–3.5cm in diameter; clutches of 4 to 107; multiple clutches common.
DIET: Chiefly carnivorous, some omnivorous; insects, crustaceans, and fish commonly eaten.
CONSERVATION STATUS: 3 species are Critically Endangered, 5 Endangered, and 6 Vulnerable.

Big-headed turtle
Family Platysternidae
Platysternon megacephalum

1 species, sometimes included as a subfamily in the family Chelydridae. S China, Indochina, Thailand, SE Myanmar. Cool mountain streams; active at night, with exceptional climbing ability.
SIZE: Length to 20cm
COLOR: Carapace olive-brown; plastron cream, yellow, orange, or reddish, with or without dark markings; head brown above, sometimes with red to orange spotting or pale stripe behind eyes.
BODY FORM: Distinguished by long, muscular tail, powerful hooked jaws, large plastron, and large, completely roofed skull that cannot be retracted into the flattened shell.
EGGS: Elongate (3.7 x 2.2cm), apparently brittle; 1–2 eggs per clutch.
DIET: Carnivorous
CONSERVATION STATUS: Endangered

SUBORDER PLEURODIRA

Austro-American side-necked turtles
Family Chelidae

53 species in 13 genera. Tropical to temperate S America, Australia, New Guinea. Family includes aquatic and semiaquatic species. Species include: **Australian snake-necked turtle** (*Chelodina longicollis*), **Fitzroy River turtle** (*Rheodytes leukops*), **Broad-shelled snake-necked turtle** (*Chelodina expansa*), **Common toad-headed turtle** (*Mesoclemmys nasutus*), **Gibba turtle** (*Mesoclemmys gibbus*), **Matamata** (*Chelus fimbriatus*), **New Guinea snapping turtle** (*Elseya novaeguineae*), **Twist-necked turtle** (*Platemys platycephala*),
Vanderhaege's toad-headed turtle (*Mesoclemmys vanderhaegei*), **Western swamp turtle** (*Pseudemydura umbrina*).
SIZE: Length from 14cm in Western swamp turtle to 48cm in Broad-shelled snake-necked turtle.
COLOR: Variable; predominantly brown, olive, and black above, with yellow, red, and orange markings in some species; plastron commonly yellow or white, sometimes red or orange, often with dark markings.
BODY FORM: Distinguished by biconvex 5th and 8th cervical vertebrae; shell without mesoplastral bones; cervical scute present; intergular scute seldom contacting plastral rim; skull and mandible without quadratojugal and splenial bones.
EGGS: Elongate to spherical, 3–6cm in greatest diameter; clutches of 1–25; 2 or more clutches per year in some species.
DIET: Omnivorous to totally carnivorous.
CONSERVATION STATUS: 3 species are Critically Endangered, 4 Endangered, and 6 Vulnerable.

African side-necked turtles
Family Pelomedusidae

19 species in 2 genera. Africa S of Sahara, Madagascar, Seychelles. Aquatic and semiaquatic. Species include: **African dwarf mud turtle** (*Pelusios nanus*), **East African serrated mud turtle** (*P. sinuatus*), **West African mud turtle** (*P. castaneus*), **Helmeted turtle** (*Pelomedusa subrufa*).
SIZE: Length from 12cm in African dwarf mud turtle to 47cm in East African serrated mud turtle.
COLOR: Brown, olive, or black above; head olive, brown, gray, or black, with yellow to cream vermiculations often present; plastron commonly yellow, gray, or brown.
BODY FORM: Shell with mesoplastral bones; cervical scute absent; intergular scute touches back rim of plastron; second cervical vertebra biconvex; quadratojugal bone present in skull and splenial bone in mandible; five claws on hind foot.
EGGS: Elongate, with leathery shells; 3–4cm long; 6 to 48 per clutch.
DIET: Mainly carnivorous
CONSERVATION STATUS: 1 species is Vulnerable.

NOTES — Length = maximum straight-line shell length.
Approximate nonmetric equivalents: 10cm = 4in. 1kg = 2.2lb.

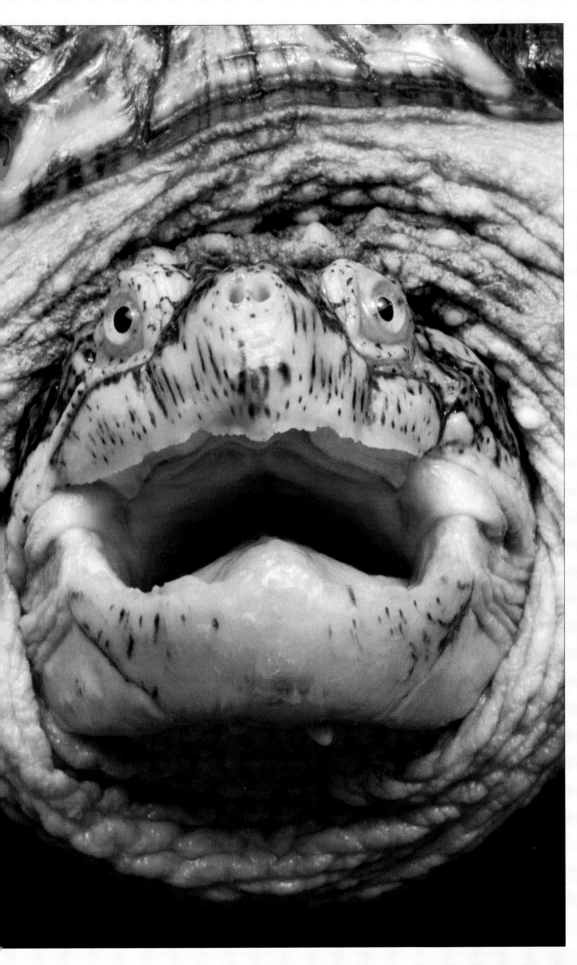

American side-necked river turtles and Madagascan big-headed turtle
Family Podocnemididae

8 species in 3 genera. Tropical S America and W Madagascar. Aquatic, mainly riverine. Species and genera include: **Big-headed Amazon river turtle** (*Peltocephalus dumerilianus*), **Madagascan big-headed turtle** (*Erymnochelys madagascariensis*), **South American river turtle** (*Podocnemis expansa*), **Red-headed river turtle** (*P. erythrocephala*).

SIZE: Length from 25cm in Red-headed river turtle to 107cm in South American river turtle.

COLOR: Brown, olive, gray, or black above; head commonly with yellow, orange, or red markings; plastron yellow, gray, or brown, occasionally with yellow, orange, pink, or red markings.

BODY FORM: Shell with mesoplastral bones; cervical scute absent; intergular scute touches back rim of plastron; second cervical vertebra biconvex; quadratojugal bone present in skull and splenial bone in mandible; 4 claws on hind foot.

EGGS: Spherical to elongate, with leathery shells, 3.2–6.1cm long; 5–136 eggs per clutch; multiple clutches common.

DIET: Mainly herbivorous, but opportunistically carnivorous.

CONSERVATION STATUS: 2 species are Endangered and 4 are Vulnerable.

◖ *Left* Aggressive and short-tempered, and capable of delivering a severe bite, the Snapping turtle (Chelydra serpentina) of eastern North America commands respect from all who know it. Reaching up to 49cm (19.5in) in length, it includes other turtles in its varied diet.

THE ASIAN TURTLE CRISIS

How international trafficking is threatening some species' survival

◁ Left The Chinese softshell turtle formed the basis of an enormous growth in the turtle farming industry in East Asia during the 1980s and 1990s.

CHELONIANS PLAY A SIGNIFICANT ROLE IN EAST Asian culture. The Tortoise is the one of the four Spiritual Creatures, associated with the element water, the color black, and the direction north. The animals are symbols of longevity, strength, good fortune, and endurance. Live tortoises were presented as gifts to Chinese emperors, and freshwater turtles are popular inhabitants of temple ponds as well as subjects for classical paintings. Divination rituals using tortoise shells date back over 4,000 years. The appreciation of turtles extends beyond symbolism and aesthetics to practical use, both in traditional medicine and as a foodstuff whose perceived benefits go beyond mere protein and calories – turtle meat is considered a "hot" food that increases circulation and warms the body. Turtle bones, preferably the plastron, are used in traditional medicinal preparations, while turtle jelly has recently become very popular in the Hong Kong region as a detoxifying substance and a claimed cure for cancer.

Traditionally the collection and exploitation of Asian tortoises and freshwater turtles were matters of subsistence use and modest regional trade, largely constrained by economic factors. In regions where trade was relatively unrestrained and where consumers had significant purchasing power, such as Taiwan and Hong Kong, dwindling local supplies of native turtles were augmented by imports. The relaxation of restrictions on private enterprise and the introduction of a convertible currency as part of Deng Xiaoping's economic reforms in the late 1980s also opened the door for the mass importation of turtles into mainland China.

Initially these imports originated from Vietnam and from Bangladesh, a country that actively promoted the export of turtles as a source of revenue. As trade flows mounted to tens and hundreds of tons per year, populations in these countries became exhausted and new sources were explored. Vietnamese traders began acquiring supplies from adjacent Laos and Cambodia, and Bangladeshi exports included turtles collected from other regions of the Gangetic plain. New, direct trade routes were developed overland between China's Yunnan province and Myanmar, and by air to Indonesia and Malaysia, as well as to Thailand with its developing aquaculture production.

Much of the trade involved turtles collected from the wild either by professional hunters or by plantation workers. The chain then continued to an extensive and well-organized network of collection depots, middlemen traders, and exporters.

As exploitation of wild populations increased, enterprising individuals also revived techniques for farming the Chinese softshell turtle (*Pelodiscus sinensis*). Pioneered in Japan in the 1860s, these practices were refined there and in Taiwan, and were adapted to tropical conditions in Singapore, Thailand, and Malaysia in the 1980s. In tropical conditions, producers enjoy fast growth and maturation rates, along with a year-round production of eggs and offspring. As apparently insatiable markets opened up in East Asia, turtle farming grew rapidly to become a significant component of the aquaculture industry. Production reached levels of about 6 million animals in Thailand in

1996 and about 1.4 million in Malaysia in 1999. Attempts were also made to farm other species such as the large native Asiatic softshell turtle (*Amyda cartilaginea*), but these proved economically less rewarding. As turtle farming expanded in China itself a few years later, exports of farmed turtles from Thailand and Malaysia declined, and farms reduced production and lost profitability.

A small but highly visible and relatively valuable component of the Asian turtle traffic was the international pet trade, supplying a wide diversity of species to hobbyists in Europe, America, Japan, and elsewhere. For ease of shipping, pet exporters generally selected small juveniles from the mass food trade and shipped these to overseas importers. Much of this trade involved locally common and well-known species of modest value, but a few traders specialized in sourcing and supplying unusual turtles from little-explored areas. When several of these unfamiliar turtles turned out to be new species, demand soared, leading to exceptional prices in the order of thousands of dollars for a single individual. Such windfalls further fueled the trade and provided incentives for wholesalers to explore new source regions, expanding the mass trade also to areas that had previously been unaffected.

With their generally slow rates of growth, late maturity, and low annual reproductive output, turtles are inappropriate animals for sustainable mass exploitation, particularly when this traffic targets the mature, reproducing adults that suffer very low natural mortality rates. The intensive collection of turtles for export brought local populations to collapse within a few years. As the trade shifted to new areas when the established sources declined, whole countries were progressively "vacuumed" of their turtle populations, as well as their pangolins, snakes, slow lorises, and anything else of value to the wildlife trade.

Concerns about the impact of the trade, which began to be voiced in the early 1990s, became widespread by the end of the decade. A workshop held in Phnom Penh, Cambodia, in December 1999 assembled disparate snippets of information to create a recognizable, if incomplete, picture: Asia's wild turtle populations were declining fast across the continent, and several species were in grave danger of imminent extinction. The trade had reached a volume of some 8,000 tons in 1999, and the survival of 67 of 90 Asian species of tortoises and freshwater turtles

was considered threatened, 18 of these being classed as Critically Endangered. These and other findings were publicized as widely as the subject permitted (Asian turtles are not very newsworthy in comparison with elephants, oil spills, or domestic politics).

Action was taken as concerns mounted, including listing the Painted terrapin (*Callagur borneoensis*) on Appendix II of the Convention on International Trade in Endangered Species of Wild Fauna and Flora (CITES) in 1997 and the inclusion of the entire *Cuora* genus of Asian box turtles on CITES Appendix II in the year 2000. Domestic wildlife protection laws and trade regulations were updated, quotas were imposed or refined, and enforcement of regulations increased. Illegal trade shipments have been confiscated. In addition, China set up a reserve in Xinjiang province specifically for the protection of the Central Asian tortoise (*Testudo horsfieldii*), and numerous national parks, wildlife sanctuaries, and other protected areas throughout Asia now provide more or less effective protection for many, though not all, Asian turtle species.

In view of the often overwhelming exploitation pressures, some people believe that captive breeding is necessary for the survival of some turtle species, and many more believe that such programs can provide "insurance populations" that act as a back-up to other measures. A workshop in Fort Worth, Texas, in January 2001, brought together the bodies responsible for numerous individual, often isolated, efforts at conservation breeding, uniting them to form the Turtle Survival Alliance.

At the time of writing, significant progress has been made, yet the challenges ahead remain daunting. A total elimination of the trade in turtles, whether for food, medicine, or pets, is probably impossible and may not even be desirable. The challenge is to reduce the removal of turtles from the wild to quantities that are sustainable, by meeting a large proportion of market demand with turtles bred in captivity and by realistic management of wild populations, combined with effective protection of the habitats in which they flourish. Tortoises and freshwater turtles have been part of Asian ecosystems for 200 million years, and they deserve to continue on their extraordinary evolutionary pathway. PPvD

◁ **Left** A Chinese vendor displays livestock – turtles and tortoises for food and traditional remedies – at a market in the southern city of Guangzhou. Economic liberalization in the People's Republic stimulated unsustainable exploitation of wild chelonian populations.

LEATHERBACKS:
BIRTH ON THE BEACH

1 *Leatherback turtles (Dermochelys coriacea) are an ancient species now under threat of extinction; the IUCN, which lists the animals as Critically Endangered, estimates that there has been a 70 percent decline in the global population in less than a generation, and total numbers worldwide may now be as low as 20–30,000. Adults spend their lives at sea, where many are killed as* *accidental bycatch by fishing fleets, but pregnant females return briefly to nesting beaches to lay their eggs, typically returning five or six times each season, with a nine- or ten-day interval between visits. Here, a female at a beach near the village of Yalimapo in French Guiana uses her hind feet to burrow out a hole in the sand; nests may be as much as 80cm (30in) deep.*

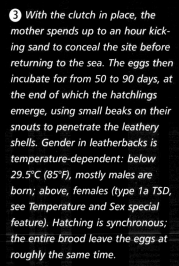

2 *When the flask-shaped nest is ready, the turtle lays from 60 to 120 eggs. Curiously, about a third are undersized and contain no yolk; infertile, they may serve as spacers or help regulate moisture. White as billiard balls, the eggs are spherical, like those of all turtle species with large clutches; the shape evidently maximizes space use while minimizing the risk of dehydration.*

3 *With the clutch in place, the mother spends up to an hour kicking sand to conceal the site before returning to the sea. The eggs then incubate for from 50 to 90 days, at the end of which the hatchlings emerge, using small beaks on their snouts to penetrate the leathery shells. Gender in leatherbacks is temperature-dependent: below 29.5°C (85°F), mostly males are born; above, females (type 1a TSD, see Temperature and Sex special feature). Hatching is synchronous; the entire brood leave the eggs at roughly the same time.*

4 *Instinct leads the brood to dig up toward the surface, a process that takes four or five days, during which the hatchlings' shells harden. Shortly before emergence the baby turtles pause, waiting for the lowering of the temperature that signals nightfall. Exposure to the full heat of the sun could be fatal at this stage, and they would also be at greater risk from predators in daylight.*

5 *As twilight falls the hatchlings erupt onto the open beach and head for the sea as fast as they can scuttle. They are never more exposed than at this time, when dogs, raccoons, and various birds of prey may attack them. Studies indicate that turtle nests laid higher up the beach suffer the greatest mortality from predation; those closer to the surfline, however, are more at risk from wave and tide damage.*

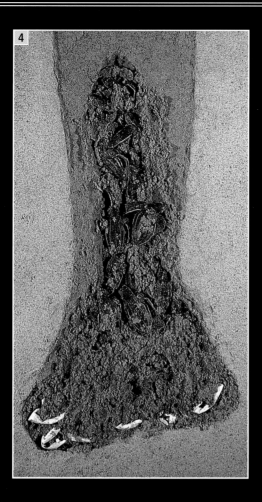

6 *The dangers facing the hatchlings do not end when they reach the waterline; a fresh range of marine predators like this sea catfish (family Ariidae) are waiting in the coastal shallows. The threat from predation is so great that estimates suggest that fewer than one hatchling in 1,000 survives to adulthood. Now that the adults too are at risk from modern deep-sea fishing techniques, the future looks bleak for this giant species.*

Lizards

a MONG GLOBAL VERTEBRATE FAUNA, LIZARDS *are almost ubiquitous features. Although most are tropical, many occur in temperate climates. In the New World, they are found as far north as southern Canada, and as far south as Tierra del Fuego, at the tip of South America. In the Old World, one species, the Viviparous lizard, occurs above the Arctic Circle in Norway. Others are found as far south as Stewart Island, New Zealand. Species occur from sea level up to 5,000m (16,500ft).*

Lizards are the most successful of the reptiles as measured by sheer number of species. They also exceed other reptiles in anatomical, behavioral, and reproductive diversity and in the extent of their geographic distribution. Whereas all of the species of turtles and snakes have adopted highly specialized body forms, lizards as a group have kept a generalized tetrapod build that has subsequently been modified somewhat differently in each of the 28 living families. Thus, while most lizards are diurnal, there are also major radiations of nocturnal and crepuscular families. Although lizards typically prefer relatively warm temperatures, they have adopted behavioral means of thermoregulation and seasonal activity patterns that allow them to succeed in all but the most severe climates.

Most lizards are, like us, dwellers on the land and active by day. Even nocturnal geckos are a familiar part of many people's lives: in tropical countries certain of these insect-eaters, with their bulging, lidless eyes, are common and welcome members of the household. Brook's half-toed gecko of West Africa, for example, advertises through its transparent belly how many captured flies have earned it the right to scurry, on its faultlessly gripping toes, up and down walls and windows and even upside-down across ceilings.

Agile and Thick-skinned
FORM AND FUNCTION

One of the most distinctive features of lizards is their skin, which is folded into scales. The outer layer is filled with keratin, a tough, nonsoluble protein that greatly reduces water loss and allows many lizards to occupy even the driest deserts. Scales vary greatly in appearance, from small and granular to large and platelike. They may touch one another (in other words, be juxtaposed) or overlap (imbricate); they can be smooth or possess one or more ridges (keels). The scaly skin of lizards is usually thick and tough and is not easily torn. Certain scales have been modified into sharp spines that can dissuade attackers. Some are fortified with internal bony plates called osteoderms.

A curious feature of lizards is their development of the pineal body. The 17th-century French philosopher René Descartes claimed that the pineal body in the human brain was the point at which mind and body interact. He saw the pineal as an eye for the immortal soul, communicating to it the sensory intake of the material body. In lizards at least, it seems that this part of the brain is indeed connected to a sort of "third eye," but it connects the animal only with the physical world. A cranial bone is perforated at this point to allow an extension of nervous tissue from the brain to a light-sensitive, transparent disk on top of the head. Research suggests that the pineal may be involved in regulating the animal's "biological clock," influenced by the recurring pattern of day and night.

The skulls of lizards are kinetic (mobile). To varying degrees lizards are capable of moving their muzzles with respect to the brain case, allowing for both a wide gape and a pincherlike action of the jaws. The two halves of the lizard's lower jaw are, in adults, firmly united at the front, which limits the size of food items that can be swallowed to something less than a head's width. The tongue is well-developed in all lizards and attached at the back of the oral cavity. There are invariably teeth bordering the cavity, and there may be additional teeth on the palate. Lizard teeth are usually pleurodont: in other words, they have elongated roots that are weakly attached to the inside margins of the jaws, with the bases of the roots not fused with the jaw. A few groups have acrodont teeth; these are usually short and very firmly attached to

◗ **Right** *Green iguanas* (Iguana iguana) *have small, granular scales and a dorsal crest running the full length of the body and tail. The prominent dewlap beneath the chin shows this young individual to be a male.*

◗ **Below** *Skin shedding is common among reptiles and amphibians. Here, a Carpet chameleon* (Furcifer lateralis) *sloughs its skin in large pieces, a pattern characteristic of lizards (snakes shed in a single piece).*

FACTFILE

LIZARDS

Order: Squamata *

5,015 species in 457 genera and 28 families

DISTRIBUTION In the New World from S Canada to Tierra del Fuego; in the Old World from N Norway to Stewart Island, New Zealand; also islands in the Atlantic, Pacific, and Indian Oceans.

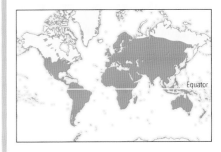

HABITAT Terrestrial, rock-dwelling, tree-dwelling, burrowing, or semiaquatic.

SIZE Length (snout to vent) 1.5–150cm (0.6–59in), but most are 6–20cm (2.5–8in).

COLOR Highly variable, including green, brown, black, and some brightly-colored species.

REPRODUCTION Fertilization internal; most lay eggs on land, but in some the eggs develop to an advanced stage inside female; some species have a true placenta.

LONGEVITY 50 years+ in captivity in some species.

CONSERVATION STATUS 43 species are listed as Critically Endangered, 53 as Endangered, and 105 are listed as Vulnerable.

* The designations Lacertilia or Sauria are no longer applied by systematists to lizards, since they do not represent a natural evolutionary group. Snakes and amphisbaenians evolved from lizards, so the term 'lizards' merely describes those squamate reptiles that lack the specialized traits of either snakes or amphisbaenians.

the jaws by either their bases, sides, or both. In most species the teeth are replaced frequently.

The external ear opening is usually visible, and there is a movable eyelid in most species. Typically, lizards have well-developed paired lungs, a urinary bladder, and the subclavian artery arising from the systemic artery. The paired kidneys are usually symmetrical in general appearance and lie in the rear part of the body cavity. The anal opening joins with the urogenital tract in a common chamber, the cloaca, which exits via a transverse slit. Male lizards have paired intromittent organs called hemipenes (singular: "hemipenis").

○ **Right** *A gecko's toes have sophisticated modifications that enable this family of lizards to climb and adhere to even the smoothest of surfaces. Seen here is the most agile of all, the Tokay gecko (Gekko gecko).*

Scurrying and Burrowing

LOCOMOTION

The familiar scurrying of lizards is not possible for some species, for in some surface-dwellers and many burrowers the limbs are reduced or absent. Female blind lizards, some anguids, and some skinks are completely legless, with only traces remaining of the internal, bony girdles associated with the missing limbs. In flap-footed lizards, male blind lizards, and a variety of girdled and plated lizards, there are no forelimbs and only vestigial hind limbs, although in one microteiid it is only the hind limbs that are lost entirely. Limb reduction is a special advantage in habitats that have narrow openings, such as dense vegetation or broken earth and rock. In these species, movement is achieved by snakelike sideways undulation, with the vestigial limbs being held close to the body.

Most species, however, have four legs, each with five toes. These saurians exhibit the most typical pattern of lizard locomotion: limbs moving in a symmetrical gait on either side of a sprawling, undulating body. The body itself can be cylindrical, depressed (flattened against the ground), or compressed (flattened vertically). The legs can be short or long, stout or slender. Burrowing lizards tend to be cylindrical, whereas crevice-dwellers are usually depressed. A compressed body form is typical of aquatic and arboreal species. Stout, long-legged species like the Australian Gould's monitor are usually runners that live in open grasslands and deserts. Arboreal species like the anoles and iguanas usually have long, slender legs, useful in jumping between perches or stretching from one branch to another. Other factors, such as digging and fighting, have probably also been important in limb evolution.

Some lizards are at least partly aquatic. Most of these, like the Crocodile lizard of South America,

have powerful, laterally-compressed tails that help to propel the animal though the water. The Bornean earless lizard is in a family by itself (the Lanthanotidae). It is a good swimmer, propelling itself with small forelimbs and snakelike sideways undulations of the body. The Marine iguana, which has webbed feet, is one of the only lizards specialized for life in the ocean. Basilisks have a fold of skin along the side of the toes that increases the surface area of the foot and helps them run across the surfaces of ponds and small streams. The Fringe-toed lizard and fringe-toed lacertids have toe modifications for running on loose sand, while the Web-footed gecko has feet that are fully webbed to serve as sand scoops.

Tree-dwelling lizards display some of the most impressive adaptations for locomotion. The Flying dragons of southeast Asia glide from tree to tree on a rib-supported membrane between the fore- and hind limbs. Other modifications for gliding or

Left *Basilisks, such as this Double-crested basilisk (Basiliscus plumifrons) have an unorthodox mode of locomotion when fleeing danger, being able to run bipedally and cross the surface of ponds and streams. Their capacity to "walk on water" has caused them to be referred to colloquially as the Jesus Christ lizard.*

Below *Using lateral membranes stiffened by lengthened ribs, the Common flying dragon (Draco volans) of Southeast Asia can glide for distances of around 60m (c.200ft) in its rainforest habitat. Fine adjustments of the tail and membrane allow this lizard to control its descent with considerable accuracy.*

parachuting occur in some geckos and lacertids. Like birds, chameleons improve their grip in the trees with feet designed for perching, with some toes facing forward and some back; specifically, the inner three and outer two digits of the forefeet are bound together and oppose one another, while the inner two and outer three of the hind feet are opposed – a condition referred to as zygodactyly. The toe modifications that enable many geckos and anoles to scale sheer surfaces can be exceedingly elaborate.

Several species of lizards have prehensile tails

that they wrap around vegetation to steady themselves as they move through their environment; in effect, they have five "limbs." Chameleons are the lizards best known for this adaptation; however, it is of much wider occurrence, among both arboreal and terrestrial lizards of many families. Some species of geckos have the scales on the undersurface of their prehensile tails modified like those on their toes, so that they can get a firm grip on vegetation. The scale at the very tip of the tail even looks like a claw in some species of lizards, and it may function as such.

THE VERSATILE LIZARD TONGUE

All lizards possess a well-developed tongue that can be extended. Many species constantly "sense" the world around them with this organ, sampling molecules from the environment and retrieving them into the mouth as chemical information on food, mates, territories, and predators.

The site of chemical sensation is called the vomeronasal, or Jacobson's, organ. It occurs in a variety of vertebrates including amphibians, lizards, snakes, and some mammals. In lizards it consists of small, paired cavities lined with sensory cells located within the roof of the mouth near the snout **1**. The molecules gathered on the lizard's tongue are transported to narrow ducts leading to these cells, and the information recorded is sent to the brain by way of the vomeronasal nerve. This type of chemosensation is different from smell and taste, of which lizards are also capable.

Lizards with long, forked tongues, such as the beaded lizards, whiptails, and monitors **2**, have better developed Jacobson's organs than do their short, broad-tongued relatives such as the Common bloodsucker **3**, which use their tongues more for capturing prey than for chemosensation.

The tongue has other functions. In many species of geckos and their relatives, it is often used to clean the spectacles of the lidless eyes (BELOW *Uroplatus*

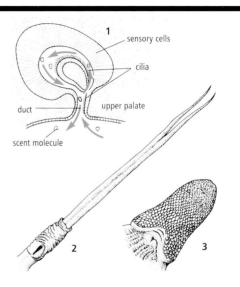

henklei), while in the Great Plains skink the mother regularly licks her eggs. In several species of Australian blue-tongued skinks, the brightly-colored tongue is probably used to frighten predatory birds and mammals. The extremely long, sticky, and highly extensible tongues of chameleons are modified versions of the fleshy tongues of chisel-teeth lizards and are used to capture distant insects.   AGK/GS

THE EXPENDABLE TAIL

Confronted by a snake or other predator, a lizard will sometimes voluntarily shed (autotomize) its tail in one or more pieces. Prior to autotomization, some lizards such as the Texas banded gecko slowly undulate the vertically erect tail from side to side. This movement is believed to cause the predator to shift the focus of its attack from the more vulnerable head and trunk. The shed tail itself wriggles convulsively for several minutes, distracting the predator and improving the tailless lizard's chance of escaping unharmed. The ability to autotomize occurs in lizards of most families, but is absent in chisel-teeth lizards, chameleons, monitors, beaded lizards, xenosaurs, and the Bornean earless monitor.

Species capable of this feat (BELOW a Cyan-tailed skink, *Emoia cyanura*) have a "fragile" tail, with fracture planes in one or more of the vertebrae. A wall of connective tissue passes through each such vertebra, making a weak point where muscles and blood vessels are also modified to allow an easy break. The lizard slowly grows a new tail, never quite the same as the original. It does not have bony vertebrae with fracture planes, but apart from autotomy has all the functions of the original, assisting in running, swimming, balancing or climbing, camouflage, courtship, mating, and fat-storage. If the tail breaks at a point other than a fracture plane, regeneration is slight.

Although losing a tail may save its owner's life, autotomy has its costs. For example, the fat stored in the tail is normally broken down and used for growth and maintenance when food is scarce or not available, especially in winter or during drought. One species of gecko, the Australian marbled gecko, is known to live longer when it has a tail. Moreover, the fat stored in a female lizard's tail appears to be important in yolk production; individuals lacking tails produce eggs with lower mass and less energy, and the offspring have less chance of survival.

Tail regeneration itself may use up energy that could otherwise be used for reproduction, perhaps making larger eggs. In the Texas banded gecko, reproduction is known to have energetic priority over tail regeneration. The same may be true for other short-lived species, especially when the probability of producing offspring is low.

A complement to tail autotomy called regional integumentary loss occurs in some geckos and possibly a few skinks. Here there are preformed zones of weakness in the body skin of the lizards. When grasped by predators, such lizards can escape by causing their own skin to rip away. The physiological and energetic cost to the lizard is almost certainly very high, but some geckos survive the loss of up to 40 percent of their dorsal skin.   AMB/AGK/GS

Strategies for Security
DEFENSE

A much more characteristic tail modification in lizards results in the remarkable phenomenon of autotomy or voluntary tail-loss (see The Expendable Tail), but camouflage, or crypsis, is by far the most effective way to avoid predation. Many lizards exhibit patterns and coloration that blend with their background. Chisel-teeth lizards and chameleons can change in a few seconds from display to camouflage coloring by movement and expansion of pigment in the skin. Crypsis is further enhanced if the lizard remains motionless, but when directly confronted by predators, whether mammals, birds, or other reptiles, several other behavioral and physical defense strategies may be employed.

Lizards are well known for their agility and swiftness. When under attack most species

attempt to escape, although the slow and clumsy blue-tongued skinks of Australia almost invariably threaten the predator with a gaping mouth, strong jaws, bright blue tongue, and bouts of hisses. The beaded lizards, venomous and generally slow-moving, also gape and hiss. Such behavior seems to be very effective, and adults of these species have few enemies.

The Australian Frilled lizard displays a large collar of loose skin and inflates its body when startled by a predator. Monitors use their speed, powerful jaws, limbs, and whiplike tail to deter attackers. The basilisks employ their long, powerful hind limbs and expanded toes to run on the surface of water to escape landbound predators; they are also good swimmers, and can stay submerged for very long periods of time. Skin and skeletal modifications allow flying dragons to glide from tree to tree or to a distant spot on the ground; however, they are incapable of the powered flight found in birds, bats, and insects. Many species of lizards escape predators by climbing trees, rocks, or man-made structures. Some, including many girdled lizards, wedge themselves in crevices or, like chuckwallas, may inflate the body to prevent predators from pulling them out. The special feet and tail modifications of geckos and anoles permit them to climb almost any surface. Several species of geckos move with ease on almost all surfaces, even upside down.

Many lizards live much of their lives underground, and thereby avoid predators. Underground species are found in such diverse families as the skinks, blind lizards, flap-footed lizards, and anguids. Those that are active on the surface from time to time escape by rapidly diving, burrowing, or sinking into soft soil. Many are most active at night, when their potential predators are not about.

Another ploy in avoiding predation is to "play dead" or, alternatively, to become very rigid ("tonic immobility"). Some predators stop their attack when the prey becomes limp or rigid and appears lifeless. They rely on behavioral cues in performing their attack, and a "dead" lizard does not supply these.

The horned lizards of North America and the moloch of Australia are especially spiny, offering attackers an unpalatable mouthful of sharp scales. Horned lizards have also evolved a mechanism to squirt blood out of the eyes at predators. This appears to be especially effective against foxes and coyotes, which find the fluid distasteful.

◁ **Left** *If molested by coyotes, horned lizards of the Southwestern US (here, the Regal horned lizard,* Phrynosoma solare) *squirt bad-tasting blood from sinuses in their eyes, making the predator release its grip.*

▷ **Right** *An Australian frilled lizard (*Chlamydosaurus kingii*) erects a rufflike collar of skin around its neck, greatly increasing its apparent size and making it more intimidating to adversaries.*

Mostly Predatory
FEEDING BEHAVIOR

Many lizards are themselves predators. They feed principally on insects and other small terrestrial invertebrates, but larger lizards often eat mammals, birds, and other reptiles. The Komodo dragon is a scavenger–predator that subdues goats and even water buffaloes. Its teeth are laterally compressed with serrated edges, resembling those of flesh-eating sharks; the dragon literally cuts chunks of flesh from its large prey. Its highly flexible skull allows it to swallow large items. The Caiman lizard eats snails; its robust skull, powerful jaw muscles, and molarlike teeth provide a base for breaking shells.

Only about 2 percent of all known species of lizards are primarily herbivorous. The iguanas consume a wide variety of plant material, especially as adults. Herbivory reaches its zenith in the Marine iguana of the Galapagos Islands. This species dives 15m (50ft) or more underwater to feed on algae, kelp, and other marine plants that grow near the rocky shores where it lives. Lizards that eat leaves and stems usually have specialized gut structures that house bacterial symbionts that help digest plant tissues, but many geckos, skinks, lacertids, and other lizards regularly supplement a more insectivorous diet with seasonally available fruits, which are more easily digestible. Many species of lizards shift their diets with maturity and seasonal changes in the availability of food.

Mutual Grooming and Threat Displays
SOCIAL BEHAVIOR

Many lizards issue threat displays, both to their own species and to others that indicate territorial ownership or aggressive intent. Color changes, body inflation and push-ups, jaw-gaping, tail-waving, and species-specific head movements are all important signals. The colorful throat fan, or dewlap, of anoles is distended when males encounter each other or their enemies. Holding a territory is beneficial in several respects, but there are costs, such as the increased risk of being taken by a predator while repeatedly displaying from the same place. Visual predators such as snakes feed more frequently on male anoles than on the inconspicuous females.

Combat often occurs when lizards are attaining or defending a territory or a mate. Male Marine iguanas acquire territories during the onset of the

⊘ **Below** *A Veiled chameleon (Chamaeleo calyptratus)* shoots out its projectile tongue to catch a cricket; the hapless insect adheres to the tongue's sticky tip. A chameleon's tongue can extend as much as twice the snout-to-vent length of the lizard.

mating season and fiercely fight intruding males. When a territory has been repeatedly defended, neighboring males become less frequently involved in boundary disputes. Larger males usually hold bigger and better territories, and they mate more often. Courtship behavior is an important part of the mating ritual. Females of some species are also territorial and fight.

Hatchling and juvenile lizards often emerge together from the nest-hole (as, for instance, in the Common green iguana), an antipredator strategy in which many eyes are better than two and large numbers make individual capture less likely.

⬤ **Right** *A wrestling bout between two male Gould's monitors (Varanus gouldii), a species native to northern and eastern Australia. Fighting takes place in the breeding season for possession of females, and dominance is established by pushing the rival over.*

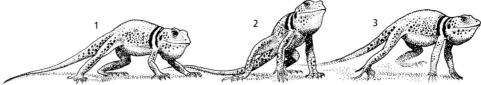

Young iguanas often remain in groups, and one of them may temporarily behave as leader. They engage in mutual tongue-licking and grooming, dewlap extension, and body and chin rubbing. At night they often sleep together on branches.

Social communication in lizards sometimes involves chemicals. Although the lizard's skin completely lacks mucus glands, other types of glands may be found on the belly, the undersurfaces of the thighs (femoral glands), and in the region of the cloaca (precloacal glands). These glands tend to be larger in reproductively mature males. Their secretions are believed to attract

females and mark territories. In species of some families, females also have them.

Basking, or voluntary exposure to the sun, is very commonly observed in lizards, although nocturnal or highly secretive lizards often obtain heat indirectly from the substrates on which they live, a strategy referred to as thigmothermy.

⬤ **Above** *Lizard aggression: preparing for battle with a rival, a male Collared lizard (Crotaphytus collaris)* **1** *catches sight of his opponent;* **2** *bobs up and down vigorously, his feet leaving the ground, and* **3** *begins to charge.*

Copulation and After

REPRODUCTION

A few lizards in each of several families are unisexual (see Unisexuality: The Redundant Male?). In species with both males and females, there is copulation, and fertilization is internal. Males achieve intromission into the female's reproductive tract with one of their two hemipenes, each part of a separate reproductive tract with a single testis, which is located within or near the middle of the body cavity. A hemipenis is not structurally comparable to the mammalian penis, being a membranous pouch (often ornamented with species-specific spines, whorls, and folds) that is turned outward through the male's cloacal opening during copulation. One hemipenis may be used three or four times before the lizard switches to the other, probably because the supply of mature sperm to the first is depleted.

Right *Double-crested basilisks hatching, after an incubation of 2–3 months. This genus of lizard produces clutches of 8–18 eggs and breeds year-round.*

Below *A Parson's chameleon (*Calumma parsonii*) carrying its young. This, the largest chameleon species, is native to the rainforests of eastern Madagascar, where its survival is threatened by habitat loss and export for the pet trade. It is CITES-listed, and legal commercial exports were suspended in 1995.*

Fertilization of one or more eggs occurs inside the female's oviducts, and the embryo achieves some maturity prior to laying. In a few species, among them Jackson's chameleon, sperm is stored in the oviducts for long periods of time, thereby making paternity difficult to establish. A female typically lays her eggs beneath a log or rock where the humidity is relatively high. Other species retain their eggs until the embryos are well-developed, and the young emerge fully developed from thick membranes. This mode of reproduction is common in groups such as xenosaurs, anguids, night lizards, girdled lizards, and skinks, and occurs more rarely in a few other families.

In the Brazilian skink, the young develop inside the oviduct and possess a placental connection with the mother. The lizard placenta, like that of mammals, is important in providing nourishment and removing waste products. Little or no nutrition is obtained from egg yolk, as is typical of most reptiles.

The number of eggs or babies is fixed in some families, such as blind lizards (1 egg) and geckos (1–2 eggs or babies), but may vary widely among other groups. The largest clutches of eggs (up to 50 in some species) are found in monitors. In general, the very largest lizards have larger clutches and litters than smaller species.

Most lizards exhibit little maternal care, except for finding and excavating a suitable site for egg-laying. There are, however, several species in which the female broods or guards her eggs. Some clean and rotate their eggs. In a very few species, including the Great Plains skink, the Desert night lizard, and some alligator lizards, the mother may assist hatchlings from the fetal membranes or egg and defend them from predators.

Some lizards have a well-defined period of annual reproduction, whereas others reproduce throughout the seasons. Most species in temperate climates are cyclic, whereas many tropical forms are continuous breeders. Environmental conditions such as temperature, rainfall and humidity, food availability, and light cycles are very important factors in all species.

LIFE IN THE LITTER

Alice, in her adventures in Wonderland, encountered a bottle with the label "Drink Me" that made her small. Fortunately for her, she did not have to deal with the consequences of small size permanently. The Jaragua sphaero (*Sphaerodactylus ariasae*) is not so lucky. This tiny gecko was described in 2001 from specimens collected in leaf litter on Isla Beata, a small island just off the coast of the Dominican Republic. Adults average only 16mm (0.6in) in length (maximum 18mm/0.7in), with a tail of similar length and a weight of about 0.13g (0.005oz), qualifying this gecko as the smallest lizard in the world. At this scale, spiders and other predatory arthropods are potential predators rather than potential prey, yet the greatest threat to the gecko's existence may be desiccation. Because of its large surface area relative to its body volume, the Jaragua sphaero tends to lose water rapidly, forcing it to conduct its life within the humid confines of the leaf litter. As well as having to contend with the problems of dwarfism, this species is also threatened by the loss of its habitat to deforestation, even within national parks.

To place the Jaragua sphaero in context, it is less than 7 percent of the length (and about 0.05 percent of the weight) of the largest member of its own family, the Gekkonidae. But geckos are not particularly large lizards as a whole. Compared to the largest living lizard, the Komodo dragon, the Jaragua sphaero is only 1 percent as long and 0.0001 percent of the weight! Interestingly, both the largest and the smallest lizards are island species; both gigantism and dwarfism are common by-products of evolution on islands, where factors such as an absence of predators and/or the limitation of food resources can promote selection for size change.   AMB

◗ Right *A fully grown Jaragua sphaero, placed on a dime for scale. It is not only the smallest lizard, but also the smallest of all reptiles, birds, or mammals.*

Populations at Risk

CONSERVATION AND ENVIRONMENT

Folklore and fables concerning lizards abound. Many harmless species, especially geckos and skinks, are feared by uninformed people who firmly believe them to be venomous. Although the beaded lizards are indeed venomous, the beliefs that they possess a poisonous tail and are capable of spitting poison are false. One ancient superstition is that lizards bite the shadows of unsuspecting people, who are then doomed to die. Lizard tails were part of the recipe for a love potion among the Salish Indians of North America, and some Asian cultures hold that a long and prosperous life is guaranteed if a gecko calls from the bedroom of newlyweds.

Humankind has relentlessly exploited lizards for food and for its social needs. Throughout the tropics of the New World, the Common green iguana is killed for its flesh and eggs. West Indian rock iguanas, many of which are severely endangered, suffer a similar fate. South American tegus and both Nile and Common Asiatic monitors are harvested by the millions annually as sources of leather. Live monitors are used in India in fertility rites, or serpent festivals. There is little doubt that this ritualized use harms them.

The pet trade has probably had only a minor effect on the decline of most lizard populations, but certain species have been depleted by this industry. Rare or unusual forms like the Solomon Islands prehensile-tailed skink and Madagascan leaf-tailed geckos are eagerly sought by collectors.

The greatest threat to lizard populations is permanent habitat alteration or destruction, especially in complex and poorly understood regions of the tropics and subtropics. In southern Florida, for example, there has been an ecological near-collapse that is due entirely to human actions. Although the introduction of exotic species of lizards to the region has been blamed for the decline of some native lizards, there is no firm evidence to substantiate these claims (exotics tend to occupy manmade environments, new niches that native forms are unable to shift into).

There is no doubt, however, that nonreptilian introductions are very harmful to many forms of lizards in any habitat. Among others, the rock iguanas of the West Indies have suffered from the presence of mongooses, feral dogs, cats, goats, and cattle. On the Canary Islands, the Critically Endangered Gomeran giant lizard was nearly exterminated by introduced predators, and all known individuals are now kept in captivity to ensure their survival. The discovery, in tiny remnant forest patches, of new species of geckos from Madagascar and anguids from Middle America highlights the disturbing truth that the wholesale destruction of certain habitat types is almost certainly causing the extinction of some lizard species before their existence is even known.

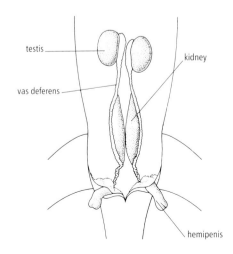

◖ Left *Reproductive and urinary organs. A male lizard's hemipenes are pouches that are turned outward during copulation. Only one hemipenis is used at each mating.*

testis
kidney
vas deferens
hemipenis

NATURAL DESERT DWELLERS
The adaptation of reptiles to arid climates

REPTILES ARE USUALLY THOUGHT TO REPRESENT the acme of desert adaptation among vertebrate animals. All of the world's deserts possess diverse reptilian faunas, and many species are well-known for their ability to survive in such inhospitable habitats – for example, the chuckwalla (*Sauromalus obesus*) in the Mojave desert in the United States, the fouette-queue or dob (*Uromastyx acanthinura*) in the Sahara, and the Netted dragon lizard (*Ctenophorus nuchalis*) in the central Australian desert. These animals are all capable of surviving in regions where daily temperatures exceed 40°C (104°F) and rainfall is rare.

Numerous studies of the behavior and physiology of such species over the past few decades have revealed many of the secrets of their survival. Firstly, as ectotherms – that is, animals which depend on external sources of heat to regulate their body temperatures – they take full advantage of the heat provided by the sun, and can maintain a remarkably constant body temperature throughout the day. Small variations in posture and the angling of their body either towards or away from the sun enable them to raise or lower their body temperature as required. Raising the body away from the hot substrate and standing vertically at midday (stilting) is also a very effective thermoregulatory strategy, as is climbing up on twigs and branches during the hottest part of the day. Then again, their low rates of metabolism – about one-tenth those of birds or mammals of similar size – mean that their material needs are very limited. Rates of turnover of water and essential nutrients such as protein and carbohydrate are very low, and this frugality enables them to survive in regions where the lack of adequate food excludes high-energy consumers such as birds and mammals.

Another helpful feature is reptiles' extreme tolerance of perturbations of their internal environment. All vertebrate animals regulate the concentration of vital elements such as sodium and potassium in their blood, but reptiles, and particularly lizards, are able to tolerate large deviations from the norm. Sodium levels may thus rise as high as 300 mmol.l-1 (millimoles per litre) and potassium to 12 mmol.l-1 in summer, concentrations that would be instantly lethal for a bird or a mammal. Lizards such as the Ornate dragon (*Ctenophorus ornatus*) in Western Australia tolerate such perturbations for many months before being relieved by thunderstorms, during which they drink the raindrops and rapidly restore their normal electrolyte balance.

Reptiles also conserve large amounts of water by excreting uric acid, rather than urea, as the end product of protein catabolism. Uric acid has a very

♦ *Above* In Australia, a Bearded dragon lizard (Pogona barbata) *stretches out full length on the topmost branches of a tree. Tree-climbing in the heat of the day allows lizards to escape the hot substrate and also take advantage of any passing breeze.*

◁ **Left** *Against the backdrop of the Grand Canyon, a chuckwalla basks in the sun. Like most lizards, this species basks each morning; once its preferred body temperature of 38°C (100°F) is reached, it begins searching for the leaves, buds, and flowers that make up its herbivorous diet.*

spend months of inactivity in summer in deep rock crevices, waiting for the onset of rain and more favorable conditions. There is also evidence that unfavorable conditions, such as dehydration and salt loading, may "reset" the animals' thermal set-point, so that they maintain a body temperature 1–2°C (1.8–3.6°F) lower than normal when active. This adjustment effectively results in a reduction in the rate at which vital body water is lost to evaporation, and thus can help extend their survival.

All of these factors underpin the successful invasion and longterm utilization of desert habitats by reptiles. When, however, we speak of "adaptations," we usually mean heritable modifications of an animal's phenotype (morphology, behavior, or physiology) that have arisen as a result of natural selection in that environment. To answer the question whether all the features seen in desert reptiles have arisen as a result of natural selection – in other words, through the differential survival of young that are so endowed in desert habitats – we need to make comparisons with reptiles that live in temperate and tropical regions and are never exposed to the desert's rigors. Surprisingly, doing so reveals little difference in the abilities of reptiles to tolerate the extremes found in deserts, regardless of whether they occur there or not. This fact leads on to the conclusion that reptiles appear to be ideally suited, or "pre-adapted," to survive in deserts because of the overall nature of their basic physiology and behavior. We thus say that they are "exapted" – "exaptation" being an adaptation that has arisen in some environment other than the one in which it is utilized. SDB

◁ **Left** *On hot sand dunes in the Namib Desert, a Shovel-snouted lizard (*Merioles anchietae*) practices "thermal dancing," raising its limbs to keep cool.*

▷ **Right** *A Namaqua chameleon (*Chamaeleo namaquensis*) demonstrates "stilting" – lifting the body as far as possible from the hot substrate.*

low solubility in water, and is precipitated in the kidney tubules and cloaca, to be finally voided as a solid white mass. Large amounts of water are reabsorbed in the cloaca and the hind part of the intestine (colon) as the uric acid is precipitated; in lizards such as the varanid *Varanus gouldii*, laboratory studies show that as much as 90 percent of the fluid entering the cloaca from the kidneys is reabsorbed into the body.

Another noticeable feature of desert-dwelling reptiles is their ability to curtail all vital activities when exposed to extreme temperatures or water deprivation. Lizards such as the chuckwalla may

Iguanas
ONE FAMILY OR EIGHT?

The lizards commonly known as "iguanas," formerly family Iguanidae, are now considered to be divided between eight separate families, although some authorities still regard them as subfamilies. Collectively they occur in North and South America, including the West Indies and the Galápagos islands, and where they are the most conspicuous group of lizards, on Madagascar and two Pacific islands, Tonga and Fiji.

Basilisks and Relatives
FAMILY CORYTOPHANIDAE

Three genera from Middle America (southern Mexico to northern South America) make up this family. They are mostly arboreal lizards with slender bodies and long hind limbs and tails. Several are bright green and have crests on their backs and the top of their heads. Males are more colourful and have larger crests than females. The four species of basilisks (genus *Basiliscus*) forage on the ground near trees and bushes and are fast-moving species that raise the front of their body off the ground as they gather speed, using only their long, powerful hind limbs (bipedal locomotion). They are well known for their ability to run across the surface of water and are sometimes called "Jesus Christ lizards". The other two genera (*Corytophanes* and *Laemanctus*) move more slowly, and are more closely associated with trees, rarely com-

ing down to the ground except to lay their eggs. All members of the family apart from one are oviparous, laying eggs over an extended nesting season, the exception being the keeled helmeted basilisk (*Corytophanes pericarinata*).

Collared and Leopard Lizards
FAMILY CROTAPHYTIDAE

Two genera of long-legged, fast-moving terrestrial species from deserts and other arid habitats in the American southwest and northern Mexico. They have stout bodies, long tails, and large heads. They are equipped with powerful jaws which they use to crush their food, often smaller lizards, such as side-blotched lizards, in which they specialize, although they also eat large numbers of invertebrates. Collared lizards in particular (genus *Crotaphytus*), occupy prominent rocks or stumps from which they can survey their surroundings and launch attacks on potential prey or rivals. Collared lizards may be green, blue, brown, or orange and have prominent black and white bands around their necks. Leopard lizards (genus *Gambelia*) are duller, with darker spots covering their backs. All the species in this family are egg-layers, with clutches of around 3–8 eggs.

Dwarf Iguanas and Weapontails
FAMILY HOPLOCERCIDAE

These are a poorly known group of 11 lizards from Panama and northern South America to Central Brazil. The dwarf iguanas (genera *Enyalioides* and *Morunasaurus*) are slow-moving, cryptic, arboreal species occurring at low densities and rarely seen. Males have clusters of long pointed scales on their neck and head and a dorsal crest. The weapontail

(*Hoplocercus spinosus*) is very different; it is terrestrial and has a stocky body, short head and a short, thick, spiny tail, which it uses to defend itself if caught in the open and to block the entrance to tree-holes in which it hides. The reproductive mode of the members of this family is unknown.

Iguanas
FAMILY IGUANIDAE

Thirty-six large lizards are retained in the family Iguanidae and are among the most familiar species. They include the green iguana (*Iguana iguana*), a very large, green or greenish-brown arboreal species with a wide range from Mexico to Paraguay, and Florida, where it has been introduced. This herbivorous species is a familiar sight perched at the top of forest trees, making only brief forays to the ground for egg-laying. One species from the Galápagos islands, the marine iguana (*Amblyrhynchus cristatus*), is the world's only ocean-going lizard and feeds exclusively on sea lettuce. Two other iguanas on the Galápagos (genus *Conolophus*), are terrestrial and feed largely on the pads and fruits of prickly pear cactus. The West Indian ground iguanas (genus *Cyclura*) are all endangered or vulnerable, having been hunted for centuries and, more recently, having lost much of their habitat to development and agriculture. These large, prehistoric-looking lizards include the rhinoceros iguana (*C. cornuta*) of Haiti and the Cayman Island ground iguana (*C. lewisi*), of which only 15–20 individuals remain in the wild and which is expected to become extinct in the next five years. Six species of chuckwallas (genus *Sauromalus*) live among rock outcrops in the American southwest and Mexico, where several are restricted to small islands in the Gulf of California. They are wary, heavily-built herbivores that

⊙ Below *On the Galapagos Islands, Marine iguanas* (Amblyrhynchus cristatus) *aggregate in large numbers in order to bask and raise their body temperature before plunging into the cool sea to feed.*

down. They are terrestrial, sometimes living among rocks, and include oviparous and viviparous species. Tree lizards (genus *Urosaurus*) and side-blotched lizards (genus *Uta*) are ubiquitous over much of the southwestern United States, and have speciated widely in Baja California and its small offshore islands; all are insectivorous egg-layers. Fringe-toed lizards (genus *Uma*) live in sandy habitats where their specialized toes enable them to run across loose, windblown sand. If pursued, they dive head-first into the dune and disappear rapidly below the surface. The same species, as well as zebra-tailed lizards (*Callisaurus draconoides*) and earless lizards (genera *Cophosaurus* and *Holbrookia*) have black and white bars on the undersides of their tails, which they raise and curl over their bodies as a form of visual communication. These are all insectivores that chase their prey with sudden rushes, and all are egg-layers.

Horned lizards (genus *Phrynosoma*) are the most distinctive members of the family. Short and wide, with short limbs and a stubby tails, these species rely on camouflage rather than speed; their shape gives them the vernacular name of "horny toads" as well as their scientific name, which means "toad-bodied". Their intricate markings of blotches and spots match the substrate on which they live and can vary from place to place, even within a single species. Irregular large spiny scales on the dorsal surface, and a fringe of pointed scales around edge of their body, break up their outline. The backs of their heads are armed with large spines which make them less attractive to

can jam themselves into crevices by inflating their bodies. Other members of the family include the spiny-tailed iguanas (genus *Ctenosaura*) of North and Central America and the relatively small desert iguana (*Dipsosaurus dorsalis*). The Fiji iguanas (genus *Brachylophus*) are restricted to the Pacific region and presumably spread westward from the South or Central American mainland by rafting on vegetation caught up in ocean currents. Most iguanids are exclusively herbivorous, feeding on leaves and flowers, with the exception of one or two of the smaller species, and young of some of the larger species, which sometimes also eat invertebrates. All are oviparous, laying up to 80 eggs in the case of the green iguana.

Madagascan Spiny-tailed Lizards
FAMILY OPLURIDAE

The seven species that make up this family are restricted to Madagascar and the neighboring Comoros Islands. They are superficially similar to iguanids from the New World. They are small to medium-sized lizards from dry, rocky, or sandy habitats. Six Madagascan collared lizards (genus *Oplurus*) are dull-colored, fast-moving, day-active species that live among rocks and climb into bushes and scale tree trunks in open forests. They have pointed scales, and some have spiny tails

◑ Above *One of the most extraordinary lizards is the agamid species* Moloch horridus, *the Thorny devil, which inhabits the dry Australian interior. It is covered in spiny, warty protuberances. Narrow channels between the scales act to draw precious droplets of dew or rain to its mouth by capillary action.*

and males become more brightly colored during the breeding season. The remaining species, the four-eyed lizard (*Chalarodon madagascariensis*) is smaller and lives in sandy places in the drier parts of the island. It lays small clutches of eggs—usually two—whereas the other *Oplurus* species lay four to six eggs.

Horned Lizards and Spiny Lizards
FAMILY PHRYNOSOMATIDAE

The Phrynosomatidae is a large assemblage of mostly small to medium-sized diurnal, terrestrial lizards found in North and Central America. Although they are most closely associated with deserts and arid scrublands, a few species occur in forests and humid environments. The spiny lizards (genus *Sceloporus*) are the most numerous genus with over 86 species found over most of the family's range. Males are often colorful, especially in the breeding season, and some have bright blue chest patches that they display by bobbing up and

◑ Above *The Central bearded dragon (*Pogona vitticeps*) is named for the fringe of soft spines around its neck. An inflatable throat pouch raises the spines for threat or courtship displays. This species lives in groups with a relatively complex social structure.*

151

mus occur throughout the southern Andes in Chile and Argentina, many of them at high altitude, and the genus is roughly equally divided between egg-layers (lowland) and live-bearers (highland) forms. *L. magellanicus* from Tierra del Fuego occurs further south than any other lizard. The 27 *Leiocephalus* species are sometimes known as 'curly-tail' lizards because they often coil their tails above their backs when resting. As with the North American zebra-tailed lizards and others, this acts as a signal to predators that may be stalking them. The 14 species of montane lizards belong to the genus *Phymaturus* are stout, flattened lizards with thick, swollen tails. They are herbivorous and give birth to live young. One species, the Patagonian mountain lizard (*P. patagonicus*) may be parthenogenic or, if not, it can retain sperm for longer than any other known squamate.

Chisel-teeth Lizards
FAMILY AGAMIDAE

Chisel-teeth lizards or agamids are the Old World counterparts of the iguanas. Like them, they are day-active, fully-limbed, and often have spines, crests, or flaps on their heads and backs. They differ in tooth attachment, however. The teeth of most lizards are rather loosely attached along the inner margins of the jaws (the pleurodont condition), but in chisel-teeth lizards (and their close relatives, the chameleons) they are firmly mounted on the top edges of the jaw bones (the acrodont condition). In most members of the family the teeth at the front of the head are compressed and may be fused to one another. They may be fanglike or more chisel-like, similar to the incisors of mammals.

Most species are insectivorous or carnivorous, but the large-bodied Egyptian spiny-tailed lizard and its relatives are herbivores. The Thorny devil or moloch of Australia is a specialist feeder on ants. Its wide body is adapted to accommodate thousands of ants at a time, which it laps up with its tongue. It is covered with spines that protect it from preda-

predators. Horned lizards' strategy is to rest motionless in the hope of being overlooked; the strategy often works. Under extreme provocation, some species can squirt droplets of blood from their eyeballs, with a range of several feet. Horned lizards feed almost exclusively on ants, often eating several hundred in a single sitting by simply positioning themselves next to an ant trail and picking off the insects as they march by. The 14 species of horned lizards are found in a variety of mostly dry habitats, including sandy or stony deserts, chaparral, conifer forests, and mountain slopes. The genus is equally divided between egg-layers and live-bearers, the latter being high altitude species. Clutches and litters are large compared with other species of phynosomatid lizards, a result of their bulky body shape and sedentary lifestyle.

Anoles and Relatives
FAMILY POLYCHROTIDAE

Numbers in this large family are bolstered by the inclusion of two large genera of anoles, *Anolis* and *Norops*, with 208 and 152 species respectively. These slender-bodies, long-tailed, diurnal lizards occur throughout the West Indies as well as the North, Central and northern South American mainland, and are often the most conspicuous sign of reptilian life. Most are green or brown, and nearly all are arboreal, although some are terrestrial and a few live in caves. Many adapt to human disturbance and live in gardens and on buildings, and a number have been introduced to Florida

and other places through accidental transport in produce, or through the pet trade. Their feet are specialized for climbing, with small pads (scansors) on each toe. Males are typically larger than females, and all species are oviparous. Each female lays one egg every two to three weeks throughout the season. Other members of the family include a number of other arboreal genera from tropical forests. Several are slow-moving, cryptic species, and all are egg-layers.

Lava Lizards and South American Swifts
FAMILY TROPIDURIDAE

Lava lizards (genera *Microlophus* and *Tropidurus*) and swifts (genus *Liolaemus*) are similar in many respects to the North American spiny lizards and have similar lifestyles; they live in open situations, often around rocks but sometimes in lightly wooded places, and are diurnal and mostly terrestrial. Lava lizards on the Galápagos Islands (genus *Microlophus*) have speciated on the various islands or island groups and at least seven species are recognised there. There are 183 species of *Liolae-*

◐ **Above** *Two striking features of chameleons are exemplified by this Panther chameleon (Furcifer pardalis) – independently swiveling eyes for all-round vision, and spectacular color change. The bold salmon-pink and red hues signal the animal's readiness to mate.*

◑ **Right** *Two male Jackson's chameleons (Chamaeleo jacksonii) confront each other in the Aberdare forest, Kenya. In common with some other species, Jackson's males are distinguished by their prominent "horns."*

tors. Its skin is also creased by tiny channels that direct water towards the corners of the mouth— a handy specialization to make the most of the infrequent rains of its desert home.

Although, with 420 species, there are not as many chisel-teeth lizards as iguanas, they are even more structurally diverse. They are especially numerous in Australia and tropical Asia, but they also widespread in the Middle East and arid Central Asia, as well as in Africa, except for Madagascar. In the forests of Asia many of the arboreal chisel-teeth lizards are green or brown, enabling them to blend in with the trees on which they perch; many such lizards bear impressive spines and crests. One forest-dweller, the Sail-fin lizard, reaches lengths of 35cm (14in), and in addition to a large crest on its back and tail bears wide flaps on the toes that help it swim in streams and rivers.

The strangest of the tropical Asian chisel-teeth lizards are the flying dragons. These lizards have extremely elongate ribs that can be folded back or extended to the side. The ribs support a thin "flight" membrane of skin that runs between the fore and hind limbs. Flying dragons launch themselves from tree trunks or branches and spread these skin flaps in order to glide to new perches. The long tail serves as a rudder to steer the lizard. The flaps of skin are colored and patterned differently in each species of dragon and proba-

bly serve as cues in sexual and social encounters. Terrestrial butterfly lizards from Southeast Asia also have elongated ribs that support brightly-colored flaps of skin, but these are not used in locomotion.

In Australia semiaquatic chisel-teeth lizards are represented by the Eastern water dragon, which has adapted to the human presence and thrives along streams, even in large cities like Sydney. When threatened, the Australian frilled lizard raises its front legs off the ground and runs on its powerful hind limbs, using its long tail as a counterbalance. Elongate modifications of the skeleton of the tongue can be manipulated to erect a large frill that nearly surrounds the head, presenting an imposing image to

would-be predators. Smaller frills under the chin or at the corners of the mouth are seen in some other Australian species as well as in one of the the Asian toad-headed lizards.

The Common bloodsucker of India and South Asia is inappropriately named. The bright red head of the male has nothing to do with its diet; rather, it is a conspicuous social cue used to signal other members of its species. Male chisel-teeth lizards usually perform complex courtship rituals to obtain matings with females. Rival males may

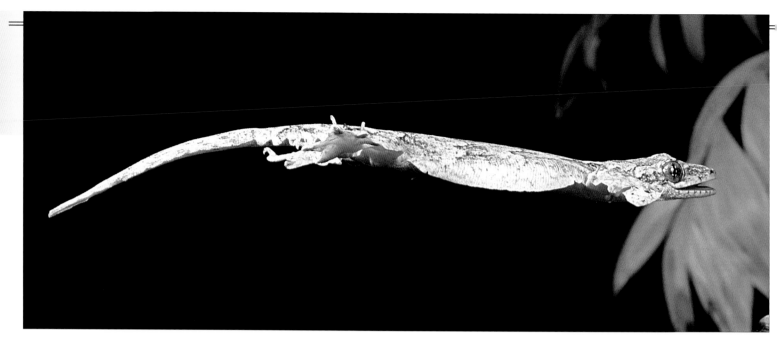

○ **Above** *Southeast Asian geckos of the genus* Pty-chozoon *(here* Ptychozoon kuhlii) *are often referred to as "flying geckos," but are more properly termed "parachuting geckos"; the skin flaps fringing their body and tail slow their fall, but do not allow them to fly or glide.*

be intimidated by one another's displays, but confrontations sometimes lead to combat, in which the enlarged front teeth may be employed. Toad-headed lizards use their tails in social encounters, coiling them up over the back as part of a threat display that also includes gaping. The tails of all chisel-teeth lizards lack fracture planes, so they rarely break. When breaks do occur, however, as in encounters with predators, the tail does not regrow, and a clublike expansion may form at the broken tip.

Almost all chisel-teeth lizards are egg-layers. As many as 35 eggs may be laid in a clutch, but smaller clutch sizes of 4–10 eggs are typical of most species; incubation times are often in the range of 6–8 weeks. Live birth appears to have evolved in two very different lineages: the toad-headed lizards of temperate Asia and the Sri Lankan prehensile-tailed lizards. Three species of butterfly lizards from Southeast Asia have all-female populations and reproduce by parthenogenesis (see Unisexuality – The Redundant Male?).

Many chisel-teeth lizards are tropical, but some occur in the mountains and on the steppes and high plateaus of Central Asia. These lizards are active for only short periods of the year and spend the remainder of their time inactive, deep within crevices or burrows to avoid the harsh conditions on the surface.

Chameleons
FAMILY CHAMAELEONIDAE
Chameleons are among the most distinctive of all lizards, possessing many traits for moving and feeding in trees and bushes. Their strongly compressed bodies, spindly limbs with grasping feet, prehensile tails, and independently mobile eyes

set them apart from all other lizards, but their firmly attached teeth show their close relationship to the chisel-teeth lizards. Like them, chameleons are strictly limited to the Old World. They are especially numerous in sub-Saharan Africa and in Madagascar, but they also extend north to southern Europe and east to India and Sri Lanka.

Most chameleons make their living in the trees, walking slowly along slender branches and twigs. On each foot opposing bundles of two- and three-fused (zygodactylous) toes ensure a firm grip on these narrow perches. The compressed body, combined with long limbs and highly mobile shoulder joints, helps to keep the animal's weight centered above the perch for stability, and the prehensile tail provides an additional anchor point when the chameleon must release its grip to reach another branch.

The slow speed of the chameleon not only gives it stability when moving but makes it difficult for both predators and prey to detect. The shape of the body, often with a high arch of the back, adds to the chameleon's camouflage, giving it a leaflike appearance. The cryptic effect is enhanced in some species such as the tiny stump-tailed chameleons of Madagascar, which move back and forth like leaves in the breeze when disturbed. The most famous of the chameleon's defenses, however, is its ability to change color. By contracting and expanding pigment-containing cells in the skin, most chameleons can alter their color and pattern, sometimes dramatically. In addition to background matching, color change is also used in social displays.

Despite their slow speed, chameleons are adept at feeding on a variety of prey, from small insects and spiders to birds and small mammals, at least in the larger species. The turretlike eyes of the chameleon can be moved independently of one another and are used to scan the foliage for potential prey. Visual information from both eyes is integrated to give the chameleon binocular vision and the ability to judge distance with great precision.

The chameleon uses its excellent depth perception to assist in acquiring prey with its amazing tongue. The chameleon tongue may be as long or longer than the body of the animal itself. At rest, the muscles of the tongue are coiled round the tongue skeleton. When the tongue is protruded, the muscles contract and the tongue is propelled forward like a wet bar of soap being squeezed. The fleshy tongue tip is covered with sticky mucus. When it contacts a prey item, muscles in the tongue tip help to create a suction cup which, with the mucus, ensures a firm grasp on the prey. The lightning projection speed of the tongue more than makes up for the chameleon's slow body movements.

Several chameleon groups have adopted a terrestrial lifestyle but have retained most of the features of their arboreal relatives. The stump-tailed chameleons of Madagascar are mostly ground-dwelling, although some occur on mossy trees or on the low branches of bushes. These animals have short, thick tails, and are among the smallest of all lizards. Some species lay eggs as small as 2.5 by 1.5mm (0.1 × 0.06in), and reach body lengths of only 18mm (0.7in) as adults. Another terrestrial chameleon is the Namaqua chameleon of southern Africa. This much larger species walks slowly on its zygodactylous feet across the stones and sand of the desert, feeding on insects, scorpions, spiders, and even mice (see Natural Desert Dwellers).

Most chameleons are sexually dimorphic. Males often have horns (as in Jackson's and Johnston's chameleons), nasal projections (as in Meller's chameleon), elaborate neck flaps, or enlarged, bony casques on the head (as in the Veiled chameleon). These ornaments help females recognize potential mates and are also used in combat between males, which are territorial. In species lacking these structures, color may be more important as a social signal. Most species of the dwarf chameleons of southern Africa are rather similar in appearance. When courting a female or facing a rival, however, males take on a unique color pattern, often consisting of blues, oranges, or pinks, contrasting with the green color of the animal at rest.

Chameleons are normally solitary. Mating takes place several times a year in many species. Courtship often involves ritualistic behavior by the male, who in some species, such as the Common chameleon, bites the female while copulating. Most chameleons lay eggs in clutches of 4–40, but the very large Meller's chameleon lays up to 70 eggs at a time. Live birth has evolved in several groups of chameleons, including all of the dwarf chameleons. Most live-bearing chameleons live in temperate climates or in alpine or subalpine habitats, and retention of the young is regarded as an adaptation for these cold conditions.

Because most chameleons are dependent on trees, the destruction of forests is a major threat to them. In Madagascar the widespread deforestation of many areas has endangered many chameleon species. The Common chameleon in Europe is threatened by habitat loss, over-collection, and road traffic. All chameleons are protected under international law.

Geckos
FAMILY GEKKONIDAE

Geckos are mostly relatively small, insectivorous, and night-active, and many species are noteworthy for their ability to vocalize and to climb. They are represented by 964 species distributed almost worldwide from southern South America to southern Siberia, although they are most numerous and diverse in the tropics. They are especially widespread on oceanic islands, which they have colonized very successfully. Although many geckos are arboreal, they are also numerous in arid areas where rock-living, terrestrial, and burrowing species predominate.

The most striking feature of many geckos is their feet. Almost every one of the 76 genera has a unique design of the toes that is related to its particular mode of locomotion. In some terrestrial forms the toes are narrow and unmodified underneath. In the burrowing Web-footed gecko of the Namib Desert, the toes are connected by skin supported by small bones to form scoops for excavating sand. The Barking gecko of southern Africa is one of several geckos to use toe fringes to assist in digging and walking on sand. In climbing geckos, the toes are expanded into pads. The undersides of these pads are divided into broad, overlapping scales called scansors. Each scansor in turn bears tens, or even hundreds of thousands, of tiny, hairlike projections (setae), each about 10–150 micrometers long. In the Tokay gecko, each seta may have more than 100 branches, each ending in a flat, spatulalike tip about 0.2 micrometers across. These tips interact with the substrate on which the animal climbs, forming weak, temporary molecular bonds. Although individually

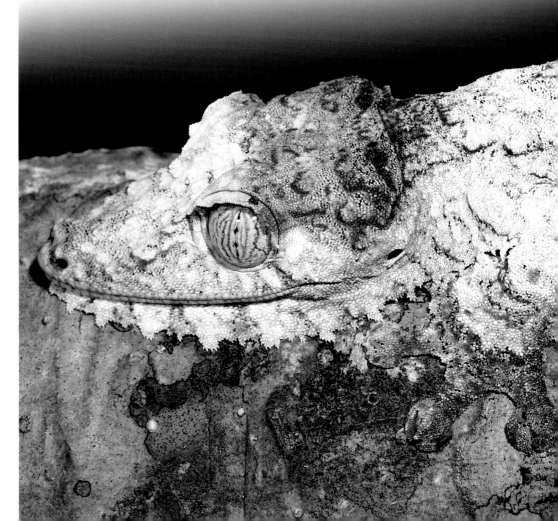

⊙ Below In addition to its highly cryptic coloration camouflaging it from predators, the Common leaf-tailed gecko (Uroplatus fimbriatus) is equipped with a fringe of skin along its head and body that helps it blend into tree trunks.

⊙ Above The diet of geckos is mostly made up of insects, spiders, and small invertebrates. Some, however, supplement this with pollen and nectar; here, a Duvaucel's gecko (Hoplodactylus duvaucelii), a diplo-dactylid from New Zealand, feeds on flax nectar.

minute, these forces, summed across all of the toes of the gecko, provide a tremendous adhesive ability – sufficient to allow the lizard to climb even on glass.

In some geckos, the tail tips may also bear setae and be used as a "fifth limb" in climbing. The tails of other geckos may be broad and flat, as in the leaf-tailed geckos of Madagascar (genus *Uroplatus*). All true geckos have fracture planes in the tails, which often break in encounters with predators or rivals.

Geckos have depressed, soft-skinned bodies and large heads. They often have loose flaps of skin on their flanks, which in the parachute geck-os (genus *Ptychozoon*) have been modified into "wings" for gliding. Gecko eyes are covered by a transparent covering (the spectacle), and eyelids are lacking; when dust or debris adheres to the eye, geckos use their mobile tongues to wipe it away. In nocturnal species, the eyes are usually large. Geckos can regulate the amount of light entering the eye by changing the shape of the

pupil from a series of pinholes or a narrow vertical slit in sunlight to a nearly round opening in total darkness. Nocturnal geckos are usually rather drab in color. In contrast, diurnal geckos such as the Day geckos of Madagascar may be bright green with blue, red, or yellow markings; in these species the eyes are smaller and the pupil may be round, even in direct light.

Geckos locate food through a combination of visual and chemical cues. The vast majority feed on insects, arachnids, and other small inverte-brates. The large Tokay gecko, however, is capable of eating other lizards, small snakes, birds, and mammals as well. Day geckos supplement insect prey with fruit and pollen or nectar from flowers.

Many geckos have a well-developed larynx and vocal cords and can produce a diversity of modu-lated chirps, clicks, growls, and barks. Barking geckos even form large choruses, with each male declaring its territory and attempting to attract mates. Visual signals are also used, especially in

diurnal species, and chemical communication may also play some role in courtship.

Most geckos lay two eggs with hard, calcareous shells, but some smaller species produce only one egg per clutch. Eggs may be laid in shallow pits, under bark, or on plant or rock surfaces. Some species, such as the Fan-footed gecko, lay eggs communally, and many individuals may place their clutches in a few favorable laying sites. Many of the gecko species that have successfully colo-nized small, distant oceanic islands are all-female forms that reproduce clonally.

Some geckos, especially island species, are threatened by habitat loss. At least one giant day gecko, *Phelsuma edwardnewtoni*, has become extinct on Round Island in the Indian Ocean.

Southwest Pacific Geckos

FAMILY DIPLODACTYLIDAE
Southwest Pacific geckos are a group of more than 140 species that occur only in Australia, New

Left *Representative species of five lizard fami-lies:* **1** *Western collared lizard (Crotaphytus collaris); Crotophytidae.* **2** *Green anole (Anolis carolinensis); Polychrotidae.* **3** *Texas horned lizard (Phrynosoma cornutum); Iguanidae.* **4** *Fringe-toed lizard (Uma notata); Phrynosomatidae.* **5** *Arabian toad-headed lizard (Phrynocephalus arabicus); Agamidae.* **6** *Spiny-tailed lizard (Uromastyx acanthinurus); Agamidae.* **7** *Common chameleon (Chamaeleo chamaeleon); Chamaeleonidae.*

Zealand, and New Caledonia. They resemble true geckos but have less complex climbing structures, and they lay leathery-shelled eggs – or, in the case of the New Zealand green geckos, give live birth to twins. Among the members of this family are the knob-tailed geckos of Australia, which have very large heads and short tails. Some species have the tails so reduced that no fracture planes are present, and the animals cannot shed the tail as a predator-escape mechanism.

The Spiny-tailed gecko is one of several species that can eject a viscous fluid from its tail. This defensive substance is thought to have evolved as a deterrent against small predators, such as lycosid spiders, by sticking to and temporarily entrapping them.

Some Southwest Pacific geckos are quite large, and many of the largest species can (and do) feed on vertebrate prey. Both New Zealand and New Caledonian species eat fruits and flower parts, and may be important pollinators and seed dispersers.

The largest living species is the New Caledonian giant gecko, which reaches more than 25cm (10in) in length and weighs 300g (10.5oz). A much larger New Zealand species, Delcourt's giant gecko (*Hoplodactylus delcourti*; 36cm/14in), now appears to be extinct.

Eyelid Geckos
FAMILY EUBLEPHARIDAE
Eyelid geckos are the smallest of the gecko groups, with only 23 species scattered across parts of North and Central America, Africa, and both tem-perate and tropical Asia. Like the Southwest Pacif-ic geckos, they lay two leathery-shelled eggs. They differ from all other geckos, however, in having movable eyelids rather than transparent specta-cles. They also always lack the climbing pads seen in other groups. Most species are terrestrial, but the Cat gecko of Southeast Asia is arboreal and has a prehensile tail. The Leopard gecko, which is typical of the eyelid geckos in appearance and

is one of the most popular of all lizard pets, has temperature-dependent sex determination. Some Southeast Asian eyelid geckos (genus *Goniuro-saurus*) are exploited for medicinal uses in China and Vietnam.

Flap-footed Lizards
FAMILY PYGOPODIDAE
Although they are strikingly different in appear-ance, flap-footed lizards, which are restricted to Australia and New Guinea, are closely related to the Southwest Pacific geckos. All species have elongate bodies with very long tails, no front limbs, and much-reduced hind limbs. Like most geckos, they have a transparent spectacle over the eye, paired egg teeth, are chiefly nocturnal, and are capable of making vocalizations. Unlike most geckos, which typically have granular or tubercu-late scales, flap-footed lizards have overlapping scales, more similar to those of snakes or skinks.

Flap-footed lizards may be chiefly burrowing or

surface-active. Delmas "swim" through spinifex grass, supporting and propelling themselves by pressing bends in their body against the stiff grass blades, often well above the ground surface. Many flap-footed lizards use their long tails to "jump" as part of an escape behavior.

Most flap-footed lizards are strictly insectivorous, but the Sharp-snouted snake lizards preferentially feed on other lizards, including geckos and skinks. They have hinged teeth that fold backward, helping to force hard-bodied prey back to the throat, where they are swallowed whole.

Social interactions in flap-footed lizards are poorly known, but males in particular are known to use their hindlimb flaps in aspects of mating and threat display. Flap-footed lizards lay two (or, rarely, one) leathery-shelled eggs. It has been suggested that some species with black heads, such as the Western scalyfoot, may be mimics of certain dangerously venomous Australian elapid snakes.

Night Lizards
FAMILY XANTUSIIDAE

Night lizards are a small group of terrestrial or saxicolous lizards with a discontinuous distribution in the southwestern United States, Mexico and Central America, and eastern Cuba. They are characterized by large head shields, granular back scales, and large, rectangular belly plates. A transparent spectacle covers the eye, which has a vertical pupil. Although called night lizards, some species are mostly dawn- and dusk-active (crepuscular), and others are diurnal.

Night lizards are secretive, hiding by day in caves or rock crevices, or else under stones or vegetation. Some species are habitat specialists, occupying only limestone or granite. Such rock-living specialists typically have flattened bodies and moderately long limbs for climbing. The Cuban

night lizard has very small limbs and uses snake-like undulations to move. The Desert night lizard is a small species most often found beneath fallen Joshua trees or under other plant debris in the American southwest. It feeds on ants and other insects, and may be active by day or night in its hiding places, depending on the temperature. The much larger Island night lizard occurs only on the Channel Islands off the coast of California. Its diet includes insects as well as seeds and flowers. Smith's tropical night lizard (*Lepidophyma smithii*) feeds largely on figs.

Night lizards all give live birth, with maternal nutrition being provided in those species that have been investigated. Following about 90 days' gestation, Desert night lizards give birth to a single litter of one to three offspring, although in years when environmental conditions are unfavorable they may not reproduce at all. The Island night lizard produces two to nine young. The tropical Middle American night lizard (*Lepidophyma flavimaculatum*) has some all-female populations that reproduce parthenogenetically. Although night lizards have a low reproductive rate, their populations often reach high densities; in preferred habitats there can be more than 3,200 Island night lizards per hectare (1,300 per acre).

Blind Lizards
FAMILY DIBAMIDAE

Blind lizards are among the most highly specialized for a subterranean existence. Fourteen species occur in tropical Asia, from southern China to New Guinea, and one additional species is restricted to eastern Mexico. Blind lizards have elongate, snakelike bodies with no front limbs and small, paddle-shaped hind limbs that are only present in males. The small eyes are covered by skin, and there are no external ear openings.

⭕ **Below** *Burton's snake lizard* (Lialis burtonis) *of Australia and New Guinea has the characteristic limbless, serpentine body shape of the pygopodids. Growing up to 60cm/24in in total length, it preys on small geckos, skinks, and snakes.*

Blind lizards have been found inside rotten logs and under stones as well as in the soil itself, and in addition have occasionally been located above the ground, under the bark of trees and in birds' nests. In Asia they are found in moist tropical forests, whereas in Mexico they occupy drier pine–oak upland habitats. All blind lizards are insectivorous, and in those species that have been studied a single egg with a hard, calcareous shell is laid. Males use their small hind limbs to grasp females during copulation.

Wall and Sand Lizards
FAMILY LACERTIDAE

Wall and sand lizards, sometimes called true lizards, are small to moderately sized. Diurnal, they are among the most active and obvious reptiles in the habitats where they occur. They are found throughout much of Europe, Asia, and Africa from tropical forests to above the arctic circle, but are most diverse in the Mediterranean Basin and in the arid and semiarid regions of north and southern Africa and the Near and Middle East. Most species are terrestrial or saxicolous, but there are specialized grass-swimmers, burrowers, and arboreal species.

Typical wall and sand lizards have somewhat depressed bodies with long limbs, long toes, and often very long tails. They are covered above by small, granular scales, and below by wide, rectangular plates. The head shields are large and conspicuous.

Most wall and sand lizards rely on speed to escape predators, running from cover object to cover object. The Bushveld lizard of southern Africa (*Heliobolus lugubris*), however, uses mimicry as a defense. The juveniles of this species are black with white markings. When threatened, they arch their backs and scuttle along the ground, resembling in appearance and motion the Oogpister beetle, a large local insect that can squirt an acidic fluid as a defense. The Gliding lizard of Africa flattens its body and uses its slightly expanded tail to parachute between perches. Flattened bodies are also seen in some of the many species that occupy rock crevices.

Sand lizards are common in most deserts of Africa and Asia. Many such species have toe fringes to help them run on loose sand, but several species are true sand-dune specialists. The Shovel-snouted lizard of the Namib Desert has long, fringed toes and a countersunk lower jaw. It forages on the slipfaces of dunes, and dives into the sand to avoid both predators and surface temperatures in excess of 44°C (111°F). When active on the dune face, this lizard often raises one front leg and the opposite hind leg to remove them from the hot sand surface and allow cooling breezes to blow between its body and the substrate.

The grass lizards of eastern Asia are among the most divergent of the true lizards. These animals have very lithe, elongate bodies, with tails up to

fewer eggs, although a few larger species, such as the large Green lizard of Europe, have clutches of up to 20 or more. Live birth occurs in the Viviparous lizard, the northernmost lizard in the world. As its name implies, this species gives birth to litters of from four to eleven young. Some egg-laying populations formerly regarded as belonging to this species have recently been recognized as a separate species.

Some species, such as the Wall lizard, coexist well with humans. These lizards are common on the retaining walls in hillside vineyards in Europe, and other species frequent other manmade structures. Among the most endangered members of the family are the giant Canary Island lizards (genus *Gallotia*). One species was thought to be extinct for more than half a century before a small population was discovered on the island of El Hierro in 1975. An even rarer species has recently been discovered on the neighboring island of La Gomera. The Gomeran giant lizard, *Gallotia gomerana*, was previously known only from subfossil remains. The rediscovered living population is known only from six individuals from an area less than 1ha (2.5 acres) in size, making this perhaps the rarest lizard in the world.

four or more times their head-and-body length. In contrast to the browns and reds of most members of the family, they are green, to match the grasses and bushes in which they live. The gliding lizards and a few other tropical African groups are among the only truly arboreal members of the family. Virtually all sand and wall lizards are insectivores that forage widely, searching actively for arthropod prey, but larger species will often eat both vertebrate prey and plant material.

Many species of true lizards, especially those studied in Europe, have complex social systems. Males usually have larger heads than the females, and they may develop bright colors during the mating season. Males often perform species-specific courtship rituals to win mates. Male combat is common, but is usually preceded by stereotypic threat displays in which the head is tilted downward, the throat inflated, and the body compressed. The flattened body is then turned broadside toward the opponent, maximizing the apparent size of the lizard and exposing the bright markings. Like many lizards, male sand and wall lizards typically possess femoral glands on the underside of the legs. These are used to mark territories and to communicate chemically with potential mates and rivals.Some species also use vocal communication, producing simple sounds that may help to maintain territories and dominance rankings. Although the great majority of true lizards reproduce sexually, the first unisexual lizards to be identified by herpetologists belonged to this family, and several of these all-female species occur in the Caucasus Mountains of Eurasia.

Most wall and sand lizards lay clutches of 10 or

🔵 *Above* Mating Sand lizards (Lacerta agilis). The female mates with several males, but lays just one clutch of 4–15 eggs per season. Note the sexual dichromatism of this species: the male's green coloration is at its most vivid during the breeding season.

🔵 *Below* The widespread Viviparous lizard (Zootoca vivipara) retains its incubating eggs within the oviduct and gives birth to live young. By protecting its developing embryos in this way, it has been able to colonize much colder habitats than oviparous species.

Whiptails and Racerunners
FAMILY TEIIDAE

Whiptails and racerunners are fast-running lizards that occupy most of the tropical and some temperate habitats of the Americas. They range from the deserts of the American southwest to the tropical forests of Amazonia. Most species are terrestrial, but some are semiaquatic. In general appearance they strongly resemble the wall and sand lizards of the Old World.

Whiptail lizards are most numerous in arid and semiarid habitats of North America. They include some of the fastest known lizards. Several species have been recorded to run at speeds in excess of 25 km/h (15mph). They also have great endurance, which is needed for their wide-ranging foraging. Whiptails have some of the highest preferred body temperatures of any lizards; some species are not active unless their temperatures exceed 36°C (97°F). The activity of whiptails is thus limited by environmental factors; many northern species are surface-active for only a few hours a day, and may spend more than six months of the year inactive in retreats. Such activity patterns are sufficient for whiptails to obtain enough food, but decrease their exposure time and risk of predation.

Smaller whiptails and racerunners eat mostly insects, but larger species are mainly carnivores. The Chilean false tegu (*Callopistes maculatus*) is a specialist lizard-feeder, while the large tegus eat eggs, mammals, birds, and reptiles in addition to all types of invertebrate prey. Tegus also include plant matter in their diet, and even a few of the smaller racerunners and junglerunners are herbivorous. The Dragon lizard and Caiman lizard of South America are both very large, semiaquatic forest-dwellers that have large, keeled scales on their backs. The Dragon lizard is the most specialized feeder in the family, with large, molarlike teeth for crushing the snails that make up the bulk of its diet.

Many jungle runners are brightly colored, and in some the males take on seasonal breeding colors. In comparison with other lizards, however, members of this family show little sexual dimorphism, and territoriality is absent or weakly developed in a majority of species. All whiptails and their relatives lay eggs. Most whiptails lay only one to seven eggs per clutch, but some larger species, such as the tegus, may produce more than 30 eggs at a time. Tegus may lay their eggs in soil or rip open termite mounds and deposit their eggs inside. The termites respond to the damage to their nest by repairing the opening, sealing the tegu eggs inside. Thus protected from predators and kept at a warm temperature by the termite mound, the eggs develop in 90–120 days, when the hatchlings dig themselves free of the nest. Many of the whiptails are all-female species. These forms originated through the hybridization of two different parental species and reproduce parthenogenetically (see Unisexuality: The Redundant Male?).

Tegus, which occupy many South American habitats, are heavily exploited by humans. More than 1 million (over 3 million in some years) are killed annually for their skins, which are used to make leather goods, especially cowboy boots. The largest members of the family, they also provide meat and medicinal products for local people throughout their range. Although tegus remain common in many areas, regional governments have now taken steps to control the trade at more sustainable levels. Two species of racerunners from the West Indies, the Guadeloupe giant junglerunner (*Ameiva cineracea*) and the Martinique giant junglerunner (*A. major*), appear to have become extinct in historical times.

Microteiid Lizards
FAMILY GYMNOPHTHALMIDAE

Microteiid lizards are close relatives of the whiptails and racerunners, but nearly all species are quite small. Instead of being conspicuous components of the fauna, most species are secretive. They vary greatly in scalation, but the scales of the back are often large and strongly differentiated from those of the flanks. Microteiids are essentially tropical, extending from southern Central America to central Argentina, with some species occurring on islands of the Caribbean, but they also occur at high elevations in the Andes, where temperatures become quite cool. Many species are

Above *Microteiid lizards are terrestrial, foraging among leaf litter in their Central and South American rainforest habitat. This Bromeliad lizard* (Anadia ocellata) *is consuming a woodlouse.*

active in leaf litter or under logs, and some are burrowers or arboreal, or even swamp-dwelling. Typical microteiids are dull-colored and have short limbs. One species of reduced-limbed microteiid is the only lizard to lack hind limbs entirely, but to retain the forelimbs.

All microteiids appear to be generalist insectivores. They are active foragers, moving frequently in search of prey. A few species have been reported to be active at night, but those species that are well studied are day-active, although they often prefer complete or partial shade to direct sun.

Most microteiids escape threats by running rapidly for cover in the base of plants or in the leaf litter. Those species inhabiting streamside forest, such as *Arthrosaura reticulata*, may dive into the water, remaining for some time at the bottom until the danger has passed.

All microteiids are oviparous, and the few forms for which information is available produce clutches of two leathery-shelled eggs. At least some species, however, can produce multiple clutches, often in rapid succession, for extended periods of the year. At least two groups are represented by all-female, asexually-reproducing forms.

Girdled Lizards

FAMILY CORDYLIDAE

Girdled lizards are the only lizard family restricted in their distribution to mainland Africa. They are most diverse in rocky habitats of South Africa, but extend northward to Ethiopia. All girdle-tailed lizards are day-active and chiefly insectivorous. Their name derives from the whorls of spiny scales that encircle their tails and sometimes their bodies also. Both the back and belly of many of these lizards are covered by regular, rectangular scales; those of the back contain osteoderms.

Most of the girdle-tailed lizards are rock-living and have bodies that are somewhat to very depressed, to allow them to enter narrow crevices. Flat lizards are extreme in this regard. They are less heavily armored than other girdled lizards, and their long legs help them to run quickly across the boulder faces where they live.

The Armadillo girdled lizard is covered with

◐ **Left** *The Dwarf plated lizard (Cordylosaurus sub-tessellatus) is a rock-dwelling species from the arid regions of Namibia in Southwestern Africa. Its most striking feature is its electric-blue tail, which it rapidly autotomizes if caught by a predator.*

◑ **Below** *To defend itself against attack, the Armadillo girdled lizard (Cordylus cataphractus) bites and clings onto its tail, thereby rolling itself into a loop and presenting its armored scales to the assailant. In such a posture, it is almost impossible to swallow.*

very large plates and lives in family groups in rock crevices, emerging for short distances to capture insect prey. Although they cannot run quickly, these lizards have a very effective defense: if intercepted outside of their retreats, they curl into a ball and bite their own tails, presenting a formidable, spiny ball to any predator, while protecting their more vulnerable belly.

The sungazer is the largest girdled lizard (20.5cm/8in) and is not a rock-dweller. It lives in colonies in upland areas of eastern South Africa and excavates burrows about 40cm (16in) deep and 1.8m (6ft) long. Its name comes from its habit of basking in the sun at the entrance to its burrow opening. The most distinctive girdled lizards are the sweepslangs, grassland specialists with tiny reduced limbs and very long tails, up to four times their body length.

Most girdled lizards are drab in color, but both the flat and crag lizards, which exhibit significant sexual dimorphism, are exceptions. In flat lizards, the much larger males may be bright red, orange, yellow, green, or blue, or some combination of these colors, whereas the females are usually black with a series of pale stripes on the back. Flat lizards have complex social behavior and form dense colonies, with males defending territories during the breeding season. Male Common girdled lizards (*Cordylus cordylus*) are also aggressive. Although many animals may share a single rocky retreat, a distinct dominance order is maintained by threats and fights.

Most girdled lizards give birth to litters of between one and four quite large offspring. The size of the litter is often limited by the flattened shape of the body, and the largest litters (up to 12) occur in the elongate sweepslangs. Flat lizards are the only oviparous girdled lizards; they lay clutches of two very elongate eggs.

Most girdled lizards are protected under international law. Because many have very restricted distributions, they are vulnerable to local habitat destruction. The sungazer is especially imperiled by the agricultural conversion of its grassland habitats in South Africa.

Plated Lizards

FAMILY GERRHOSAURIDAE

Plated lizards, which are closely allied to girdled lizards, inhabit both sub-Saharan Africa and Madagascar. Like the girdled lizards they are day-active and share the presence of osteoderms in the skin. They usually lack the spiny scales of the girdled lizards, however, and possess a prominent fold at the edge of the belly. Most have cylindrical bodies with small but well-formed legs, but limb reduction has occurred in the seps. These are small, elongate, grassland species, some of which exhibit complete loss of the forelimbs. Most plated lizards are relatively large, but the Dwarf plated lizard of Namibia reaches only about 50mm (2in). It has a bright blue tail that is used to distract

predators from attacking the more vulnerable head and body.

The African species include both rock-dwellers and savanna species that occupy burrows, including those abandoned by mammals. The Giant Malagasy plated lizard (*Zonosaurus maximus*) is semiaquatic, and some other Madagascan forms are dry-forest dwellers.

The Namib plated lizard is highly specialized for life in the dunes. It has toe fringes for movement on loose sand and a countersunk jaw that excludes sand from the mouth. It uses its whole body to dive deeply into the sand to avoid predators and escape high midday temperatures. Small colonial groups often spread out on a dune face to bask and forage. They are very wary, and will dig down out of sight when any potential threat appears. Namib plated lizards are opportunistic feeders, but the dunes where they live provide little predictable animal prey. They feed mostly on beetles and seeds and other plant material that blows in from the better-vegetated inland. Most other plated lizards are chiefly insectivorous, but many larger species eat small vertebrates and plant material as well.

Most plated lizards are solitary, and they may become aggressive around other members of their species. There can be changes in color pattern during maturation, especially among Madagascan species, and adult males are often more brightly colored than the females. All plated lizards are oviparous, and females typically produce small clutches of 2–12 eggs.

Eastwood's seps (*Tetradactylus eastwoodae*) from South Africa is almost certainly extinct. It has not been seen since it was described in 1913, and its habitat has since been converted to exotic pine plantations.

Skinks
FAMILY SCINCIDAE

The 1,400 skink species occupy almost all types of habitats from southern Canada and northeastern Asia to southern New Zealand, and they have also successfully colonized many of the tropical islands of the world. Most surface-active skinks are diurnal, but species that live in burrows or leaf litter may be active at night or at different times throughout the day. Most skinks are covered by smooth or weakly keeled overlapping scales and have roughly cylindrical bodies with short limbs. The limbs have been reduced or lost at least 25 times in the history of the group. Reduced-limbed skinks usually spend much of their time below ground or under cover objects, but some are mostly surface-active.

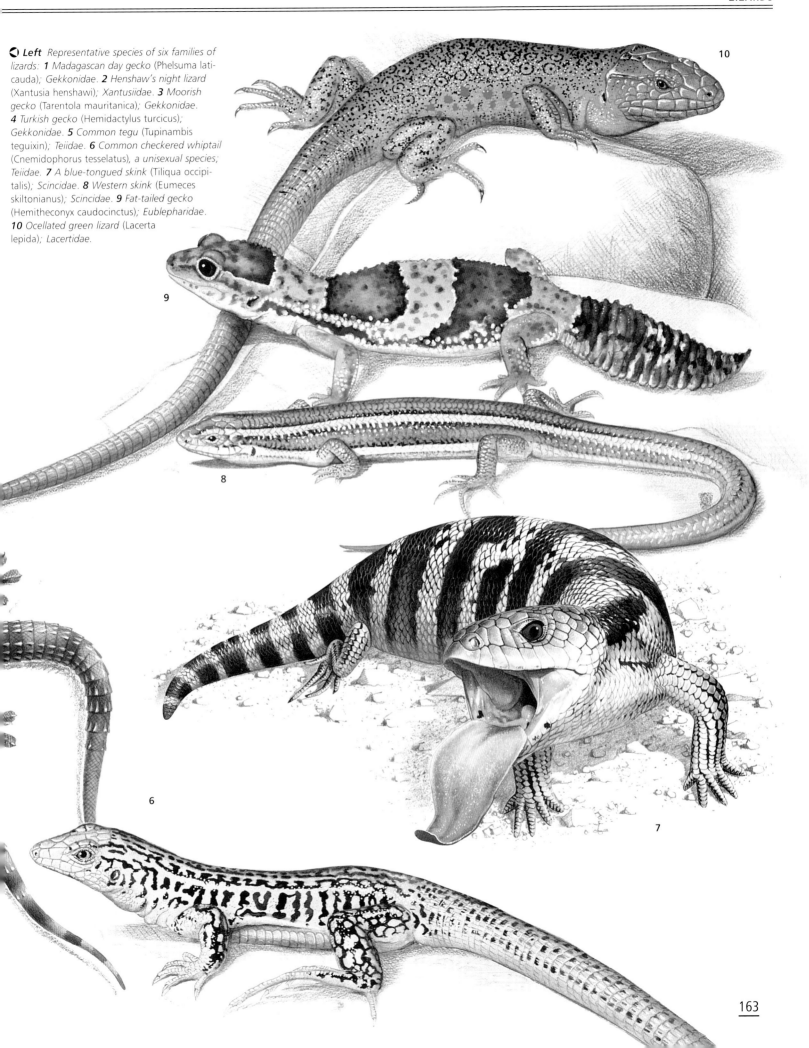

◑ **Left** *Representative species of six families of lizards:* **1** *Madagascan day gecko (Phelsuma lati-cauda); Gekkonidae.* **2** *Henshaw's night lizard (Xantusia henshawi); Xantusiidae.* **3** *Moorish gecko (Tarentola mauritanica); Gekkonidae.* **4** *Turkish gecko (Hemidactylus turcicus); Gekkonidae.* **5** *Common tegu (Tupinambis teguixin); Teiidae.* **6** *Common checkered whiptail (Cnemidophorus tesselatus), a unisexual species; Teiidae.* **7** *A blue-tongued skink (Tiliqua occipitalis); Scincidae.* **8** *Western skink (Eumeces skiltonianus); Scincidae.* **9** *Fat-tailed gecko (Hemitheconyx caudocinctus); Eublepharidae.* **10** *Ocellated green lizard (Lacerta lepida); Lacertidae.*

The most extreme cases, such as the blind burrowing skinks, have no external traces of limbs, and both the eyes and the ears may be covered over by scales. These species spend their entire lives underground or beneath leaf litter or rocks, where they feed on termites and other small, patchily distributed insects. The sandfish represents a different solution to burrowing; this species from North Africa and Arabia retains its limbs and uses its fringed toes and wedge-shaped snout to move through the desert sand.

The majority of skinks are active foragers that seek out prey. The diet of skinks consists mostly of arthropods, and small skinks are almost exclusively insectivorous, although somewhat larger skinks occasionally supplement this fare with plant material. Those that eat fruit may serve as important seed dispersers, especially on islands. The very large Australian blue-tongued skinks and their relatives are omnivorous, eating fruits, flowers, and other plant parts in addition to snails, birds' eggs, arthropods, and the occasional small vertebrates. The most extreme herbivory is seen in the prehensile-tailed Solomon Islands skink. This large, arboreal skink has a diet composed almost entirely of plant matter. After processing its food, the lizard will often ingest its own feces, in order to recycle the bacterial gut symbionts that the lizard needs to break down the cell walls of the plant tissue and to provide an opportunity for further digestion.

The Solomon Islands skink is unusual in other ways too. It is nocturnal and, after a gestation period of about 39 weeks, gives birth to a single offspring that is nearly one-third the size of the mother. Another strange arboreal skink is the Green-blooded skink of New Guinea. Its green blood pigment, biliverdin, is related to bile and causes the scales, tongue, mouth lining, muscles, bones, and even eggs of the species to be green. This skink has toe pads similar to those of geckos and anoles, but somewhat less complex and less effective.

Skinks rely mostly on a combination of chemical and visual cues to communicate. Bright colors are rare, but occur on the heads of males of some terrestrial species such as Five-lined skinks and Great Plains skinks during the breeding season. Most skinks are not territorial, but some will defend burrows or basking sites. During the breeding season, males become aggressive, and head bobbing, lateral body displays, and actual combat are common. Males often lick females prior to mating and typically hold the female in a mating grip, biting her on the neck, limbs, or forebody before and during copulation.

The shingleback or Sleepy lizard (*Tiliqua rugosa*), which can live for 20 years or more, has a complex social structure. Males and females form pair bonds that may last at least 14 years. The pair reunite to mate each spring but spend winters apart. Infidelities occasionally occur,

but most bonds last from year to year.

Most skinks lay eggs, but live birth has also evolved independently many times. Some species are known to have some egg-laying populations and others that retain the eggs internally for the duration of development but have live birth. The embryos of these skinks are nourished by large yolk deposits. Several groups, however, have evolved a form of viviparity that is much more like that seen in mammals. In the Brazilian skink and a few other species each egg is minute and contains virtually no yolk. As the embryo develops, it obtains its nutrition directly from the mother via a placenta. Gestation in the Brazilian skink lasts almost a year (as compared to 9–39 weeks for species that rely on yolk for nutrition). Egg-laying skinks can produce clutches ranging from 1 to more than 30 eggs, about the same range of variation as seen in live-bearing forms. Most lay their eggs in protected crevices or cavities with high humidity and then abandon them, but many skinks of the genus *Eumeces* brood their eggs. This behavior may take the form of turning, licking, moving, and guarding them, as well as sometimes actively fighting egg predators.

Several insular, ground-dwelling skinks appear to have become extinct in recent times. One of the largest skinks to have lived was the Cape Verde giant skink, *Macroscincus coctei*, which reached a length of 32cm (12.5in). This species was relatively common in zoos in the 19th century, but has not been seen in the wild for almost 100 years and is presumed to be extinct. Several Indian Ocean island skinks appear to have been wiped out shortly after European ships arrived, bringing rats, cats, and other mammalian predators.

Anguids
FAMILY ANGUIDAE

Anguids occur throughout most of the tropical and some temperate parts of the northern hemisphere; in the New World they extend south to central Argentina. Like plated lizards, they have relatively uniform, usually rectangular scales with osteoderms embedded inside. Many species also have a fold along the margin of the flank and belly. Anguids occupy a diversity of habitats from dry deserts through cool pine and oak forests to tropical cloud forests. Some are secretive and either nocturnal – for example, the galliwasps – or, if active by day, then only in deep shade or under cover objects. Others are day-active and relatively sun-loving.

The alligator lizards of North and Central America are fully limbed. One group, the arboreal alligator lizards, are denizens of the tropical forest, often living high in the canopy among the epiphytic plants that crowd the tree branches. One species has been found as much as 40m (130ft) above the ground in bromeliads. These lizards possess prehensile tails and are often brightly colored. The related Southern alligator lizard of the western United States, a mostly terrestrial diurnal

lizard, has also been observed to use its tail as a grasping organ to help it reach birds' nests to feed on eggs.

Many anguids have reduced limbs and some, such as the slowworm and California legless lizard, are completely limbless, but all such species retain moveable eyelids, and all except the California legless lizard have external ear openings. The glass lizards of North America are so called because of the ease with which their long tails break. The Latin name of the slowworm – *Anguis fragilis* – similarly refers to the fragile nature of the tail in comparison to those of snakes, with which they are sometimes confused. Tail autotomy is a common defense mechanism employed by most anguids, but larger species, especially alligator lizards, may gape and lunge if disturbed and can inflict painful bites.

Most reduced-limbed anguids are surface-active and diurnal for much of the year, but the California legless lizards are true burrowers found in sand or loose soil. They locate prey on the surface by chemical and vibrational cues and emerge to grasp insects or spiders before returning underground. In addition to a wide range of insects, anguids are known to eat lizards, small mammals, eggs and nestling birds, tadpoles, earthworms, spiders, scorpions, and sow bugs. Slowworms and their much larger relatives, the scheltopusiks, are particularly effective predators of snails and slugs.

A gradation from robust-limbed to virtually limbless forms occurs among the tropical American galliwasps, many of which strongly resemble certain skinks. One species of Brazilian galliwasp has brightly-banded juveniles but rather drab adults. The similarity in size and color of the young to a toxic millipede has led to the suggestion that the galliwasp is a mimic that gains protection through its resemblance to its noxious arthropod model. On tiny, rocky Malpelo Island off the Pacific coast of Colombia, Malpelo galliwasps (*Diploglossus millipunctatus*) overcome a lack of arthropod prey by subsisting on seafood. The galliwasps mob seabirds returning to their nests with food for their young, and survive largely on regurgitated fish. One of the largest members of this group, the Jamaican giant galliwasp (30.5cm/12in) appears to be extinct. The limited information about this species suggests that it lived in swampy areas, where it fed on both fish and fruit.

Anguids use chemical cues to locate both prey and mates. Although they do not maintain territories as do many other lizards, males regularly engage in combat and bite one another. They also typically bite the head of the female during copulation. Although they are mostly solitary, anguids

◐ **Above** *The tree-dwelling Solomon Islands skink* (Corucia zebrata) *is large and stoutly built, but is equipped with a strong prehensile tail that it uses to support itself as it forages for fruit and leaves.*

◑ **Left** *With its wedge-shaped nose and streamlined body, the* Sandfish (Scincus scincus) *is extremely adept at moving through its desert environment, literally "swimming" through loose dune sand.*

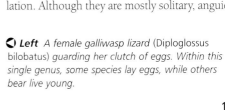

◑ **Left** *A female galliwasp lizard* (Diploglossus bilobatus) *guarding her clutch of eggs. Within this single genus, some species lay eggs, while others bear live young.*

in cooler climates may share winter retreat sites; 30 slowworms were found together in one such location.

The majority of anguids lay eggs in small clutches of 2–12, but some alligator lizards have clutches of up to 40 eggs. Glass lizards frequently lay their eggs in abandoned mammal burrows, and may brood their clutches, wrapping themselves around the eggs and providing them with protection. Live birth occurs in the slowworm, which produces 4–28 (usually up to 12) offspring in late summer after 8–12 weeks' gestation. Alligator lizards and galliwasps include both egg-laying and live-bearing species; litters of up to 27 have been recorded for some of the largest live-bearing galliwasps. Many anguids reproduce only every other year, and this fact, combined with their often small clutches and litters, means that the reproductive output for many species is quite low, a handicap compensated for by a long reproductive life. Many anguids are known to live for at least 10 years, and the slowworm has been known to survive for 54 years in captivity, a record for any lizard.

Some of the arboreal alligator lizards that have very restricted ranges in the mountains of Central America are endangered due to deforestation. It is likely that some species have been wiped out before even becoming known to science. The fearsome appearance of some of the alligator lizards has given rise to a widespread popular misconception that they are venomous; in fact, they are completely harmless.

Legless Lizards
FAMILY ANNIELLIDAE
These two species of small legless lizards (genus *Anniella*) are sometimes included in the family Anguidae. They live in sand or sandy soil in California and Baja California and come to the surface at night to feed on small invertebrates, especially around the bases of bushes. These legless lizards move through the soil by

"sand-swimming" and leave characteristic tracks. Both species give birth to one or two live young.

Xenosaurs
FAMILY XENOSAURIDAE
Xenosaurs are a small group of lizards consisting of five species in tropical Middle America and one in China. The Chinese species is sometimes placed in a separate family of its own (the Shinisauridae). Xenosaurs have large, keeled tubercles on the back that contain osteoderms. The tails may also have a crest of such bony plates. The New World species are scattered in southern Mexico and Guatemala and are mostly terrestrial. They occur from dry scrub to cloud forest in volcanic or limestone mountains, but usually seek out moist microhabitats. They are secretive, and are often encountered in rock crevices, hollow logs, and tree holes, but are also found beneath stones in shallow streams or standing water. They may climb up to 2m (6.5ft) into vegetation, but are not truly arboreal. All species have somewhat depressed heads and bodies.

The Chinese xenosaur, or Chinese crocodile lizard, readily climbs vegetation, often overhanging streams. It is semiaquatic in its habits and is never found far from pools of slow-flowing water. The head and tail are both laterally compressed, and the lizard is a strong swimmer. It feeds on fish, tadpoles, frogs, crabs, and insects. In contrast, the American species

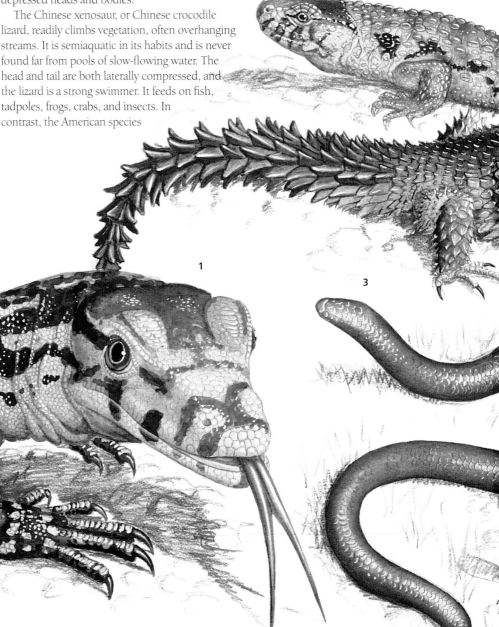

Below *Representative species of seven families of lizards:* **1** *Common Asiatic monitor (Varanus salvator); Varanidae.* **2** *Chinese xenosaur (Shinisaurus crocodilurus); Xenosauridae.* **3** *An Asian blind lizard (Dibamus novaeguineae); Dibamidae.* **4** *Southern alligator lizard (Elgaria multicarinata); Anguidae.* **5** *Sungazer (Cordylus giganteus); Cordylidae.* **6** *Bornean earless lizard (Lanthanotus borneensis); Lanthanotidae.* **7** *Gila monster (Heloderma suspectum); Helodermatidae.*

are primarily insectivorous, although they also occasionally consume millipedes and small lizards. Most xenosaurs appear to be principally diurnal, but are mostly active in shaded conditions. American xenosaurs are for the most part solitary animals, normally coming together only to mate. Adults are aggressive toward one another, and will fight and bite vigorously. When threatened by potential predators, they gape widely and will bite.

All xenosaurs are live-bearing. The New World species give birth to 1–8 young from spring to midsummer, following about 1 year of gestation. In the Chinese crocodile lizard, mating may last for up to 1.5 hours, and litters of as many as 15 offspring are born after a gestation period of 8–14 months. Breeding in xenosaurs usually takes place every other year. In at least one Mexican species,

Newmans' xenosaur (*Xenosaura newmanorum*), adult females may share retreats and provide protection to their offspring for periods of up to several months.

In Latin America, some people erroneously believe that xenosaurs are venomous. The geographically restricted Chinese species is considered endangered as a result of destruction of its forest habitat and excessive collection in the 1980s for the herpetocultural trade. Only about 2,500 animals remain in the wild, although the species is frequently kept and bred in captivity.

Beaded Lizards
FAMILY HELODERMATIDAE

The two species of beaded lizards are the world's only venomous lizards. The Gila monster lives in

arid areas of the American southwest and western Mexico, while its larger relative, the Mexican beaded lizard, extends as far south as Guatemala. Both species have cylindrical, heavy bodies with short legs, a relatively thick tail, and a blunt head. The venom glands are modified salivary glands and are located in the lower jaws. Venom is drawn into narrow grooves in the teeth by capillary action and is injected as the lizard chews on its victim. The venom acts on the nervous system and results in cardiac failure and respiratory paralysis. Beaded-lizard bites in humans are painful but rare, and are generally not fatal for healthy adults.

Beaded lizards avoid temperature extremes and are usually active at dusk or during the night. They may use rocky crevices or abandoned mammal burrows as retreats. In the colder parts of their

range Gila monsters are inactive for several months each year. Beaded lizards are chiefly terrestrial, but they are also powerful diggers and adept climbers and may be found 5–7m (16–23ft) up in trees. Both species have diverse diets including small mammals, birds, lizards, frogs, reptile and bird eggs, insects, earthworms, myriapods, and carrion. Prey are detected by chemical cues using the tongue and vomeronasal organ. The tail serves as a fat-storage organ and changes in diameter according to the nutritional state of the lizard.

Mating takes place in spring and copulation lasts for up to an hour. Females lay 3–13 elongate, leathery-shelled eggs and bury them in shallow (12cm/5in) nests dug in the ground in sunny spots. In Gila monsters, incubation lasts up to 10 months.

Bornean Earless Monitor
FAMILY LANTHANOTIDAE

The Bornean earless monitor is in a family of its own. It is most closely related to the true monitors, but resembles the beaded lizards in body form. It differs from these in having the nostrils placed high on the snout and in lacking an external ear opening, a fold of skin across the throat, and any trace of venom apparatus. It is known only from southern Sarawak in the Malaysian portion of Borneo.

Earless monitors are chiefly nocturnal, and appear to be both burrowing and semiaquatic. Their limbs are short, and their bodies are elongate and covered by enlarged, osteoderm-containing scales. Earless monitors are accomplished swimmers, but move awkwardly on land.

In captivity the earless monitor eats fish, earthworms, and both turtle and bird eggs, but their diet in the wild is unknown. It is believed that they have a low metabolic rate, and captive specimens generally appear sluggish. Earless monitors lay small clutches of 2–6 eggs.

Monitor Lizards
FAMILY VARANIDAE

Monitors, also known as goannas in Australia and leguaans in South Africa, are the largest of all living lizards, with the Komodo dragon of Indonesia reaching 3.13m (over 10ft) in total length. Extinct monitors were even bigger; the Australian *Megalania prisca*, which lived during the Pleistocene, reached total lengths of 7m (23ft). Monitors are also related to mosasaurs – fossil marine lizards that were larger still – and to snakes. The Komodo is the top predator on the small islands where it occurs. It kills and eats deer, pigs, and goats, and has been known to bring down a 590kg (1,300lb) water buffalo. A 46kg (100lb) dragon has been known to devour a 41kg (90lb) pig in a single meal. It has been known to attack humans, and a few deaths are attributable to it. Although not venomous, the saliva of Komodo dragons contains a

variety of bacteria that can lead to sepsis and death if untreated. At the other end of the scale are small species such as the Short-tailed monitor of Australia, which feeds on insects and small lizards and reaches a maximum body size of 12cm ((less than 5in) and a weight of less than 20g (0.7oz).

Monitors occur throughout Africa, southwest and southeast Asia, and Australia. They are especially numerous in Australia and Indonesia. Large, carnivorous mammals are absent from much of the eastern range of monitors, and to some extent these lizards may fill this niche. Most larger monitors are carnivorous, eating small mammals, birds, eggs, lizards, snakes, fish, and crabs; Nile monitors are especially important predators on crocodile eggs. Larger prey may be ambushed. Like snakes, monitors swallow their prey whole and usually head first. Insects are important in the diets of small and even some larger species. Carrion of all types is readily taken by many monitors. In some species the teeth become modified for crushing to accommodate a diet of snails. Gray's monitor (*Varanus olivaceus*) of the Philippines eats mostly snails and crabs when young, but switches to a diet rich in fruits as it matures. Although invertebrate prey are still consumed by the adults,

◑ Above *The Emerald tree monitor* (Varanus prasinus) *has many specialized adaptations for life in the rainforest canopy: a slender build; long digits with sharp claws; a prehensile tail; and effective green and black camouflage coloration.*

the digestive tract of this species is specialized for processing plant material.

Most monitors share a common body shape. The body is elongate and the limbs are well developed, with strong claws on all digits. The neck is long and the tail is muscular and slightly to highly laterally compressed. Most monitors are terrestrial and occur in sandy deserts, savannas, or forests, but many smaller species, including the Emerald tree monitor of New Guinea, are agile climbers; even juvenile Komodo dragons spend most of their time in trees. Others are excellent swimmers; monitors swim using their tails, with the limbs pressed against the sides of the body, so these species often have very strongly compressed tails. The Common Asiatic monitor occasionally swims far out to sea, where it regularly dives beneath the surface to escape predators and can remain submerged for up to an hour.

Other common defenses include lashing with

COURTING LIZARDS

Like all vertebrates, lizards have courtship rituals that help them to find suitable partners and win their cooperation in mating. In lizards it is always the male that courts, but he can only progress if his partner provides him with the right stimuli. She must behave like a female, and like one of his own species; with a female of the wrong species his reproductive investment would yield sterile hybrids, if any offspring at all. She must also signal that she is receptive, with mature ova ready for fertilization.

Many lizards use distinctive visual signals in courtship. In the Green anole, the male begins by bobbing his head and expanding the dewlap under his throat. Often, he will approach the female with a stiff-legged walk. If receptive, she remains still and then arches her neck when he is near; if unreceptive, she simply runs away. In the Alligator lizard, the male makes side-to-side movements of the head. Receptive females approach displaying males and nudge and walk over them, performing head-jerks. Unreceptive females retreat, waving their tails.

Visual signals are often important in gecko courtship – in the Western banded gecko, males approach females in a prostrate position, undulating their tails – but in some species such as the Barking gecko the most important signals are vocalizations, produced by males as they sit at the entrances to their burrows.

Monitors use very few auditory or visual signals. The male Komodo dragon, for example, approaches a female, presses his snout to her body **1**, and flicks her with his long, forked tongue to obtain chemical information about her receptivity. He then scratches her back with his long claws **2**, making a ratchetlike noise. If unreceptive, she raises and inflates her neck and hisses loudly.

Males often bite, scratch, or lick females that have signaled their receptivity. Anguid species such as the slowworm and Alligator lizard grasp the female's neck or head in the mouth while attempting copulation. The Green anole parallels the female's body and, as he firmly grasps her neck with his mouth, his hind limbs typically clasp the base of her tail. The Western banded gecko pushes against his mate with his snout and either licks her or bites her tail and flanks. Eventually he obtains a bite-hold on her neck and aligns his body and his fore and hind limbs with hers. In the Komodo dragon the male crawls onto the female's back **3** and, flicking his tongue on her head, rubs the base of her tail with his hind limbs to stimulate her to raise it. AGK/GS

❂ **Below** *A male Flap-necked chameleon (Chamaeleo dilepis) pursues a female during courtship. His elaborate body color is designed to attract a mate.*

the powerful tail, clawing, and hissing with the neck inflated and body compressed to maximize apparent size. Gould's monitor may stand on its hind legs in response to threats, but this posture is also useful for scanning the surroundings for mates and potential prey.

All monitors are day-active, and most terrestrial and arboreal species have preferred body temperatures of 35°–40°C (95°–104°F). They bask to raise their temperatures, and retreat to burrows or other shady spots when temperatures rise too high. More aquatic species typically prefer temperatures below 33°C (91.4°F). Species living in temperate areas, such as White-throated and Desert monitors (*Varanus griseus*), regularly become dormant in winter.

Monitors have long tongues that are used to pick up chemical cues in the air. Prey are often located this way, as are mates. In the African White-throated monitor, males travel up to 4km (2.5mi) per day to locate females. Males and females differ little from one another in size, color, or appearance. Male monitors are territorial, and will fight rivals for access to mates. Combatants stand on their hind legs, grasp each other with their forelegs, and try to push their opponent to

the ground; in some species the victor bites its vanquished foe. In Rosenberg's monitor (*Varanus rosenbergi*) of Australia, males and females form pair bonds. Courtship involves the male licking and nuzzling the female as well as multiple copulations over a period of several days. All monitors lay eggs, which are most often deposited in burrows, tree hollows, or termite mounds. Clutches range from 7 to 51 eggs, with the larger clutches characterizing larger-bodied species.

Many species of monitors are hunted for their skins and their meat. In central Australia, these lizards make up an important part of the traditional diet of the indigenous people, while in Asia and Africa monitors are exploited for use as both food and medicinal products. More than 1 million lizards, especially Nile and Common Asiatic monitors, are killed for their skins, which are exported for use in the manufacture of lizard-skin products such as shoes, belts, and purses. AMB/AGK/GS

Lizard Families

LIZARD CLASSIFICATION IS ACTIVELY BEING researched, but the interrelationships of many families remain uncertain. Most available data support the recognition of the major groupings shown in the phylogenetic hypothesis opposite (iguanians, gekkotans, scincomorphs, anguimorphs). The placement of some families, however, including the Xantusiidae and Dibamidae (both of which are sometimes regarded as related to the Gekkota), is uncertain. In alternative classifications the families Eublepharidae and Diplodactylidae are sometimes included within the Gekkonidae, and the Gerrhosauridae are regarded as part of the Cordylidae. Both the fossil marine mosasaurs and the snakes are believed to be derived from within the varanoids.

Chisel-teeth lizards
Family: Agamidae

399 species in 54 genera. Throughout Africa, Asia, and Australia, and all areas between, N to 52°, except Madagascar; up to more than 3,600m. Species and genera include: **Australian frilled lizard** (*Chlamydosaurus kingii*), **Butterfly lizard** (*Leiolepis belliana*), **Common bloodsucker** (*Calotes versicolor*), **Eastern water dragon** (*Physignatus lesueurii*), **Egyptian spiny-tailed lizard** (*Uromastyx aegyptius*), **flying dragons** (genus *Draco*), **moloch** or **Thorny devil** (*Moloch horridus*), **Sri Lanka prehensile-tail lizard** (*Cophotis ceylanica*), **Sail-fin lizard** (*Hydrosaurus pustulatus*), **toad-headed lizards** (genus *Phrynocephalus*).
LENGTH: 4–35cm
COLOR: Generally brown, gray, or black, but green in some forest forms; many species sexually dichromatic; some capable of rapidly changing color.
SCALES: Most keeled or drawn out into spines; in many species males and females have different head ornaments, throat fans, and neck frills, and folds, crests, or enlarged spines on back and tail.
BODY FORM: Head large and set off from neck; body cylindrical, compressed or depressed; 5th toe absent or reduced; tail long and not fragile. Eyes and lids well

developed. Ears external; opening rarely covered with scales. Tongue thick and fleshy, blunt and slightly notched at tip. Teeth acrodont (fixed to tooth-bearing bones); often incisorlike at front, or long and fanglike; others truncated blunt cylinders or compressed, with irregularly wavy shearing margins. Skin glands in front of cloaca and on undersides of thighs in many species.
CONSERVATION STATUS: 2 species – the **Spineless forest lizard** and the **Leaf-nosed lizard** – are Endangered and 2 others are Vulnerable.

Chameleons
Family: Chamaeleonidae

168 species in 6 genera. Madagascar and neighboring Indian Ocean islands, Sri Lanka, India, S Arabian Peninsula, Mediterranean Europe, Africa except the Sahara; up to 4,260m. Species and genera include: **Common chameleon** (*Chamaeleo chamaeleon*), **Jackson's chameleon** (*C. jacksonii*), **Meller's chameleon** (*C. melleri*), **Veiled chameleon** (*C. calyptratus*), **dwarf chameleons** (genus *Bradypodion*), **stump-tailed chameleons** (genus *Brookesia*).
LENGTH: 2–28cm
COLOR: Predominantly brown, green, or yellow; can quickly change color, even to extremes of white and black.
SCALES: Small, juxtaposed, without bony plates.
BODY FORM: Head and body compressed; head often with horns, flaps, or crests; limbs tending to be long and slender; feet zygodactyl (with some toes bound together and opposed); tail muscular, prehensile in most species, never fragile. Eyes protruding, covered with scaly lids, independently mobile. Ears have no external opening. Tongue long and slender, can be extended more than head-body length. Teeth acrodont. Skin glands absent from belly and undersides of thighs.
CONSERVATION STATUS: **Smith's dwarf chameleon** is Critically Endangered, while **Setaro's dwarf chameleon** and **Tiger chameleon** are listed as Endangered; 4 other species are Vulnerable.

Iguania

Gekkota

Scleroglossa

Scincomorpha

Autarchoglossa

Corytophanidae
Crotaphytidae
Hoplocercidae
Iguanidae
Opluridae
Phrynosomatidae
Polychrotidae
Tropiduridae
Agamidae
Chamaeleonidae
Eublepharidae
Gekkonidae
Diplodactylidae
Pygopodidae
Dibamidae
Xantusiidae
Lacertidae
Teiidae
Gymnophthalmidae
Scincidae
Cordylidae
Gerrhosauridae
Anniellidae
Anguidae
Xenosauridae
Helodermatidae
Lanthanotidae
Varanidae

Teioidea | Scincoidea | Anguimorpha

Varanoidea

⚫ **Above** *This dendrogram gives one account of the relationships among the 28 extant families of lizards. In this account, the Corytophanidae, Crotaphytidae, Hoplocercidae, Iguanidae, Opluridae, Phrynosomatidae, Polychrotidae, and Tropiduridae are considered to be separate families. Some authorities treat them all as one family, Iguanidae.*

Basilisks and casque-headed lizards
Family: Corytophanidae

9 species in 3 genera. Central America; arboreal and terrestrial, in open forests and often close to rivers. Species include **Plumed basilisk** (*Basiliscus plumifrons*) and **Helmeted basilisk** (*Corytophanes cristatus*).
LENGTH: 9–20cm
COLOR: Brown or green.
SCALES: Small on body; bony plates absent; crests along back.
BODY FORM: Slender; long tail; long limbs and claws; large eyes; teeth pleurodont.

Collared lizards
Family: Crotaphytidae

10 species in 2 genera. Southern N America. In deserts; terrestrial and diurnal. Genera are **collared lizards** (*Crotaphytus*) and **leopard lizards** (*Gambelia*).
LENGTH: 10–15cm
COLOR: Brown or green, often with bands or spots of white or black.
SCALES: Small on body, larger on head.
BODY FORM: Stocky, with long tail, long legs, and long claws; head large and wide; gape large; eyes large.

Dwarf and spiny-tailed iguanas
Family: Hoplocercidae

11 species in 3 genera. C America and northern S America in forests. Terrestrial and arboreal; diurnal. Species include

NOTES   Length = length from snout to vent.

Approximate nonmetric equivalents: 10cm = 4in 1kg = 2.2lb

Weapon-tailed lizard (*Hoplocercus spinosus*), dwarf iguanas (genus *Enyalioides*).
LENGTH: 9–15cm
COLOR: Mostly brown, mottled.
SCALES: Small on body, variable on head.
BODY FORM: Slender, with long tail; long limbs and claws; dorsal crest often greatly expanded on top of head. Weapon-tailed lizard is stocky with a thick, heavily armored back.

True iguanas
Family: Iguanidae

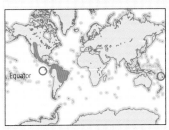

36 species in 8 genera. SW N America to S America, the Galápagos Islands, Caribbean islands, and Fiji. Terrestrial and arboreal; diurnal. Species and genera include: **Fiji banded iguana** (*Brachylophus fasciatus*), **Green iguana** (*Iguana iguana*), **Marine iguana** (*Amblyrhynchus cristatus*), **Galápagos land iguanas** (genus *Conolophus*), and **West Indian ground iguanas** (genus *Cyclura*).
LENGTH: 20–75cm
COLOR: Brown, gray, or bright green.
SCALES: Small on body; crests on top of head and back.
BODY FORM: Mostly stocky with well-developed limbs, long tail. Eyes large, with well-developed lids. Ears with external openings. Tongue thick and fleshy. Teeth pleurodont, with complex crowns.
CONSERVATION STATUS: 11 of the 36 species, including the **Fiji crested iguana** (*Brachylophus vitiensis*); several **spiny-tailed iguanas** (genus *Ctenosaura*) and most of the **West Indian ground iguanas** are Critically Endangered. Most other species are either Endangered or Vulnerable.

Madagascan iguanas
Family: Opluridae

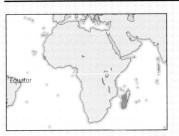

7 species in 2 genera. Madagascar and the Comoros Islands in deserts, scrub, and open forest. Arboreal and terrestrial; diurnal. Species include: **Four-eyed lizard** (*Chalarodon madagascariensis*) and **Cuvier's Madagascan iguana** (*Oplurus cuvieri*).
LENGTH: 6–15cm

COLOR: Brown or gray.
SCALES: Variable; mostly small on body; small dorsal crest of enlarged scales.
BODY FORM: Stocky with well-developed limbs and claws; moderately long tails; large eyes.

Spiny lizards, horned lizards, and relatives
Family: Phrynosomatidae

194 species in 9 genera. Southern N America, northern C America, in deserts and mountains. Mostly terrestrial and rock-dwelling; diurnal. Species and genera include: **Side-blotched lizard** (*Uta stansburiana*), **Fringe-toed lizard** (*Uma notata*), **horned lizards** (genus *Phrynosoma*), **spiny lizards** (genus *Sceloporus*).
LENGTH: 5–20cm
COLOR: Variable; usually brown and gray, sometimes marked with paler areas and often matching the substrate on which they live. Some have green markings, and others have bright colors on hidden surfaces.
SCALES: Variable; small or large, smooth or keeled and often with pointed and raised tips. Small crests on some species.
BODY FORM: Variable; slender and agile with long limbs and long tail, to short, squat and laterally depressed, with short limbs and short tail (horned lizards). Fringed toes in several.
CONSERVATION STATUS: *Sceloporus exsul* is Critically Endangered and **Coachella Valley fringe-toed lizard** (*Uma inornata*) is Endangered.

Anoles and relatives
Family: Polychrotidae

398 species in 9 genera. SE N America, C America and S America, in a wide range of habitats, mostly forests; arboreal (some terrestrial, rock- and cave-dwelling); mostly diurnal. Species and genera include: **Green anole** (*Anolis carolinensis*), **anoles** (genera *Anolis* and *Norops*); **bush anoles** (genus *Polychrus*).
LENGTH: 3–18cm
COLOR: Brown, gray, green or bluish. Some

capable of color change.
SCALES: Small, with a low dorsal crest.
BODY FORM: Mostly slender with very long limbs and a long tail; feet specialized for climbing in some species, with adhesive scansors; eyes large.
CONSERVATION STATUS: **Culebra Island giant anole** (*Anolis roosevelti*) is Critically Endangered.

Lava lizards, curly-tailed lizards, and relatives
Family: Tropiduridae

339 species in 10 genera. West Indies, S America, Galápagos Islands, in a wide range of mostly dry, open habitats and some in forests. Terrestrial; diurnal. Species and genera include: **Española lava lizard** (*Microlophus delanonis*), **curly-tailed lizards** (*Leiocephalus*), South American swifts (genus *Liolaemus*).
LENGTH: 6–12cm
COLOR: Brown; some with bright green markings.
SCALES: Variable; small and smooth or large and spiny.
BODY FORM: Stocky with moderately long limbs and long claws; triangular heads with large eyes.

Geckos
Family: Gekkonidae

964 species in 77 genera. Almost worldwide between 50°N and 47°S; up to 3,700m. Species and genera include: **Australian marbled gecko** (*Christinus marmoratus*), **Barking gecko** (*Ptenopus garrulus*), **bent-toed geckos** (genus *Cyrtodactylus*), **day geckos** (genus *Phelsuma*), **Fan-toed gecko** (*Ptyodactylus hasselquistii*), **gliding geckos** (genus *Ptychozoon*), **Indian wall gecko** (*Hemidactylus flaviviridis*), **Mourning gecko** (*Lepidodactylus lugubris*), **tokay** (*Gekko gecko*), **Web-footed gecko** (*Palmatogecko rangei*).
LENGTH: 1.5–20cm
COLOR: Usually brown or gray, but some species green or other bright colors.
SCALES: Almost always small, comprising a

soft skin; scattered keeled tubercles and spines sometimes present; without bone, rarely overlapping; enlarged, symmetrical head shields exceptional.
BODY FORM: Typically with depressed head, body, and tail; tail varying considerably in shape and ornamentation, fragile; limbs short, with 5 toes that are especially variable in shape and scalation. Eyes usually large, lids fused to form transparent covering. Ears with external opening that may be small. Tongue short, broad, slightly notched at tip. Teeth cylindrical and pointed, with blunt crowns in a few species. Skin glands in males in precloacal position and/or on underside of thighs; in most species small paired sacs and associated bones present at base of tail.
CONSERVATION STATUS: The **Paraguanan ground gecko** of Venezuela and *Phelsuma antonosy* are Critically Endangered; 3 species are Endangered, and 9 are Vulnerable. 1 species – *Phelsuma gigas* from Mauritius – is thought to have become extinct in the course of the 19th century, while *Nactus coindemerensis*, also from Mauritius, is now Extinct in the Wild.

Southwest Pacific geckos
Family: Diplodactylidae

121 species in 15 genera. New Zealand, New Caledonia, Australia except S Victoria and Tasmania; up to 2,200m. Species and genera include: **Duvaucel's gecko** (*Hoplodactylus duvaucelii*), **knob-tailed geckos** (genus *Nephrurus*), **New Caledonian giant gecko** (*Rhacodactylus leachianus*), **New Zealand green geckos** (genus *Naultinus*), **Spiny-tailed gecko** (*Strophurus ciliaris*), **velvet geckos** (genus *Oedura*).
LENGTH: 4.5–36cm
COLOR: Usually brown, gray, or purplish, often with contrasting dorsal markings; a few species green.
SCALES: Usually small and granular; some species with pronounced tubercles or caudal spines.
BODY FORM: Typically depressed, sometimes with loose folds of skin at sides; tail short to very short, fragile in all but some knob-tailed geckos; limbs short and robust to long and slender; 5 toes with or without adhesive scansors. Eyes usually large, lids fused to form transparent covering, pupils vertical. Ears have external opening exposed. Tongue short, broad, slightly notched at tip. Teeth small, pleurodont, cylindrical, pointed. Skin glands in males in precloacal position; paired cloacal sacs

and bones at base of tail in all species.
CONSERVATION STATUS: 1 species
(*Nephrurus deleani*) is Endangered, and 1
Vulnerable. **Delcourt's giant gecko** (*Hoplo-dactylus delcourti*) is extinct.

Eyelid geckos
Family: Eublepharidae

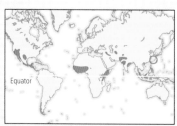

23 species in 6 genera. Scattered areas of N
and C America, E and W Africa, India, C
Asia, Borneo, Malaysia, Ryukyu Islands,
Hainan Island and adjacent China and
Vietnam; up to 2,500m. Species and genera
include: **Cat gecko** (*Aeluroscalabotes felinus*),
Leopard gecko (*Eublepharis macularius*),
Western banded gecko (*Coleonyx variegatus*).
LENGTH: 5–17cm
COLOR: Usually brown or pale pinkish,
yellowish, or purplish, frequently with contrasting markings.
SCALES: Small, granular, with scattered
enlarged tubercles in some.
BODY FORM: Trunk cylindrical or somewhat depressed; limbs short, with 5 toes;
no adhesive toe scansors; tail short, often
very thick, fragile. Eyes large, with vertical
pupils and moveable lids. Ears not covered
by scales. Tongue short, broad, weakly
notched at tip. Teeth small, conical, with
multiple cusps in some species. Small paired sacs and associated bones at base of tail.
CONSERVATION STATUS: 1 species, **Kur-
oiwa's ground gecko** (*Goniurosaurus kuroi-
wae*), is Vulnerable

Flap-footed lizards
Family: Pygopodidae

38 species in 7 genera. Aru Islands, New
Britain, New Guinea, Australia except Tasmania; up to 850m. Species and genera
include: **delmas** (genus *Delma*), **sharp-
snouted snake lizards** (genus *Lialis*),
Western scalyfoot (*Pygopus nigriceps*).
LENGTH: 6.5–31cm

COLOR: Usually brown or gray, some
striped or with dark bands on head and neck.
SCALES: Large and symmetrical on head in
almost all species; overlapping above on
trunk and tail, usually smooth; on underside, large rectangular shields; bony plates
absent.
BODY FORM: Snakelike in appearance;
forelimbs absent, hind legs persisting as
small to large flaps held tightly against
body near cloacal opening; tail very long,
fragile. Eyes modest in size, with fused lids
forming transparent covering; pupil usually
a vertical slit. Ears with external opening;
inconspicuous in a few species. Tongue
short, broad, and thick; slightly notched at
tip. Teeth conical; stout to slender. Small
paired sacs and associated bones at base of
tail, like geckos.
CONSERVATION STATUS: 6 species
are Vulnerable.

Blind lizards
Family: Dibamidae

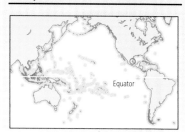

21 species in 2 genera. New Guinea to SE
China, E Mexico; up to 1,525m. Species
and genera include: **Asian blind lizards**
(genus *Dibamus*). **Mexican blind lizard**
(*Anelytropsis papillosus*).
LENGTH: 5–20cm
COLOR: Flesh-colored tones or violet
purplish-brown predominating.
SCALES: On head, large and plate-like,
especially on snout and lower jaws; on
body and tail, smooth and overlapping,
with concentric lines of growth;
without bone.
BODY FORM: Snakelike; forelimbs absent;
hind limbs small and paddle-shaped, present only in males; tail fragile, extremely
short. Eyes lidless and concealed beneath
skin. Ears have external opening covered
with scales. Tongue short, wide, not divided at tip. Teeth small and pointed, curving
inward; absent from palate. Preanal glands
in some species.

◗ **Right** *The vivid skin colors, spiny
crest, and dewlap of Boyd's forest
dragon (Hypsilurus boydii), a chisel-
tooth lizard from Australia, all serve to
enhance the visual impact of its display
postures and movements..*

Girdled lizards
Family: Cordylidae

55 species in 3 genera. E and S Africa; up
to 3,500m. Species and genera include:
Armadillo lizard (*Cordylus cataphractus*),
sungazer (*C. giganteus*), **flat lizards** (genus
Platysaurus), **snake lizards** (genus
Chamaesaura).
LENGTH: 5–20cm
COLOR: Brown, black, or brightly multicol-
ored; some species sexually dichromatic.
SCALES: on head, large, symmetrical
shields with bony plates; on body usually
in regular series, rectangular and overlapping, usually keeled on upper surfaces;
small and granular in flat lizards.
BODY FORM: stout to snakelike; typically
moderately to very depressed; limbs short,
reduced in some species; tail fragile, short
to very long, usually with whorls of spines.
Eyes well developed, with conspicuous
lids. Ears have distinct external opening.
Tongue short to moderately elongated;
pointed; may be feebly notched at tip.
Teeth cylindrical, with tapering tips or 2
cusps; present on palate. Femoral glands
with conspicuous external pores in
most species.
CONSERVATION STATUS: 4 species,
including **Armadillo lizard**, are listed as
Vulnerable.

Plated lizards
Family: Gerrhosauridae

34 species in 5 genera. Most of Africa S of
the Sahara, Madagascar; up to 3,500m.
Species and genera include: **Dwarf plated
lizard** (*Cordylosaurus subtessellatus*), **Giant
plated lizard** (*Gerrhosaurus validus*), **Malagasy plated lizards** (genus *Zonosaurus*),
Namib plated lizard (*Gerrhosaurus skoogi*),
seps (genus *Tetradactylus*).
LENGTH: 5–30cm
COLOR: Brown, gray, reddish, or black,
often with bright markings.
SCALES: On head, large, symmetrical
shields with bony plates; on body usually
regularly rectangular and overlapping.
BODY FORM: Stout to elongate; cylindrical

or depressed with conspicuous ventrolateral fold; limbs short, reduced in some
species; tail fragile, moderate to very long.
Eyes well developed, with conspicuous
lids: lower lid with transparent window in
some species. Ears have distinct external
opening. Tongue short to moderately elongated; pointed: may be feebly notched at tip.
Teeth pleurodont, cylindrical, with tapering
or cuspate tips. Femoral glands with conspicuous external pores in most species.
CONSERVATION STATUS: 1 species Vulnerable

Microteiid lizards
Family: Gymnophthalmidae

206 species in 38 genera. S Central America to C Argentina, Lesser Antilles; up to
almost 4,000m. Species and genera
include: **keeled microteiids** (genus
Arthrosaura), **Ocellated microteiid** (*Cer-
cosaura ocellata*), **reduced-limbed microteiids** (genus *Bachia*), **Schreibers's
microteiid** (*Pantodactylus schreibersii*).
LENGTH: 3.7–6.5cm
COLOR: Predominantly brown, gray,
or black.
SCALES: Large and symmetrical on head in
most species, variable from granular to
large and keeled on body, and in rows of
enlarged plates on belly; bony plates
absent.
BODY FORM: Highly variable, from relatively short to very elongate; limbs short, from
well developed to reduced to absent; tail
variable from moderate to long, fragile.
Eyes typically moderately large, with oval
pupils; lids well developed, moveable, covered with scales, or lower lids transparent.
Ears open or completely covered with
scales. Tongue long, narrow, covered above
with fleshy protuberances, deeply forked
at tip. Teeth conical; pterygoid teeth vari-
ably present. Femoral skin glands
usually present.

Whiptails and racerunners
Family: Teiidae

120 species in 9 genera. S USA through S
America and W Indies except Patagonia
and southern forests; up to 3,962m.

NOTES	Length = length from snout to vent.
	Approximate nonmetric equivalents: 10cm = 4in 1kg = 2.2lb

Species and genera include: **Caiman lizard** (*Dracaena guianensis*), **Common tegu** (*Tupinambis teguixin*), **Dragon lizard** (*Crocodilurus lacertinus*), **junglerunners** (genus *Ameiva*), **Desert grassland whiptail** (*Cnemidophorus uniparens*), **Little striped whiptail** (*C. inornatus*), **New Mexico whiptail** (*C. neomexicanus*), **St. Croix ground lizard** (*Ameiva polops*), **Tiger whiptail** (*C. tigris*).
LENGTH: 5.5–45cm
COLOR: Chiefly green, brown, or gray.
SCALES: Large and symmetrical on head, small and granular on body and in rectangular plates on belly; bony plates absent.
BODY FORM: Stout with large head and well-developed limbs, hind toes elongate; usually with pointed snout and long, fragile tail. Eyes moderately large, with oval pupils; lids well developed, moveable and covered with scales. Ears large, not covered by scales. Tongue long, narrow, covered above with fleshy protuberances, deeply forked at tip. Teeth conical at front, variable at sides. Femoral skin glands usually present.
CONSERVATION STATUS: The **St. Croix ground lizard** is Critically Endangered, and 1 other species is Vulnerable. The **Martinique giant ameiva** (*Ameiva major*) and *A. cineracea* are thought to have become extinct during the 20th century.

🔊 *Below* The African Savanna monitor (*Varanus albigularis*) *will often lay its eggs inside termite mounds.*

Wall and sand lizards
Family: Lacertidae

276 species in 31 genera. Europe, Africa, Asia, Indo-Australian Archipelago; up to 3,810m. Species and genera include: **East Asian grass lizards** (genus *Takydromus*), **fringe-toed lacertids** (genus *Acanthodactylus*), **Gliding lacertid** (*Holaspis guentheri*), **European green lizard** (*Lacerta viridis*), **Hierro giant lizard** (*Gallotia simonyi*), **racerunners** (genus *Eremias*), **Viviparous lizard** (*Zootoca vivipara*), **Common wall lizard** (*Podarcis muralis*).
LENGTH: 4–26cm
COLOR: Green, yellow, reddish, brown, gray, or black.
SCALES: May be enlarged and symmetrically arranged on head, containing thin bony plates usually fused to skull; on back, small and usually granular; belly with large, smooth, rectangular plates in well-defined rows; in whorls on tail.
BODY FORM: Head conical and distinct from body; snout pointed; usually have fold of skin across throat; body long, with well-developed limbs, typically five-toed; tail very long and fragile. Eyes moderately large, with moveable lids; some species with transparent window in lower lid. Tongue long and narrow, deeply forked at tip, covered above with small projections or folds. Teeth tend to be cylindrical; some on palate. Femoral and precloacal glands usually present.
CONSERVATION STATUS: 10 species, including 3 **fringe-toed lacertids** and 3 **giant Canary Islands lacertids** (*Gallotia*) are Critically Endangered; 12 species, including **Lilford's wall lizard** (*Podarcis lilfordi*) are Endangered; others are Vulnerable.

Skinks
Family: Scincidae

1,345 species in 136 genera. All tropical and temperate regions, especially in Africa, S and E Asia, Indo-Australian archipelago, New Guinea, Australia, New Zealand, and from S Canada to N Argentina; up to 4,800m. Species and genera include: **blind burrowing skinks** (genus *Typhlosaurus*), **blue-tongued skinks** (genus *Tiliqua*), **Brazilian skink** (*Mabuya heathi*), **casqueheads** (genus *Tribolonotus*), **Common five-lined skink** (*Eumeces fasciatus*), **Great Plains skink** (*E. obsoletus*), **Green-blooded skink** (*Prasinohaema virens*), **Marine skink** (*Emoia atrocostata*), **sandfish** (*Scincus scincus*), **Solomon Islands prehensile-tailed skink** (*Corucia zebrata*).
LENGTH: 2.3–49cm
COLOR: Brown, gray, and black predominating; some species green or blue.
SCALES: Usually enlarged on top of head to form series of symmetrical shields; on body, usually smooth, flat, and overlapping each other in regular order.
BODY FORM: Very variable; in general, elongated and somewhat depressed, with a small, flattened head and short legs, reduced or absent in many species. Eyes: moderate to small and nonfunctional; pupil usually round; lids well developed, moveable or fused; lower lids with transparent disk or covered with scales. Ears have external opening present or covered by scales. Tongue short, broad, flat, and fleshy; scaly or ridged; slightly notched at tip. Teeth very variable. Precloacal and femoral skin glands absent.
CONSERVATION STATUS: 3 species, including **Bermuda rock lizard** are Critically endangered; 8 others, including the **Adelaide pygmy blue-tongued skink**

(*Tiliqua adelaidensis*) are Endangered, and 26 are Vulnerable. In addition, 2 species, including the **Cape Verde giant skink** (*Macroscinsus coctei*) are thought to have become extinct in the course of the 20th century.

Night lizards
Family: Xantusiidae

24 species in 3 genera. E Cuba; Panama to C Mexico and SW USA; up to 2,600m. Species include: **Cuban night lizard** (*Cricosaura typica*), **Desert night lizard** (*Xantusia vigilis*), **Island night lizard** (*X. riversiana*), **Middle American night lizard** (*Lepiodophyma flavimaculatum*).
LENGTH: 3.5–12cm
COLOR: Predominantly somber shades of brown, gray, or black.
SCALES: Very large on head, with bony plates in some species; much smaller on upper surfaces of body and tail, juxtaposed; underside of body covered with large, rectangular plates.
BODY FORM: Head quite depressed in several species; body long and slightly depressed; limbs short and typically five-toed; tail long and fragile. Eyes small, with vertical pupil and transparent, fused lids. Ears with external opening. Tongue long and slightly notched at tip. Teeth cylindrical, either blunt or with 3 cusps; absent from palate. Femoral glands present, precloacal glands absent.
CONSERVATION STATUS: *Lepidophyma lipetzi* is Endangered, and *Xantusia riversiana* is considered Vulnerable.

Glass lizards, alligator lizards and relatives
Family: Anguidae

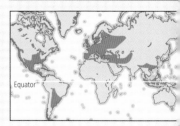

114 species in 12 genera. SW Canada and much of N USA to C Argentina, including West Indies; Britain to China, Sumatra and Borneo, between Arctic Circle and NW Africa in the west and S Asia in the east; absent from Sri Lanka and Arabian Peninsula; up to more than 4,260m. Species and genera include: **arboreal alligator lizards** (genus *Abronia*), **galliwasps** (genus

NOTES   Length = length from snout to vent.
Approximate nonmetric equivalents: 10cm = 4in 1kg = 2.2lb

Diploglossus), **glass lizards** (genus *Ophisaurus*), **scheltopusik** (*O. apodus*), **slowworm** (*Anguis fragilis*), **Southern alligator lizard** (*Elgaria multicarinata*).
LENGTH: 5.5–52cm
COLOR: Bright green, brown, gray, silver, or black.
SCALES: Smooth or moderately keeled; most with well-developed bony plates, including those on underside; large and juxtaposed on head; on body, smaller, overlapping, and disk-shaped.
BODY FORM: Most species have streamlined shape; some with prominent ventrolateral skin folds; limbs short when present: tail usually very long and fragile. Eyes functional but small to modest; lids moveable, lower ones sometimes with transparent window. External ear opening covered with scales in a few species. Tongue long, slender; portion with forked tip may be extended; less retractable fleshy portion found well back in the mouth. Teeth crushing and close-set to fanglike, inwardly curved and widely separated; absent from palate. Skin glands absent from precloacal and femoral regions.
CONSERVATION STATUS: 4 species, including the **Montserrat galliwasp** (*Diploglossus montiserrati*), are listed as Critically Endangered and 11 as Endangered.

Legless lizards
Family: Anniellidae

Equator

2 species in 1 genus: **California legless lizard** (*Aniella pulchra*) and **Baja California legless lizard** (*A. geronimensis*) from California and Mexico. Fossorial species that live in sandy deserts; nocturnal.
Length: 15–18cm
COLOR: Brown
SCALES: Smooth, shiny; small and overlapping.
BODY FORM: Legless; slender and snake-like; tails short; eyes small with well-developed eyelids.

Xenosaurs
Family: Xenosauridae

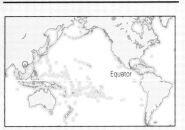
Equator

7 species in 2 genera; the Chinese species is sometimes included in a separate family, Shinisauridae. E Mexico S to Guatemala, SE China; 300–2,100m. Species and genera include: **Chinese crocodile lizard** (*Shinisaurus crocodilurus*), **New World xenosaurs** (genus *Xenosaurus*).
LENGTH: 10–15cm
COLOR: Predominantly brown, gray, or black.
SCALES: Minute to large, with small to large disks of bone; some species with a series of enlarged scutes on midline of nape and trunk, which may continue to tail.
BODY FORM: Head compressed in Chinese xenosaur, depressed in New World species; body stout, with well-developed limbs; tail moderately long, not fragile. Eyes of modest size, with round pupil. Ears have very large opening, conspicuous or inconspicuous depending on scale covering. Tongue has a long, slender portion forked at tip and at least partially retractable into more fleshy basal portion. Teeth fanglike, with slight inward curve, or blunt and cylindrical; tips of some with sharp cutting edges absent from palate. Femoral and precloacal glands absent.

Beaded lizards
Family: Helodermatidae

Tropic of Cancer

2 species in 1 genus: **Gila monster** (*Heloderma suspectum*), **Mexican beaded lizard** (*H. horridum*). SW USA through W Mexico to Guatemala; up to 2,000m.

LENGTH: 33–45.5cm
COLOR: Brown or black with pinkish or yellowish markings.
SCALES: Relatively large, raised, rounded scales on upper surfaces; underside with smaller, rectangular scales in regular rows; larger scales, including those of head, with bony disks inside.
BODY FORM: Almost cylindrical; head large, depressed, with blunt snout; distinctive fold of skin across throat; body stout, limbs short, five-toed; tail short to moderate in length, thick, blunt-tipped, not fragile. Eyes have small, round pupils and well-developed, moveable eyelids. Ears have external opening exposed. Tongue with a longer, deeply-forked portion and a basal part covered with fleshy projections. Teeth elongated, inwardly curved and sharply pointed; somewhat compressed, deeply grooved on anterior edge. Precloacal and femoral glands absent. The only venomous lizards; bite is painful but rarely fatal to humans.
CONSERVATION STATUS: **Gila monster** is Near Threatened.

Bornean earless monitor
Family: Lanthanotidae

Tropic of Capricorn

1 species: *Lanthanotus borneensis*. Borneo; elevations near sea level.
LENGTH: To 20cm
COLOR: Uniform brown or olive.
SCALES: Small and undifferentiated on head except in region of temples, where there are three rows of much enlarged tubercles; 3–5 rows of enlarged tubercles along midline of back and tail; remaining body scales small and flattened, with scattered tubercles.
BODY FORM: Relatively elongate, roughly cylindrical; head depressed, not strongly offset from neck; small nostrils high on snout; limbs short, five-toed; tail about as long as body, tapering, not fragile. Eyes: small, with round pupils and well-developed lids; transparent window in lower lid. Ears have no external openings. Tongue as in beaded lizards. Teeth recurved and fanglike; small palatal teeth.

Monitor lizards
Family: Varanidae

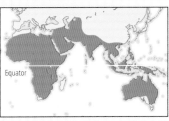
Equator

57 species in 1 genus. Africa to S Asia, Indo-Australian Archipelago, Philippines, New Guinea, Australia; up to 1,830m. Species include: **Bengal monitor** (*Varanus bengalensis*), **Cape monitor** (*V. albigularis*), **Common Asiatic monitor** (*V. salvator*), **Gould's monitor** (*V. gouldii*), **Komodo dragon** or **ora** (*V. komodoensis*), **Nile monitor** (*V. niloticus*), **Short-tailed monitor** (*V. brevicauda*).
LENGTH: 12cm–1.5m
COLOR: Largely drab brown, gray, or black.
SCALES: Mostly small, pebblelike granules that may form rings around larger, juxtaposed scales with conspicuous pits; thin disks of bone in some.
BODY FORM: Distinctive, with long neck and relatively short body; head usually very long and narrow, frequently with pointed snout and slitlike nostrils placed near eyes; skin fold across throat; limbs strongly built, five-toed; toes and curved claws long and strong; tail long and muscular, not fragile, usually compressed. Eyes large, pupils almost circular, lids moveable. Ears with external opening exposed. Tongue very long and slender, deeply forked, no expanded basal portion. Teeth usually compressed, with sharp cutting edges, fanglike; absent from palate. Femoral and precloacal glands absent.
CONSERVATION STATUS: 2 species, including the **Komodo dragon**, are considered Vulnerable.   AMB, CM

◖ **Left** *The Gila monster* (Heloderma suspectum) *and the closely related Mexican beaded lizard are the only lizards with venom glands. These are located in the lower jaw and, unlike a snake's, do not have any sophisticated mechanism for injecting the toxin.*

Worm-lizards

◐ **Left** Unlike other worm-lizards, the Mexican worm-lizard or ajolote (Bipes biporus) has prominent forelimbs. Their position well forward on its body, close to its head, has given rise to its folk name lagartija con orejas ("little lizard with ears").

WORM-LIZARDS, ALSO KNOWN AS RINGED lizards, are the only true burrowers among the reptiles. Other reptiles live underground for part of their lives, and some of these use tunnels dug by other animals, but only worm-lizards are exclusively subterranean. They generate their own tunnels, many of them driving passages through very hard soils, and their digging adaptations, among other features, are distinctive.

Worm-lizards first appear in the fossil record of the North American Paleocene 65 million years ago; other early specimens have been found in Europe (England and Belgium), while the largest known fossil forms are from Kenya. Three of the four families lack all traces of limbs and may even have lost remnants of the internal pelvic and pectoral girdles, which other legless reptiles retain. A fourth family, the Mexican worm-lizards (Bipedidae), have retained and even increased the size of their forelimbs; their body has become elongated behind the shoulder girdle, giving the impression that the hands are sited on the head. The hands help *Bipes* move above ground and also to generate the initial divot from which it burrows into the soil.

Head-bangers and Scrapers
FORM AND FUNCTION

All worm-lizards form tunnels by forceful movements of the head. The heads are formed into various kinds of digging tools, and the skulls are more solid than those of other reptiles of equivalent size. In some species, the skull is shielded by hard keratin, in others by scales with a low-friction surface that facilitates penetration of the soil. The eyelids are fused, and the eyes lie deep below the translucent skin. There are no external ear openings, and the nostrils point backward so that pressure tends

to close them, which prevents sand entering during burrowing. The upper lip extends over the lower lip, and the lower jaw normally closes with the head in such a way that it is likely to be pushed closed rather than open as the head drives into the soil at the end of the tunnel. The ears are highly specialized; they are used not only for hearing but also for detecting subterranean vibrations.

Under the ground, worm-lizards propel the body by concertina and rectilinear movements. The skin, in rings around the body, is relatively free from the trunk and can slide over it, folding like a bellows. In rectilinear motion, portions of the skin are fixed against the tunnel wall and the attaching muscles pull up the trunk, generating the force required for a penetrating stroke into the tunnel's end.

Once the head is driven into the end of the tunnel, it may be wriggled and then withdrawn slightly before the next stroke. About half of the species, however, have separated the effort of extending tunnels from that of widening them. They first ram the head into the tunnel's end and then use different muscles arranged in a complex pulley system to press the soil into the tunnel roof and floor. They thus avoid having to dump soil on the surface, and can spend most of their lives underground.

Members of the commonest family, the true worm-lizards (Amphisbaenidae) found in the Mediterranean region, Africa, and the Americas, have specialized their widening system. They either form the head into a vertical keel, swinging it to left and right after penetration, or into a horizontal spade that is lifted to compress soil onto the roof of the tunnel. Keel-headed species have the disadvantage that less than half of the neck muscles can be used for compression; the spade-headed species are larger and more numerous.

As the sole living member of a family (the Rhineuridae) with an extensive fossil record across North America, the Florida worm-lizard also shows a spade-snouted pattern. It digs in a fashion similar to that of the true worm-lizards, and its head segmentation also shows similarities. However, it differs profoundly from the pattern of the African and South American species in the nature and arrangement of the bones that form the spade, as well as in the muscular arrangement along the back, particularly of the neck.

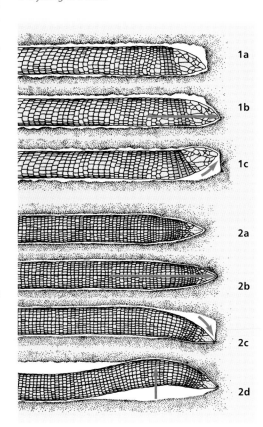

◐ **Above** The distinctively-patterned Checkered worm-lizard (Trogonophis wiegmanni) lives in the arid, sandy environment of Morocco. It is unique within its family grouping for being viviparous, giving birth to 4–5 young in a litter.

1a
1b
1c
2a
2b
2c
2d

FACTFILE

WORM-LIZARDS

Suborder: Amphisbaenia

166 species in 23 genera and 4 families: Amphis-baenidae, Trogonophidae, Bipedidae, Rhineuri-dae. Species include: **Florida worm-lizard** (*Rhineura floridana*).

DISTRIBUTION Subtropical regions of N America, W Indies, S America, into Patagonia; Africa, Iberian Peninsula, Arabia, W Asia.

HABITAT Burrowing reptiles highly specialized for digging.

 SIZE Length (snout to tip of tail) from 10–75cm (4–30in), but most are 15–35cm (6–14in).

COLOR Reddish-brown to brownish black; black and white.

REPRODUCTION Fertilization internal; some lay eggs, in others the eggs develop inside female.

LONGEVITY Generally 1–2 years in captivity, some longer.

CONSERVATION STATUS Not threatened.

Worms with Teeth

FEEDING AND REPRODUCTION

Worm-lizards are formidable predators. They use large, interlocking upper and lower teeth to bite and crush their prey, and then withdraw down the tunnel, pulling their victim behind them, so that pieces are ripped out as the prey is dragged along.

Prey (small arthropods, worms, and even verte-brates) is recognized by scent and sound. The predator's ears are modified so that sound is de-tected through a substitute eardrum on the side of the jaws, which enables worm-lizards to hear air-borne sounds transmitted down the tunnel before them. Experiments suggest that they respond to sounds made by the movements of prey. They get some moisture from their food, but certain species can apparently collect soil water, pulling it in by capillary action between the lips and tongue.

Worm-lizards reproduce like snakes and lizards, with hemipenes allowing internal fertilization. Some species lay eggs, a few in ant and termite colonies, and in some the eggs develop within the female before live birth. As in snakes, the embryos are complexly coiled within the egg, but in worm-lizards the posterior end of the embryo remains coiled into a relatively small spiral until shortly before birth.   CG

◑ **Above** *Amphisbaena fulginosa is a member of the largest family of worm-lizards. This individual has shed its tail (a feature known as caudal autotomy, used to escape predators). Yet, unlike other tail-shedding lizards, amphisbaenians cannot regenerate their tails.*

◁ **Left** *Two worm-lizard digging methods. 1 In shovel-snouted species, at the start of the stroke (a) the rings of scales behind the head are close together. During penetration (b), these rings separate, pushing the head forward. In tunnel-widening (c), the head is lifted against the tunnel roof. 2 In keel-snouted species, the tunnel-extending stroke (a,b) is similar, but the tunnel is widened either (c) near the snout by bending the head or (d) around the back of the skull by bending the body sideways.*

The Trogonophidae are a family of edge-snouted worm-lizards that burrow in a rotating, oscillatory pattern. Their face tends to be a flattened plate before the eyes. The edges of this plate form scrap-ers that shave a layer of soil off the end of the tun-nel so that the side of the trunk can, in the next movement, force it into the tunnel's wall. As the penetrating strokes produce torsional forces between body and tunnel wall, these are the only worm-lizards that have opted for a triangular or beam-shaped cross-section of the body (and the tunnel). The method is highly effective in the fri-able, sandy soils of the Arabian Peninsula and the Horn of Africa, where the most specialized trogo-nophid species is found.

Snakes

O VER 3,200 SPECIES OF SNAKES HAVE *been identified, and new ones are being added all the time. In many respects, they are all similar, having an elongated, roughly cylindrical body with a long tail at one end and a head at the other. Snakes have no limbs or other extremities to interrupt their shape, neither do they have external ear-openings or eyelids. Despite these apparent limitations, they have managed to find their way into a wide variety of habitats and have diversified into many niches. To achieve this success, they have evolved specialized methods of locomotion and of sensing the world around them. In some cases, these senses are unique; in others they are developed to a higher degree than in other animals.*

Even so, snakes are similar enough to lizards for both groups to be placed in the same order, Squamata, and taxonomically speaking they are hardly separate from them. The most obvious way in which snakes differ from the majority of lizards is their lack of limbs. Leglessness has evolved independently in several families of lizards such as the glass lizards and the skinks – the Anguidae and the Scincidae – usually in response to a burrowing or semiburrowing lifestyle. Snakes evolved from neither of these families, however, but from a branch of the lizards' family tree, now extinct, that is most closely related to the advanced families,

especially the monitors (Varanidae).

Snakes occur on every continent except Antarctica but, like most reptiles, they are more numerous in warm places, especially the tropics. They have not, however, dispersed onto small islands as successfully as lizards. The distribution of the various families (some very widespread and others very limited) is due to the way in which the global landmasses have drifted apart and rejoined through the period since snakes first appeared. Ancient families, therefore, have a wide distribution, especially in the southern hemisphere, whereas more recent families that appeared later have had limited opportunities to spread further than the coastlines of the continent on which they evolved. Yet other families appeared early on and were formerly widespread, but have since become fragmented as a result of local extinctions and the appearance of barriers such as mountain ranges and large rivers.

C *Right As adults, Emerald boas (Corallus caninus) are bright green, but their newborn young may be yellow or red. Arboreal boas gain a purchase on trees by wrapping their long, prehensile tails around branches.*

C *Below The heavy-bodied Gaboon viper (Bitis gabonica), with its distinctive spade-shaped head, is found in woodland in tropical Africa. Its disruptive, cryptic skin patterning enables it to conceal itself in leaf litter while waiting to ambush its prey.*

Life without Limbs

FORM AND FUNCTION

Within the suborder Serpentes, variations in size, shape, color, markings, and texture reflect, to a large extent, each species' lifestyle. However, characteristics reflecting lifestyle are often superimposed on those common to a family – the mainly squat body shape of vipers as opposed to the more elongated shape of most cobras and their relatives, for instance – so they can sometimes be misleading.

Snakes range in size from tiny, threadlike creatures that barely exceed 10cm (4in) in length to gigantic boas and pythons that approach 10m (33ft). Very large and very small snakes, however, are in the minority; most species measure between 30cm (12in) and 2m (6.6ft).

Species may be long and slender or short and squat, depending mainly on their feeding habits. Thus, snakes that sit and wait for their prey tend to be heavy-bodied. The reasons are twofold. First, the large body helps to anchor them to the ground, giving them a good purchase from which to launch a strike. Secondly, snakes that have

SNAKES

Order: Squamata

Over 3,200 species in 466 genera and 18 (20) families

DISTRIBUTION Worldwide except Arctic regions, Antarctica, Iceland, Ireland, New Zealand, and some small oceanic islands. No seasnakes inhabit the Atlantic Ocean, due to their inability to cross cold currents.

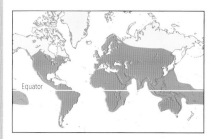

Equator

HABITAT Mostly terrestrial, but many burrowing, aquatic, marine, or tree-dwelling.

SIZE Length from snout to tip of tail from 15cm–11.4m (6in–37.5ft), but most are 25cm–1.5m (10–60in).

COLOR Mostly brown, gray, or black, but some with bright red, yellow, or green bodies, and markings that vary from spots and blotches to rings, crossbands, and stripes.

REPRODUCTION Fertilization internal by means of paired male hemipenes (only one inserted at a time); most lay eggs, but many bear live young; some have true placental connection with mother; some snakes guard eggs, and a few (notably most temperate pitvipers) give parental care to young.

LONGEVITY Even small snakes may live as long as 12 years; large species may live 40 years, perhaps longer.

CONSERVATION STATUS 13 species are Critically Endangered, 26 Endangered, and 33 Vulnerable. In addition, 3 recent species including the St. Croix racer and the Round Island burrowing boa are now considered Extinct.

given up the option of chasing their prey do not need to be as streamlined as those that do and, therefore, can afford to be stouter, with an increased capacity for large meals. Having a large body allows them to accommodate prey that might otherwise have to be rejected.

In addition, many ambushers are venomous and need wide, triangular heads to accommodate their venom glands, which are situated just behind their eyes – good examples would be the Gaboon viper and the Australian death adders (which, despite their common name, belong not to the Viperidae but to the Elapidae, a family that also includes the cobras, mambas, and coral snakes). Nonvenomous ambushers, however, may have disproportionately small heads, perhaps to allow them to strike more rapidly: for example, the Short-tailed or Blood python, *Python curtus*. All these snakes have small eyes since sight is subsidiary to smell, and in some cases to heat detection, in locating prey.

At the other extreme are those snakes that hunt actively, often poking their heads into cracks, crevices, and dense vegetation in the hope of

◗ Below *Some primitive snakes, such as pythons and boas, retain a vestigial remnant of a hindlimb in the form of a bony spur on their pelvic girdle. Shown here is a South African python* (Python natalensis).

flushing out prey, then giving chase. These species are long and slender, with narrow heads, long tails, and large eyes, since they track their prey by sight. Examples are the garter snakes of the genus *Thamnophis*, the *Masticophis* and *Coluber* whipsnakes and racers, and the African sand snakes, genus *Psammophis*.

There is also variation in the cross-sections of snakes' bodies. Burrowing species are almost per-fectly cylindrical, perhaps the starting point from which other shapes are derived. Ground crawlers such as the large vipers are often flattened from top to bottom to provide plenty of surface area in contact with the ground (like the wide tires of high-performance cars), whereas climbing snakes are flattened from side to side, so that they can hold their bodies rigid when they stretch across open spaces, like a girder.

Regardless of size and proportions, snakes' skeletons are highly modified, with a large num-ber of vertebrae – up to 500 in exceptional cases. These are quite loosely articulated and each one can rotate on its neighbors, allowing the snake to bend and coil in every direction. They are preven-ted from rotating too much, and so possibly caus-ing damage to the spinal cord, by overlapping, winglike processes on each vertebra. Each of the vertebrae in the body and neck region has a pair of ribs attached, and these are important in loco-motion. Although they have no visible limbs, some species retain the pelvic (hind) limb girdles, and may also have small vestiges of their limbs attached to them. These are the "spurs" or claws of the more primitive species such as the boas and pythons, which are often more prominent in males. Snakes' skulls are, for the most part, extremely flexible, with most bones reduced in size and weight and only loosely articulated with one another – a trait that allows the snake to open its mouth widely to accommodate meals that may be several times the diameter of its head. Blind-snakes, which have a diet of small, soft-bodied invertebrates, have more rigid skulls.

Snakes' teeth are pointed, very sharp, and directed backwards. They have evolved for grasp-ing and holding prey rather than for chewing. Although members of some of the most primitive families have few teeth, most species have a large number, arranged along the rims of the upper and lower jaws and with two additional rows on the roof of their mouth (the palatine and pterygoid teeth). Members of some families have a few teeth modified for injecting venom (see Venomous Snakes) and other groups may have modified teeth for dealing with specialized prey.

The elongation of snakes' bodies has resulted in corresponding elongations to some of their internal organs so that they can fit into the avail-able space. For the same reason, certain organs are reduced in size, whereas others are reposi-tioned. Most snakes (but not the boas, pythons, and other primitive groups) have only a single

functional lung, the right one, the left lung having reduced or disappeared altogether. The right lung is greatly enlarged by way of compensation, and there is an extra structure, the tracheal lung, derived from the windpipe, that also helps in res-piration. Snakes' stomachs are large and muscular, little more than a long, thickened region of the gut, and the intestines are not intricately coiled as they are in many other animals. The kidneys are elongated and staggered, as are the testes in male snakes. Females of a few small and very slender species have lost one of their oviducts.

○ Above *The shed skin of an Eastern hog-nosed snake* (Heterodon platirhinos). *As a snake grows, it periodically sloughs this keratinous layer. Note the large, transparent scales (brilles) that cover the snake's eyes – vital protection for an animal with no eyelids.*

Scaly Coatings
SKIN

Snakes' skin is covered in scales. Each scale is a thickened part of the skin, separated from neigh-boring scales by thinner areas of elastic, interstitial skin that allow the snake's body to be flexible. The dorsal scales are the most conspicuous; they may have either pointed or rounded, overlapping edges, like roof tiles, and they may be smooth and shiny or keeled and rough. Some snakes have granular, or beadlike, scales that do not overlap. The number, shape, arrangement, and color of the dorsal scales are very helpful in snake identifica-tion. On the underside of the body, the ventral scales of most snakes are wide and arranged in a single row. Similarly, the scales below the tail (subcaudal) may also be arranged into a single row, or they may be paired. In thoroughly aquatic

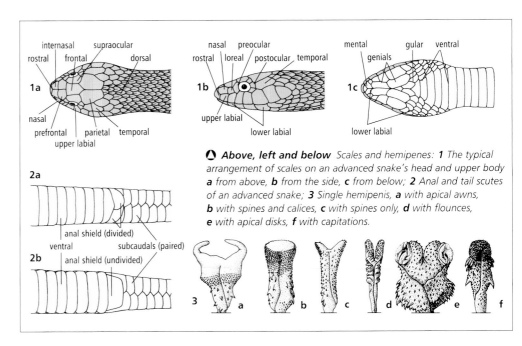

○ Above, left and below *Scales and hemipenes:* **1** *The typical arrangement of scales on an advanced snake's head and upper body* **a** *from above,* **b** *from the side,* **c** *from below;* **2** *Anal and tail scutes of an advanced snake;* **3** *Single hemipenis,* **a** *with apical awns,* **b** *with spines and calices,* **c** *with spines only,* **d** *with flounces,* **e** *with apical disks,* **f** *with capitations.*

species these ventral and subcaudal scales have become narrow and are represented by a simple keel or ridge running the length of the underside. The most primitive snakes do not have differentiated ventral scales.

The scales covering the head may be small and granular but are more commonly large and plate-like. Small head scales are found among the boas, pythons, and vipers, although there are some species in all these families with large scales, while others have head scales of an intermediate size. Conversely, members of the Colubridae and Elapidae, for example, have large head scales, although, again, there are a few exceptions. Since snakes do not have eyelids, their eyes are protected by a large circular scale, called the brille, that is periodically shed along with the rest of the skin. Scales protect the snake's skin from abrasion as it moves across the substrate and also limit water loss. In addition, pigment cells in the skin give each species its characteristic colors and markings.

Covering the scales, and the interstitial skin between them, is a layer of keratin, which provides additional protection but which is not flexible enough to allow for growth. It also becomes

scratched and damaged during the normal rough and tumble of a snake's daily activities. Consequently, the snake must shed this outer layer occasionally. Young snakes, which grow more quickly than adults, tend to shed more often, although the frequency depends on how well the snake has fed and how active it has been. All snakes apparently shed immediately after a period of hibernation (the vernal shed) as well as just before laying eggs or giving birth to young, and again just afterwards. Freshly-shed skin (strictly speaking, the epidermal layer) is moist and tacky, owing to oils that the snake secretes between its old and new surfaces to facilitate shedding. After a few hours the skin dries and becomes brittle. The segments of a rattlesnake's rattle consist of the dried epidermis of the enlarged terminal scale, retained when the rest of the skin is shed, and held loosely together by constrictions around their "waists."

⚪ **Below** *Flicking out the forked tongue is a typical action of snakes, as seen in this Salvin's rattlesnake (Crotalus scutulatus salvini). In doing so, the snake is picking up chemical signals that allow it to track prey, detect a threatening predator, or find a mate.*

Tasting the Air
SENSES AND SENSE ORGANS

Snakes' evolutionary history, which included a long spell as burrowing animals, has been the driving force behind some of the unique ways they have of sensing the world around them. Sight, for example, became largely superfluous when they lived underground, and their eyes therefore became redundant. The more primitive snakes, which still live below the surface, have small, almost nonfunctional eyes covered by thick scales. More advanced species, which made the move back to the surface, needed to reinvent the eye or a satisfactory substitute, leading to fundamental differences between their eyes and other animals' visual organs. For example, snakes' eyes focus not by changing the shape of the lens but by moving the lens backward and forward. This trait appears to be a limitation, and most snakes have poor eyesight; in particular, they are not very good at seeing stationary objects.

Snakes' eyes vary in size according to species' lifestyles, and their pupils vary in shape. Small, secretive species have small eyes with round pupils; diurnal hunters also have round pupils but

■ The Mechanics of Slithering
■ LOCOMOTION

An obvious necessity for any group of animals that has lost its limbs is to develop a new method of locomotion. Snakes have evolved several, some of which are common to most species while others overcome problems unique to certain types of habitat.

Rectilinear, or "straightline," locomotion is achieved by movement of the ventral scales. Each one is attached to a pair of ribs by means of obliquely-arranged muscles. As these muscles contract and relax, the edges of the scales hook over slight irregularities in the ground surface and pull the snake along. At any given time, several groups of scales will be pulling while alternate groups will be moving forward, so the movement is wavelike. From above, the snake seems to glide effortlessly over the ground. Rectilinear movement is fairly slow and is used by heavy-bodied snakes such as large boas, pythons, and vipers, or by other snakes when they are creeping forward slowly, for instance while stalking prey.

Serpentine locomotion, sometimes called lateral undulation, is used by most snakes when they are moving rapidly. The snake's body follows its head in a series of gentle curves. Its flanks push against many small or large irregularities simultaneously so that it moves forward smoothly. At the same time, its ventral scales may also be moving in the way described for rectilinear locomotion, increasing the amount of thrust. A similar method of locomotion is also used in swimming, where the flanks are pushed against the water. A number of semiaquatic snakes have rough, keeled scales that may help to improve the amount of thrust, while some of the more thoroughly aquatic species, notably the seasnakes, have bodies and tails that are compressed from side to side.

Most typically used by climbing species, concertina locomotion involves gripping the substrate with the rear part of the body and the tail while reaching forward with the head and front part. Once the snake gets a new grip, the rear portion is drawn up and the process starts again. Some climbing snakes, notably the rat snakes of the genera *Elaphe* and *Pantherophis* have a pronounced ridge where the flanks meet the underside, giving them a better grip, especially on bark.

Snakes living on loose, windblown sand have an unstable surface to contend with. They cope with this challenge by "sidewinding," a method of locomotion in which the head and neck are raised off the ground and thrown sideways while the rest of the body anchors the snake. Once the head and neck are back on the ground, the rest of the body

their eyes are much larger. These species frequently stop and raise their heads off the ground in order to obtain a better view, demonstrating that vision is important to them. Nocturnal snakes tend to have vertical pupils (when viewed in daylight), which contract to a narrow slit in order to protect their sensitive retinas. About 10 species of tree snakes – two in Africa and eight in Southeast Asia – have horizontal pupils. These species also have long, narrow snouts, and the "wraparound" pupils allow the snake to peer forward with both eyes, giving good binocular vision and helping it to judge distances.

Snakes have no external ear openings, again thanks to their burrowing past. Internal parts of the ear are present, however, and it seems likely that snakes can detect sound in the form of vibrations transmitted through the ground. To some extent, they can also hear some airborne sounds.

For most snakes, smell is the most important sense for hunting, avoiding predation, and finding mates. It is hardly possible for us to imagine a world landmarked by smells, some of them produced by animals that passed by several days previously. This degree of sensitivity is helped by a specialized sense organ, known as Jacobson's organ, which is used, in partnership with the tongue, to detect airborne smells. When a snake is hunting, or any time that it senses a change in its environment, it flicks out its forked tongue repeatedly. Scent particles stick to the forked tip and are brought into the snake's mouth. The two tips are then inserted into the Jacobson's organ via a pair

of small pits in the roof of the mouth. The minute traces of scent are analyzed there, and the results are relayed to the olfactory part of the snake's brain.

Some snakes, including the pitvipers and most boas and pythons, have organs that can detect small changes in temperature, such as those given off by warm-blooded animals or even by other reptiles that have raised their body temperature by basking. These organs are lined with cells containing many thermoreceptors, each connected to the brain. In boas and pythons, the heat-sensitive organs take the form of a row of shallow pits along the snake's lips. In boas they lie between adjacent scales, while in pythons they are within the scales. They vary in number from one species to another and in some are lacking altogether.

The pitvipers' system is more advanced, with a single deep pit directed forward and positioned on either side of the head, just below an imaginary line drawn from the eye to the nostril. Each pit consists of two compartments, divided by a membrane. The inner compartment is connected to the outside world through a pore, opening just in front of the eye, which serves to equalize the pressure on either side of the membrane and to measure the ambient temperature. Using their pits, pitvipers can not only detect minute rises in temperature amounting to a fraction of a degree but can also pinpoint the position and range of an object by comparing the messages received on each side of the head, allowing them to strike accurately in total darkness.

and tail follow. Before the tail has touched the ground, however, the head and neck are thrown sideways again, resulting in a continuous, looping movement across the sand, with the snake moving at about 45° to the direction in which it is pointing. All the sidewinding snakes are vipers, but they are widely distributed, being found in the deserts of Africa and Central Asia as well as North and South America.

Burrowing snakes propel themselves through the substrate in a variety of ways. Species that live in loose soil or sand often push their way through it as though they were swimming, and the substrate collapses again after they have passed. Other snakes living in firmer soil use a complex of tunnels that they make themselves. Tunneling species often have a reinforced skull and a pointed or flattened head with a ridge around the snout. Their necks are not differentiated from their bodies, and they use muscular contractions to ram their heads through the soil, sometimes making side-to-side movements to compact the soil. Species that merely root around in the soil when looking for buried prey may have a upturned snout, as in the American hognose snakes of the genus *Heterodon* and some of the *Causus* African night adders.

Contrary to popular belief, snakes do not travel particularly quickly. A typical, medium-sized snake will probably move at just 4–5km/hr (2.5–3mph), a brisk walking pace for humans, while even rapid species, such as the African mambas, can only attain speeds of around 11km/hr (7mph). Even then, snakes quickly run out of energy and can only sustain fast movement over a short distance.

◐ **Left** *Most gartersnakes are terrestrial, but some live in freshwater streams or brackish inlets. They regularly leave the water to bask or to reproduce. Here, a Two-striped gartersnake* (Thamnophis hammondii) *swims through the shallows in Baja California, Mexico.*

◑ **Below** *Peringuey's desert adder* (Bitis peringueyi) *from the Namib desert of southwestern Africa, is one of the species of viper that moves by sidewinding. This mode of locomotion is the most efficient way of traversing shifting, burningly hot sand dunes.*

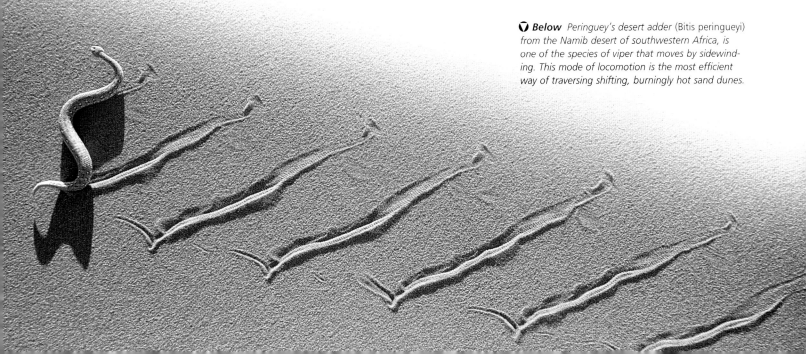

Swallowing Prey Whole

All snakes live off other animals. Among them, they eat almost every possible type of prey, from small invertebrates to large mammals. Their main limitation is the inability to dismember prey (because they have no limbs), so they must swallow it whole. Although their highly flexible skull and elastic skin allows them to engulf items that are considerably larger than their own diameter, there are limits. The type of prey therefore depends on the size of the species concerned as well as on availability.

Whereas some species are generalists, and eat more or less anything they can overpower, others are highly specialized and restrict themselves to a single type of prey. Most snakes will eat a variety of prey species within a single animal group: invertebrates, frogs, lizards, mammals, and so on. Specialists include species that eat only spiders (hook-nosed snakes, *Ficimia*) or centipedes (the Centipede snake, *Scolecophis atrocinctus*) or slugs and snails (*Dipsas* and *Sibon* in the Americas and *Pareas* in Asia) or birds' eggs (*Dasypeltis* in Africa and *Elachistodon westermanni* in India), but there are many others. Some species undergo a shift in diet as they grow, moving from small species to larger ones, for instance from lizards to mammals.

Snakes hunt in a number of ways. Some species simply live among their prey; these include the primitive thread snakes that reside inside termite colonies. Others actively search for prey, often employing strategies that give them advantages over specific prey species. For instance, nocturnal snakes may hunt diurnal lizards, moving through cracks or checking leaves and branches where

○ **Above** *The hunting strategy of juveniles of some snake species, such as the Cantil (Agkistrodon bilineatus), a Middle American pitviper, involves luring prey with a contrastingly colored tail tip. Other species to use this ploy include death adders (Acanthophis spp.).*

they are most likely to be sleeping. Others target their prey much more accurately; the cat-eyed snakes of Central and South America, for example, search for clutches of frog spawn attached to leaves overhanging small pools, while the egg-eating snakes of Africa necessarily seek out the nests of small birds.

The means by which snakes overpower and engulf their prey are closely linked with the prey type. The majority of small snakes, which feed on invertebrates and other small fare, simply grasp and begin swallowing. Snail-eating snakes have modified lower jaws that can be thrust into the snail's shell and used to winkle out the fleshy part of the mollusk. Egg-eating snakes have modified vertebrae that saw through the eggshell, allowing the contents to flow into their gullet while the empty shell is regurgitated. Frog-eating species similarly have little need for sophisticated methods of subduing their food – they simply swallow it live.

Species that eat animals – whether lizards, other snakes, birds, or mammals – that might be expected to fight back and cause damage need more advanced means of disabling prey. Snakes use two main methods of overcoming potentially harmful prey: constriction and envenomation. All boas and pythons are constrictors, as are the sunbeam snakes, dwarf boas, and several colubrids. These species kill their prey by coiling around it and preventing it from drawing breath; at the same time they restrict its circulation, sometimes hastening the outcome. Once the snake is satisfied that the prey is dead it loosens its coils slightly, searches for the head, and begins to swallow, pulling the carcass through the coils as it does so.

HOW A SNAKE SWALLOWS ITS PREY

Although some snakes have taken up a burrowing life and specialize in eating small insects and worms, many eat animals that are large in proportion to their own size. In fact, it would seem that one of the most important aspects of snake biology – and one that separates them from other animals – is their ability to capture and swallow large prey. This ability, coupled to their slow metabolism, gives snakes the advantage of not having to eat so often. Many eat less than once a week, and some only 8–10 times per year. Large boas and pythons can, if necessary, go for 12 months or more without eating.

Snakes' teeth are little more than sharply pointed, inwardly curving cones used to hold prey animals and drag them back into their esophagus. The jaw bones of all except those from the three most primitive families bear numerous teeth of this kind. In addition, the jaw bones articulate with the quadrates (movable bones on the back of the skull), so each bone can be moved individually up and down, back and forth, or from side to side. The two halves of the lower jaw are not fused at the front but are connected only by ligaments and muscles. Thus, each half of the lower jaw, as well as each of the six toothed

bones of the upper jaw and palate, can be moved independently.

The snake grasps the prey item, after having killed it in the case of venomous snakes or constrictors, and usually turns it so that it will pass down the gullet headfirst. Then the toothed bones of the jaw work alternately, "walking" the prey into the esophagus. The snake then forms a sharp curve in its neck behind the prey and pushes it down into the stomach.

The snake's body skin is flexible enough to let a large animal stretch it on the way down without tearing it. Snakes have dispensed with the pectoral girdle associated with front limbs in other vertebrates, so this does not impede the passage of food through the esophagus. Also, they have lost the sternum, which connects the front ends of the ribs in most animals. Without any obstruction, the prey can slip easily from the mouth to the stomach, with flexible skin and ribs spreading to make room for it to pass.

◁ **Left** *Specialized morphological adaptations allow the extraordinary Common egg-eater (Dasypeltis scabra) to engulf birds' eggs that are three or four times the diameter of its own head.*

Venomous snakes may strike and hold onto their prey until it dies, or they may strike and release it, preferring to go in search of the body after a short time has elapsed, thus avoiding any risk of being bitten or otherwise injured. The snake uses its tongue and Jacobson's organ to track the dying victim.

Disguise and Warning
DEFENSE

Just as snakes are predators, so they are also prey. Even the very largest boas and pythons are not immune from attack, although small to medium-sized snakes are the most vulnerable. Snake predators may be generalists, such as raccoons or crows, or specialists, like mongooses or secretary birds. Often, other snakes are their most significant predators.

The most valuable defense is to escape detection. Snakes are experts at concealing themselves, not least because they change their outline at will, from coiled to stretched out and anything in between. No other vertebrate has a parallel ability, which prevents predators from building up a search image. In addition, most snakes are the same colors as the substrate on which they live; as this may vary from place to place, the same species may occur in a number of geographically separate color forms or races. Markings, which may take the form of stripes, bands, spots, blotches, or irregular mottling and shading, break up the snake's outline when it is resting in its natural

habitat, although they sometimes seem gaudy and conspicuous when seen out of context in zoos.

Members of the Elapidae are frequently brightly colored, often with rings of red, black and white, or yellow arranged along their bodies. These are warning colors to discourage predators. When attacked, these species thrash around or move forward in short, jerky spurts, creating a kaleidoscope of color that may act as an optical illusion as well as a startle display. Species with coloration of this type occur in North, Central, and South America, in southern Africa, Southeast Asia, and Australia; most are called coral snakes.

Nonvenomous snakes may mimic these species, benefiting from the similarity. So-called

◑ Above The aquatic Green anaconda (Eunectes murinus) of South America can grow to 10m (33ft) or more. With its great girth and powerful muscles, it is able to subdue large animals, including other top predators such as caiman.

◐ Below Another formidable constrictor is the African rock python (Python sebae), which habitually feeds on ungulates, including Thomson's gazelle and even larger antelopes like the kob and impala. Because of the snake's low metabolic rate, such a huge meal means that it will not need to eat again for many weeks.

◗ Right *Among non-gregarious snakes such as certain species of rattlesnake, males compete during the short breeding season for access to females. Here, two Speckled rattlesnakes (Crotalus mitchellii) engage in a ritualized pushing contest to establish dominance. Such struggles may last for an hour or more.*

◐◗ Above and right *Since some venomous snakes' patterning appears to deter predators, mimicry by harmless species from the same habitat is widespread. For example, in North and Central America, numerous species and subspecies of milksnake (above, Nelson's milksnake* Lampropeltis triangulum nelsoni) *mimic the bold red, yellow, and black banding of venomous coral snakes (right, the Variable coral snake* Micrurus diastema). *Several other genera of colubrids have evolved the same strategy.*

"false" coral snakes are found notably in the American genus *Lampropeltis*, the king or milk snakes, as well as in several Central and South American colubrids. There do not appear to be any coral snake mimics in other parts of the world. Other harmless species that are chunky and have broad heads mimic vipers.

Once discovered, many snakes will try to make a hasty escape by sliding into crevices or dense vegetation, but, in extreme circumstances, some will turn defense into attack in the hope of deterring their enemy. These species may puff themselves up and hiss loudly, sometimes striking repeatedly. Cobras raise the front third of their body off the ground and spread a hood by rotating forward the elongated ribs in their neck region.

Rattlesnakes produce a loud, buzzing sound by raising and rapidly vibrating their tail, causing the segments of the rattle to click together. Sounds are produced in a different way by some desert vipers from North Africa and the Middle East. They have modified scales on their flanks, with serrated keels that are arranged obliquely. When threatened, they form a characteristic, horseshoe-shaped coil and, by moving one section of their body in the opposite direction to the adjacent one, the scales produce a loud rasping sound. The Common egg-eating snake has similar scales and behavior, flattening its head and making mock strikes while

rubbing its scales together. The spitting cobras of Africa and Asia spray venom through a small pore in the front of their fangs. The snake raises its head and forces the venom out at high speed and with great accuracy. The venom causes intense pain and temporary blindness if it enters the eyes.

Other snakes defend themselves by coiling into a tight ball with their head in the centre. One such, the Royal or Ball python, gets one of its names from this behavior. A small number of snakes feign death, flipping over onto their backs with their mouths agape and their tongue hanging out. This behavior may be accompanied by the production of a foul-smelling cloacal secretion, possibly suggesting decomposition.

▌ Sperm Storers
REPRODUCTION

Snakes' reproductive cycle varies according to habitat. Tropical species do not always have well-defined breeding seasons, although the onset of the rainy season may produce increased mating and courtship. Snakes from temperate climes usually mate in the spring, following a period of hibernation or reduced activity. Some species hibernate in groups, making mate-finding easier in the spring before they have dispersed. Mating in these species usually occurs immediately after the vernal shed and before feeding activity has begun,

although a few species mate later in the year. Otherwise, males use scent trails to track females and to recognize those that are receptive. In some species, including mambas, vipers, and some rattlesnakes, males compete for females by rearing up with their bodies entwined, each trying to force the other to the ground. These bouts may last for hours, with long breaks between periods of activity, and the victor will often mate immediately afterwards.

Snakes (in common with some lizards) are able to store sperm for long periods, giving them the option of mating at almost any time of the year and then delaying fertilization until a more suitable season. The gap may extend over a period of several months; for example, species that have a very short period of activity may mate in the autumn and store the sperm until the following summer. Females are also able to produce two or more fertile clutches of eggs from a single mating, an advantage for those species that are sparsely distributed over a uniform habitat and therefore have limited opportunities to meet and mate. Under typical conditions, females lay their eggs about 40 days after mating. Live-bearers have gestation periods that rarely last less than four months but that can be as long as 10 from fertilization to birth, not including the time the sperm is stored.

The factors that determine whether a species lays eggs or gives birth to live young are largely associated with its distribution, coupled with its evolutionary history. Very simply, species that live in cold places are more likely to bear live young than those that live in warm ones, because females that carry their eggs for the whole period are able to seek out the warmest places in which to bask and, therefore, to speed up the process of development. Egg-layers have to leave the development of their eggs to chance: not such a problem in warm climates, but more so at higher latitudes and altitudes.

Reproductive methods may also be influenced by family trends. All members of the Boidae except for one give birth to live young, for instance, whereas all members of the Pythonidae lay eggs. Similarly, the majority of vipers and pitvipers are live-bearers, whereas all members of the most primitive families, as far as is known, are egg-layers. On the other hand, some genera of snakes contain live-bearing as well as egg-laying species. For example, the Northern smooth snake, *Coronella austriaca*, gives birth to live young, whereas the southern species, *Coronella girondica*, lays eggs. Even more unusually, the method may even vary within a single species. The South American water snake, *Helicops angulatus*, and the African grass snake, *Psammophylax variabilis*, both have egg-laying and live-bearing populations; in every case, the live-bearers live in cooler parts of the species' range.

Clutch or litter sizes are largely independent of

whether a species is oviparous or viviparous, but are closely correlated with the species' size. Within a species, larger females produce larger clutches. The most prolific species include the two largest pythons, the Reticulated and the Burmese, which can both produce clutches approaching 100 eggs, and the African mole snake and the Puff adder, which are known to have produced clutches of over 100 live young. Most snakes, however, produce clutches or litters of about five to 20, and some of the very smallest species produce only one or two at a time. Tropical species may breed over an extended season, producing several relatively small clutches instead of a single large one. At the other extreme, species from cooler places, which can only be active for part of each year, may only breed every two or three years.

As far as is known, there is only a single species of snake that breeds without needing to mate. This is the Brahminy blindsnake, a parthenogenetic species originally from India and Southeast Asia. Every individual is a female and, as soon as it is mature, begins to lay fertile eggs that produce females that are clones of their mother. Because the snakes are small and easily overlooked, and any individual can begin to reproduce immediately on reaching maturity, this species has been accidentally introduced into many of the warmer parts of the world. It is especially prone to being transported inadvertently in the root balls and soil of potted plants and commercial crops such as rubber trees, earning it the alternative common name "Flowerpot snake."

▷ **Right** *A diver hunting seakraits off the Philippines. Tens of thousands of kraits are collected annually; their flesh is eaten, and their skins turned into goods for the tourist trade, such as shoes and bags.*

Running Out of Room
CONSERVATION AND ENVIRONMENT

The 20th century was not a good one for snakes, and indications are that the 21st will be even worse. The major problem, by far, is habitat destruction. Snakes are rarely able to move into better situations if their habitat is disturbed. Populations quickly become isolated, decimated, and, after quite short periods, eliminated. On a global scale, rainforest species are probably the worst affected, although virtually no figures are available. Those inhabiting deserts and marginal lands, however, are also losing habitat to the spread of agriculture, industry, and urbanization, as well as suffering from the knock-on effects of pollution and increased traffic spawned by these activities.

Species living on islands are especially vulnerable to habitat loss and to the introduction of predatory or competing species. The unique family of Mascarene boas, for example, were restricted to Round Island by the turn of the 20th century, having already been eliminated from neighboring Mauritius by introduced rats and possibly from other islands in the region. After the introduction of rabbits and goats caused soil erosion, the habitat on their final refuge was largely destroyed. Of the two species in the family, one, *Bolyeria multocarinata*, a burrowing snake, seems to have gone extinct around 1975, when the last specimen was seen. The other, *Casarea dussumieri*, was in danger of following it into oblivion when it was pulled back from the brink thanks to some desperate measures, including the elimination of goats and rabbits and a captive breeding program; even so, its future will not be assured for some considerable time to come. Other vulnerable island species include the Milos viper from the Greek island of the same name, the Aruba rattlesnake, and a number of West Indian racers that are either extinct or nearly so.

In addition to habitat destruction, but fairly insignificant by comparison, are the problems caused by widespread prejudice against snakes, which results in large numbers being killed for no good reason. Staggeringly high numbers are also run over by traffic each year. Additional threats, perhaps more easily controlled, include collection for the skin and pet trades.

Apart from a small element of captive breeding, measures to help snakes survive are, in the main, still at an early stage. Conservationists are still working to acquire a better understanding of their requirements, often by longterm ecological studies using radio telemetry. Natural areas in which snakes live may be preserved, but this is usually incidental to the conservation of other, more conspicuous organisms. Educational programs pointing out the esthetic and economic value of wild plants and animals, including snakes, will perhaps be the most important contribution toward their continued survival.

VENOMOUS SNAKES
Toxin evolution and delivery mechanisms

VENOMS – BIOLOGICAL TOXINS SECRETED IN specialized glands and injected into other organisms by a variety of methods – are relatively rare in the animal kingdom, but are most prevalent by far among snakes. Representatives of around 200 genera of snakes, all belonging to "advanced" families, are thought to possess venoms to at least some extent. They include all vipers, all elapids (i.e. cobras, sea snakes, and Australian venomous snakes) and several colubrids. What factors determined the evolution of these complex toxins in snakes, and what are the mechanisms by which they produce and deliver their venoms?

The cost in human deaths and injuries from snakebites (perhaps as may as 100,000 people are killed annually; see The Threat from Snakebites) has inevitably colored our perception of these creatures. However, any notion that snakes harbor some "aggressive" intent toward human victims can be dispelled forthwith; snakes are reactive, and only defend themselves by biting when they perceive a threat and can find no means of escape. Injection of venom into people is quite incidental to the primary role of this weapon in the snake's armory, and to stress the human attrition rate from interactions with snakes is to misconstrue the evolutionary purpose of venom. Biologists believe that snake venoms evolved in response to the need to subdue a wider range of prey items. Envenomation

debilitates and disorients larger, potentially dangerous animals, while hard-to-get-at prey hiding in burrows or crevices can be paralyzed and extracted. While this offensive, prey-subjugating capacity is clearly the principal function of venom, it is by no means its sole purpose.

The biochemical origins of venoms have been traced to digestive juices, pointing to another important function of venom – the predigestion of prey. This is especially true of the powerful tissue-destroying enzymes present in the venom of viperids, notably phospholipase (PL) A_2. For example, within 24 hours of a Western diamond-backed rattlesnake (*Crotalus atrox*) ingesting a prey item, its venom will typically have digested the skin, exposed the body cavity, and begun to break down the internal organs. Viper envenoming in humans often results in severe tissue necrosis around the bite site.

A defensive role, deterring the approach of predators, may also have had a strong influence on the evolution of venom in certain species. This deterrent factor has not been extensively investigated, but is well attested in the spitting cobras (*Hemachatus*, plus some *Naja*), which use their venom primarily for this purpose. It is intriguing to speculate whether the development of particularly powerful toxins in some snakes might also be linked to a defensive strategy. Certainly, the

potency of the venom of the Inland taipan (*Oxyuranus microlepidotus*), an Australian elapid, which discharges enough in a single bite to kill 200,000 mice, is out of all proportion to its habitual prey of rodents and small marsupials. If its powerful venom gives it no feeding advantage, then might it have evolved for another purpose?

Other snakes have evolved anatomical modifications such as sounds, colors, and behavior patterns that warn possible predators of their dangerous nature, and this in turn has led to Batesian mimicry of venomous snakes by nonvenomous species. This phenomenon is exemplified by the striking similarity of two Central American species, the False terciopelo (*Xenodon rabdocephalus*) and the highly venomous Terciopelo (*Bothrops asper*). In this case, one external feature – the round pupils of the former species – helps distinguish it from the latter. However, this is by no means a reliable rule-of-thumb in telling innocuous species from venomous ones; all elapids have round pupils, while many nonvenomous such as boas and pythons have elliptical pupils.

Venom production and delivery mechanisms vary considerably. Snake venoms are produced by head glands located in the upper jaw or temporal region (except in some stiletto snakes,

◁ **Left** *A Black-necked spitting cobra* (Naja nigricollis nigricollis). *Eleven cobra species with specially modified fangs have the capacity to "spit"; the venom of this species can cause permanent blindness.*

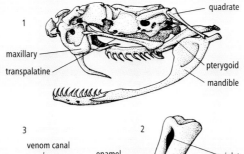
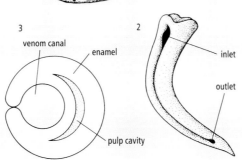

Below *Fang equipment.* **1** *Rattlesnake skull, showing specialized teeth in the upper jaw.* **2** *Fang of rattlesnake showing inlet for the poison and its outlet near the tip.* **3** *Cross-section of a cobra's fang, showing cavity through which the poison in conducted.*

Above *The primary role of snake venoms is to immobilize prey. Here, a Terciopelo subdues a Rainforest whip-tailed lizard (Ameiva festiva).*

some elapids, and some night adders, which have long, tubular glands running down the body that are extensions of the head glands). The type of gland differs according to group. True venom glands, characterized by a spacious, central venom storage chamber (the lumen) connected by a duct to a single fang, are found in all viperids and elapids, as well as some atractaspidids (Stiletto snakes). Most colubrids, on the other hand, possess Duvernoy's gland, which rarely has a lumen, and may be attached to several of the posterior teeth; however, a few of the more venomous colubrids have a mechanism that more closely resembles a true venom gland.

There are also different types of venom-conducting teeth. In the front-fanged snakes, they are located in the front part of the upper jaw, and the grooves or canals that transport the venom are closed for at least part of their length. In vipers and atractaspidids the fang can be folded against the roof of the mouth. The fangs of these snakes are proportionally longer than those of elapids, except for a few Australian venomous snakes. (The longest fangs are those of the Gaboon viper, which measure 2.9cm (1.2in) in a 1.3m (51in) snake.) In many colubrids, dental modification simply takes the form of an enlargement of one or two pairs of back teeth; in others, these teeth are equipped with grooves on the front or sides.

Venom storage capacity is greatest in vipers and some cobras. The amount of venom that can be obtained from such snakes by extraction, or "milking," may be as much as 6–7ml (0.2–0.24fl oz; with a dry weight of about 1.5g/0.05oz). Unsurprisingly, given their lack of a lumen, the colubrids produce the smallest yields, amounting to just a few microliters (with a negligible dry weight). This family, which makes up almost two-thirds of all snakes, used to be categorized as the "harmless snakes." Yet just as there is no simple correlation between venom toxicity and size of prey, so is there none between venom yield and toxicity. Research following the deaths of eminent herpetologists from accidental colubrid bites revealed that the venom of some species is far more harmful than was once supposed. Gram for gram, the venoms of the Boomslang (*Dispholidus typus*), two *Thelotornis* species (*T. capensis*, the Savanna Twigsnake, and *T. kirtlandii*, the Birdsnake), and the Yamakagashi (*Rhabdophis tigrinus*) are even more toxic than that of the Black mamba.   HWG/SAM

Left *Snake venom can benefit humans: a compound synthesized from the venom of the Jararaca (Bothrops jararaca) is used to treat high blood pressure, heart failure, and kidney disease in diabetics.*

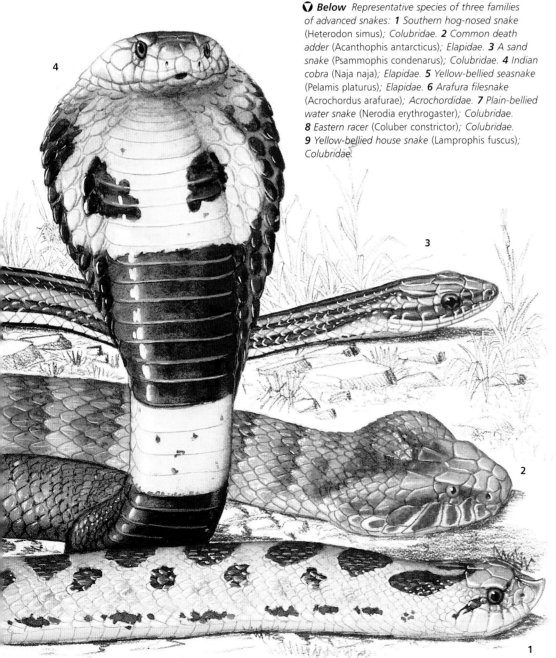

⬡ *Below* Representative species of three families of advanced snakes: *1* Southern hog-nosed snake (Heterodon simus); *Colubridae*. *2* Common death adder (Acanthophis antarcticus); *Elapidae*. *3* A sand snake (Psammophis condenarus); *Colubridae*. *4* Indian cobra (Naja naja); *Elapidae*. *5* Yellow-bellied seasnake (Pelamis platurus); *Elapidae*. *6* Arafura filesnake (Acrochordus arafurae); *Acrochordidae*. *7* Plain-bellied water snake (Nerodia erythrogaster); *Colubridae*. *8* Eastern racer (Coluber constrictor); *Colubridae*. *9* Yellow-bellied house snake (Lamprophis fuscus); *Colubridae*.

Snake Classification

TAXONOMY

The classification of snakes is not fixed, but at present most experts recognize 18 families divided into two broad groups: the Scolecophidia, or blindsnakes, and the Alethinophidia, containing all the remaining species. The Scolecophidia form a relatively homogenous group of just three families. The relationships within the Alethinophidia, however, are much less certain and there is considerable debate about them. Most taxonomists agree that the vipers, colubrids, burrowing asps, and the snakes of the cobra family (Viperidae, Atractaspididae, Colubridae, and Elapidae respectively) form a distinct subdivision, and there is a fairly well-established relationship between these four families and the filesnakes (Acrochordidae). Together, these five families are sometimes grouped as the Caenophidia or advanced snakes.

Among the other Alethinophidia families, boas and pythons form a distinct subdivision (and were placed in the same family until quite recently). The other eight families are all small, some containing only one or two species and all are very restricted in distribution. At least some appear to be intermediate between the primitive Scolecophidia and the higher snakes, but their exact evolutionary status has not been positively established. Others may be the results of evolutionary dead ends. Members of some of the families are rarely collected and are represented in museums by a mere handful of specimens. Classification is therefore tentative, and a conservative approach has been taken.

Blindsnakes

FAMILIES ANOMALEPIDIDAE, TYPHLOPIDAE, LEPTOTYPHLOPIDAE

The Scolecophidia contains three families of primitive snakes, all very similar to each other. They are the Anomalepididae, Leptotyphlopidae, and Typhlopidae, known collectively as the blindsnakes, although the names wormsnakes and

threadsnakes are also used for some or all of them. Their combined total is about 400 species, or 15 percent of all snakes, but they are poorly known, partly because their small size and secretive lives make study difficult. The Anomalepididae are confined to South and Central America, but the other two families have a much wider distribution, especially in the southern hemisphere. Only a single species from the family Typhlopidae, *Typhlops vermicularis*, is found in Europe. The Leptotyphlopidae are absent from Europe and Australasia. New species are described regularly, sometimes from just one or two specimens collected from isolated places. Most blindsnakes turn up accidentally while biologists are searching for other organisms.

Although the three families differ from each other, notably in the structure of their skulls and the arrangement of their teeth, they share several common characteristics. All have very rigid skulls that they use to push their way through soil. Members of all three families have pelvic girdles, and some have small spurs at either side of their cloaca, the rudiments of legs that have been reduced through the evolutionary process. They do not have specialized scales on their heads or undersides; every part of their body is covered in small, smooth scales with rounded edges (cycloid scales). They are mostly slender – hence the name "threadsnakes" – and cylindrical in shape, both adaptations to a burrowing lifestyle. Some species lack eyes altogether, whereas others have small eyes covered by scales. Their most important sense is that of smell. Being burrowers, they have little use for pigment, and most are pink, pinkish brown, or buff, although a few have striped or mottled markings.

One species, Schlegel's blindsnake, grows to almost 1m (3.3ft) in length, and there are a couple of others that approach this size, but they are exceptional; the majority of species are under 30cm (12in) in length, and some barely reach 10cm (4in). All the world's smallest snakes come

from these families. So far as is known, all blindsnakes lay eggs, although those of one species, Bibron's blindsnake, hatch only five or six days after laying. Another African species, Diard's blindsnake, appears to be viviparous in part of its range at least, but this fact is still unconfirmed. Clutch sizes range from one in some very slender species to 60 in the larger ones. Texas blindsnakes, and perhaps other species, stay with their eggs during incubation, although whether this occurs coincidentally, because the females lay their eggs where they happen to be living, or whether there is an element of parental care, is difficult to say. Many species feed almost exclusively on ants and termites and may live most of their lives inside termite mounds, where they protect themselves from attack by chemical secretions. They may swallow entire insects, eat only the soft abdomen, or squeeze the soft parts out before

discarding the remainder of their bodies. Their main predators are probably other snakes but, being wormlike in appearance, they must certainly be eaten by all manner of birds, lizards, frogs, and even large insects and spiders.

Pipesnakes and Shield-tailed Snakes
FAMILIES ANILIIDAE, ANOMOCHILIDAE, CYLINDROPHIIDAE, UROPELTIDAE

Three small families of snakes are popularly known as pipesnakes. They are the Aniliidae, Anomochilidae, and Cylindrophiidae. All three are small families, containing one, two, and ten species respectively. All are burrowing snakes with cylindrical bodies and smooth, shiny scales, and all retain the pelvic girdle and have small vestigial limbs in the form of spurs.

The South American or Red pipesnake, *Anilius scytale*, is the sole member of the Aniliidae. It lives

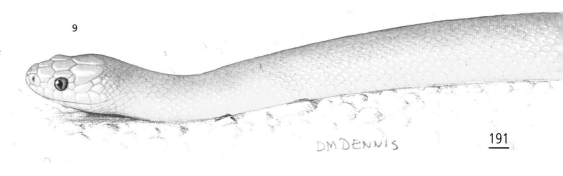

⚫ *Above* *The Red-tailed pipesnake (Cylindrophis ruf-fus)* has a blunt, stout skull appropriate to its burrow-ing lifestyle. It specializes in eating long-bodied prey such as other snakes, eels, and caecilians, sometimes of a size almost equaling its own.

in the Amazon Basin and is a false coral snake, being boldly marked with red and black rings. It has a heavily-built skull, and its eyes are small and covered with large, hexagonal head shields. It lives below the surface, in leaf-litter or damp soil, sometimes occurring in fields and other disturbed habitats. It eats other snakes and worm-lizards, and females give birth to live young.

The other two families occur in Southeast Asia. The Asian pipesnakes, Cylindrophiidae, live in tunnels in damp soil but emerge at night to hunt for food, including other snakes and eels. Their tails are flattened from top to bottom and, when threatened, they lift them up and curl them over their backs to expose the underside, which may have red or bold black-and-white markings. This behavior deflects the attention of attackers away from their head, giving the snakes vital seconds in which to escape. Their dorsal surfaces are brown, and some species have black markings. They give birth to live young.

The Anomochilidae, or dwarf pipesnakes, are among the most enigmatic of all snakes. Although two species are known, from West Malaysia, Borneo, and Sumatra, less than 10 specimens have been found altogether. They are small, with blunt heads and short tails, and are almost certainly burrowers. They have small mouths, and probably feed on earthworms or other soft-bodied inverte-brates; they are thought to lay eggs. Anatomically, they seem to be intermediate between the blind-snakes and the more advanced snakes, although this may be purely coincidental. Until relatively

recently, the two species in the family were includ-ed within the Cylindrophiidae.

The shield-tailed snakes, Uropeltidae, seem to be closely related to the pipe snakes and share with them a burrowing lifestyle and all the adapta-tions that go with it; they lack a pelvic girdle, how-ever. There are 47 species altogether, unevenly shared among nine genera; two genera account for 35 species, and the other seven contain one to four species each. The snakes are found in Sri Lanka and near the very southern tip of India, especially in the Western Ghats, and they range in size from about 15 to 40cm (6–16in).

Most species are poorly known, although some are apparently common. They live in cool, forest-ed areas, or where the soil is loose enough for them to burrow down to avoid the hot surface lay-ers during the day. They are sometimes found when land is plowed or when heavy rains flood their burrows, forcing them to the surface; other-wise they are rarely seen. Like the Asian pipe-snakes, shield-tails raise the tip of their tail when they are molested, while hiding their head among their coils. The neck and front part of the body are very muscular. When burrowing, they brace the body against the sides of the tunnel and use the neck muscles to ram the thick, rigid skull through the soil. Once the head has reached the end of its travel, it is moved sharply from side to side to enlarge the tunnel and compact the sides. Then the head, in turn, is braced against the tunnel walls so that the body can be dragged up behind it. Once the snake has begun to burrow, it is rela-tively safe from attack because its tail ends in a modified scale, or shield, that effectively blocks the tunnel behind it. The shape of the shield varies among species, but it is invariably large and has a rough surface. In some species it is an oblique, disklike scale ringed with spines, whereas

others have a thick, conical tip to their tail. Shield-tails are thought to eat earthworms; most species – possibly all – give birth to live young.

Sunbeam Snakes

FAMILIES XENOPELTIDAE, LOXOCEMIDAE

Two small families, the Xenopeltidae and the Lox-ocemidae, are commonly called sunbeam snakes because their scales are highly iridescent. They are superficially similar to each other even though they are well separated geographically.

There are two Asian sunbeam snakes, *Xenopeltis unicolor* and the more recently described *X. hainan-ensis*. Both grow to just over 1m (3.3ft), and have

⚫ *Above* As its common name indicates, the *Madagascan tree boa (Boa manditra)* is an arboreal species, maneuvering its relatively slender body through the forest canopy to capture lemurs.

◗ *Right* Brazilian rainbow boas (Epicrates cenchria cenchria) *display a beautiful iridescent sheen on their scales. Other species with this trait are Xenopeltis sun-beam snakes and Ringed pythons (Bothrochilus boa).*

large, smooth, highly polished scales and a flattened head covered with large platelike scales. They have wide ventral scales but no pelvic girdles or vestigial hind limbs. They are dark gray above, white below. They live in burrows in moist soil, often beneath logs, palm fronds, or, in disturbed habitats, human rubbish. They eat almost anything of a suitable size, including lizards, snakes, frogs, and small mammals, and they lay eggs.

The Loxocemidae contains a single species, the Mexican burrowing snake, *Loxocemus bicolor*. It is about the same size as *Xenopeltis*, and it too is dark gray in color, often with irregular, scattered patches of white scales. Unlike the Asian species, it has vestigial hind limbs, and its scales are smaller and less glossy. It lives in forests, possibly in burrows or otherwise hidden, emerging to forage at night. Its food includes small mammals and reptiles, but it is also know to raid the nests of iguanas and even sea turtles. It lays small clutches of large eggs.

Boas, Pythons, and their Relatives
Families Boidae, Pythonidae, Tropidophiidae, Bolyeriidae

In the past, the boas, pythons, and two smaller groups, the tropidophiids and the Mascarene boas, were often regarded as a single family. Now they are split into four: the Boidae, Pythonidae, Tropidophiidae, and Bolyeriidae. They share several characteristics, including pelvic girdles and vestigial hind limbs. Most have a functional left lung (not found in more advanced snakes), and many have heat-sensitive pits in the jaws. They are all powerful constrictors and, although there are semi-burrowing species, they are mostly terrestrial, with a few species that have become arboreal. Between them, they contain the world's six largest snakes.

The first two families are by far the best known. The Boidae consist of 36 species in 11 genera, living in North and South America, Africa, Madagascar, Europe, and the Pacific region. Although some, such as the anaconda, are enormous, there are many small to medium-sized boas as well; most species are less than 2m (6.6ft) in length. The family is divided into two subfamilies, which are quite distinct. The "true" boas of the Boinae include the largest species, such as the anaconda and the Common boa. Anacondas are semiaquatic, living in thickly-vegetated swamps and flooded forests, but the other species are ground-dwellers or climbers. The tree boas (*Corallus*) are slender with long, prehensile tails, and their bodies are compressed from side to side, like iron girders, so that they can support themselves when they reach across from one branch to another. They typically rest on horizontal boughs in a characteristic coil, and hunt by hanging the heads down and making a single S-shaped coil in the necks so that they can strike rapidly. Their teeth are very long and curved backward for gripping, and they eat mainly roosting birds.

Several other boas, including the Common boa and the anaconda, also climb, but they are not as highly adapted as the tree boas. There are 11 boas belonging to the genus *Epicrates* living on various West Indian islands, some of which are rare due to habitat disturbance and the introduction of predators. They vary greatly in shape and size, ranging from large, heavy-bodied forms such as the Cuban boa, which can reach 3m (10ft) in length, to the slender Vine boa from Haiti. All "true" boas give birth to live young, and the litter size in some of the larger species can be over 50. They feed mostly on mammals and birds, killing their prey by constriction. Several appear to specialize in catching bats in their roosting caves, and some also eat lizards. A number of species switch diet from frogs and lizards to mammals as they grow. The Green anaconda also eats caiman and turtles. Many boines have heat-sensitive pits in the scales bordering their mouths, although these are absent in members of the genera *Eunectes* (anacondas), *Boa*, and *Candoia* (Pacific boas). They are most obvious in the tree boas.

The other subfamily, the Erycinae, contains the Rosy and Rubber boas from North America, the Burrowing boa from West Africa, and the sand boas of Africa, Asia, and Europe. They are all small, with thick bodies, short heads, and blunt

◐ **Above** Boas and pythons are often mistakenly thought of as solely equatorial rainforest dwellers. Many species, however, inhabit dry areas of the world; this Bredl's carpet python (Morelia bredli) sits high above a gorge in the arid interior of Australia.

◐ **Below** The Reticulated python (Python reticulatus) of Southeast Asia can grow to more than 10m (33ft). These giant snakes have been documented as occasionally devouring humans. Note the large infrared pits along the snake's upper lip.

COURTSHIP AND AGGRESSION IN SNAKES

Snakes normally appear to be oblivious to others of their kind but when the mating season arrives their attitude changes. Male snakes become aggressive toward one another, and the scent of an attractive female may send them into a frenzy. Within a species, not every female is conditioned to breed in any given year but, if food has been abundant, perhaps half the adult females will have mature eggs in their oviducts. These "ripe" females exude a chemical secretion, or pheromone, signaling that they are ready. They are more attractive to males just after they have shed their skins. In gregarious species, such a female may attract numerous courting males. In species that live as lone individuals, however, her scent trail may be picked up by only a single tongue-flicking suitor.

In either case the male, upon finding the female, moves forward to bring his chin onto the nape of her neck, with his body overlapping or alongside hers, pressing against her along his entire length. The tongue flicks become more rapid, and are nearly continuous as his body approaches hers. He rubs his chin over her nape and twitches his body, while trying to force his tail under hers. The female may also respond by twitching if she finds him suitable and will raise her tail and open her cloaca for the insertion of one of the male's two hemipenes. After it is inserted, the pair may remain mated for between 10 minutes and 24 hours, depending on the species.

In boas and pythons, another element is introduced, with the male using his "spurs" (rudimentary pelvic limbs) to scratch the female in the cloacal region as an additional stimulus. The males of some of the advanced snakes, which lack spurs, may also grasp the female's neck in their jaws.

In several species of snake, both large and small, many males may compete to mate with a single female. Sometimes they are so numerous as to form a "ball" with their intertwined bodies (BELOW a breeding ball of Green anacondas). Male aggression in these species amounts to no more than attempting to push one another from the mating position. In natricine colubrids, the male leaves a "plug" of waxy, solidified fluid in the female's cloaca after mating so she cannot be mated by additional males, whose sperm might otherwise displace that of the original male. Then he is free to go off in search for more females with which to mate.   HGD

tails, and none of them have heat pits. Erycine boas are mostly burrowing species and many live in dry areas, including sandy deserts. The Rubber boa, the smallest species, is an exception because it lives in the cool montane forests of the Pacific northwest. The West African ground boa is also a forest species, and is unusual in laying small clutches of very large eggs. It has recently been suggested that the Arabian sand boa also lays eggs, but otherwise erycine boas give birth to live young, usually numbering between three and ten.

Pythons, the Pythonidae, are only found in the Old World, with four species in Africa, six in Asia, and the remaining 20 in New Guinea and Australia. The largest species is the Reticulated python, which may reach 10m (33ft) or more in length. Next come the African, Indian, Amethystine, and Oenpelli pythons; these species can potentially reach 5m (16ft) in length, although they rarely do so. The smallest species is the Anthill python, *Antaresia perthensis*, which rarely grows to more than 30cm (12in).

The Green tree python, which is superficially similar to the Emerald boa, is exclusively arboreal, but most pythons are generalists. Some species are restricted to forest habitats whereas others are found in grasslands, open or lightly wooded country, wetlands, or, in the case of some Australian species, deserts. As its name suggests, the Anthill python sometimes turns up in ant and termite nests, probably searching for geckos. Otherwise, pythons feed mainly on small mammals and birds, with juveniles also eating lizards. Like the boas, they are all powerful constrictors. Most pythons have heat-sensitive pits bordering their mouths, but two Australian species of the genus *Aspidites* lack them.

Pythons are all egg-layers, and females of all the species that have been studied (which is nearly all of them) coil around their eggs throughout the incubation period. At least one species, the Indian python, can raise the temperature of her clutch, if necessary, by making a series of rhythmic twitching movements. Clutch sizes vary in rough proportion to the adults' size; the Reticulated python may lay up to 100 eggs, whereas the Anthill python rarely lays more than two or three.

Closely related to the boas and pythons, the Mascarene boas are placed in the family Bolyeriidae. Two species are known from Round Island, north of Mauritius, but one of them, *Bolyeria multocarinata*, is probably extinct, none having been seen since 1975, and the other, *Casarea dussumieri*, is also rare. This species lays eggs.

The Tropidophiidae, or dwarf boas, comprise 21 species divided into two subfamilies, but the differences between species are slight. These are mostly small snakes, although one species can reach 1m (3.3ft) in length. Most species are placed in the genus *Tropidophis*. They occur primarily in the Caribbean region, especially on Cuba, although three species live in South America. They inhabit forest floors and disturbed habitats such as fields, and are mainly active at night. Other members include two eyelash boas and two "banana boas," so named because they occasionally turn up in shipments of bananas. The remaining species is very rare and lives in cloud forests in Central Mexico. Ttopidophiids apparently eat most types of prey, including invertebrates, and are constrictors. They give birth to very small live young, except for *Trachyboa gularis*, which is an egg-layer.

Filesnakes
FAMILY ACROCHORDIDAE

There are only three species of filesnakes, and they are among the most peculiar of all snakes. They are totally aquatic, living in freshwater lakes and rivers, estuaries, and coastal waters, and have several physical and behavioral adaptations suiting them to this lifestyle. They measure from 1 to 3m (3.3–10ft) in length, depending on species and sex; males are significantly smaller than females. Their skin is loose and baggy, especially when they are taken out of the water. They are

◗ **Right** *A mating pair of Asian green vinesnakes (Ahaetulla prasina). With well-camouflaged bodies little thicker than a pencil and keen eyesight, these agile colubrids lie in wait among creepers and vines to ambush skinks and young birds.*

covered with numerous small, wartlike scales that do not overlap and that are covered in tiny bristles, making them rough to the touch. Their heads are also covered with granular scales, and their eyes are small and beadlike.

The Little filesnake has dull black-and-white markings, but the other two are almost uniformly gray or grayish-brown in color; the Javan filesnake is sometimes called the "elephant trunk" snake. The smallest species may eat marine crustaceans, but otherwise they eat fish, including eels. The bristly scales help them grasp slimy prey in their coils. They all give birth to live young – an essential adaptation for entirely aquatic species – but they have a remarkably slow rate of reproduction, appearing to breed only every 8–10 years.

Stiletto Snakes and their Relatives
FAMILY ATRACTASPIDIDAE

The 70 or so species now included in the Atractaspididae have been a thorn in the side of classifiers for many years. Some or all of them have, at various times, been placed in the Colubridae, Elapidae, and Viperidae. Now they are tentatively combined into a family of their own. Because various species and genera have been linked together and separated again several times, it is hardly surprising that common names have proliferated. Many, such as "mole vipers" and "burrowing asps," are taxonomically misleading but widely accepted.

Atractaspidids, as currently understood, include snakes with very different fang arrangements – venomous species with long, hinged front fangs, or short, fixed front fangs, or grooved rear fangs, and some species have no venom-delivering fangs at all. These diverse systems have likely evolved in response to the specialized diets of various species.

With a single exception from the Middle East, the atractaspidids occur in Africa. They live underground, beneath rocks, in loose soil, leaf-litter, or sand, or in burrows made by other animals or of their own construction. In keeping with their burrowing habits, they have smooth, shiny scales and long, cylindrical bodies. They eat a range of prey, including earthworms, centipedes, other burrowing reptiles, and small mammals. Among the more interesting feeding habits are those of the stiletto snakes, *Atractaspis*, which have two long, hollow, venom-delivering fangs on their upper jaw. The snakes can bring these into play by rotating them sideways so that they emerge from the side of its mouth. They strike by jabbing one of the fangs backward into prey, and can strike successfully in the confines of a burrow. As far as is known, all the atractaspidids lay eggs with the sole exception of *Aparallactus jacksoni*, although the breeding habits of several species are not known at present.

Colubrids
FAMILY COLUBRIDAE

The Colubridae is a massive and confusing family of over 1,900 species (comprising over 60 percent of all snakes). Such a large group defies any attempt to describe it in general terms, and there have been many attempts to break it down into smaller, more meaningful units. Although some genera have obvious close affinities with one another, and several distinct subfamilies are recognized, the relationships of others are unclear. At present, then, the family is divided into a number of subfamilies, some (or possibly all) of which will eventually be promoted to full family status; some may even give rise to more than one full family in time. To further complicate the situation, none of the subfamilies are easily assigned common names.

As presently understood, the "typical" colubrids of the subfamily Colubrinae include over 650 species with a worldwide distribution, among them many of the more familiar species from North America and Europe, such as the kingsnakes and milksnakes (*Lampropeltis*), the rat snakes (*Elaphe, Pantherophis*, etc.), and the whipsnakes and racers (*Masticophis, Coluber,* etc.). They range in size from less than 20cm (8in) to over 3m (10ft) and include long, slender "racers" that chase down lizards and other snakes as well as species with muscular bodies that stalk and constrict birds and mammals. Some species have very specialized diets; these include the African egg-eaters (*Dasypeltis*), the hook-nosed snakes (*Gyalopion*), which feed largely on spiders, and the

Central American centipede-eating *Scolecophis*.

In their habitat preferences they are equally diverse; they occur almost everywhere, from deserts to wetlands. A few species venture into brackish estuaries and mangrove forests, but there are no marine species. They may be burrowers, sand-swimmers, ground-dwellers, or tree-climbers. There is, however, some consistency in their reproductive habitats: the vast majority of species in this subfamily lay eggs, ranging from one to 40 or more.

Two groupings currently included in the Colubrinae are sometimes considered as separate subfamilies. The 50-odd reed snakes ("Calamarinae") are small, burrowing Asian species with smooth scales, which eat mostly earthworms. The sand snakes ("Psammophiinae") are slender, whiplike, fast-moving, diurnal hunters from Africa and Europe, with rear fangs. This group includes about 35 species.

The homalopsine water snakes (Homalopsiinae) form a well-defined group of 35–40 species from southern and Southeast Asia and Australasia. They are all well adapted to an aquatic lifestyle, having crescent-shaped nostrils that they can close and upwardly-directed eyes on top of their heads. Most species belong to the genus *Enhydris*, small snakes with shiny scales that live in well-vegetated freshwater lakes, swamps, and flooded fields. The Tentacled snake, *Erpeton tentaculatus*, has unique, fleshy "tentacles" on its snout that it may use to navigate in murky water, although their exact function is not known. Its scales are heavily keeled and its body is flattened from top to bottom and almost rectangular in cross-section. All homalopsines have enlarged venom fangs at the back of their mouths with which they grasp and envenomate their prey, consisting mainly of fish and frogs. The majority probably lie in wait for their prey, but the White-bellied mangrove snake (*Fordonia leucobalia*) from Australia eats crabs that it catches on exposed mud flats at night. It apparently pins large crabs down with its body, then tears off their limbs. This makes it perhaps unique among snakes in not swallowing its prey whole. All the homalopsines give birth to live young.

The Asian slug- and snail-eating snakes form another well-defined group and are placed in the subfamily Pareatinae. There are 15 species altogether, 11 of them in the genus *Pareas*. They are all very slender, with wide heads and large eyes; the Montane slug-eater has deep red eyes. Several species live in trees and shrubs, where they feed on tree snails, and all are nocturnal, hunting apparently by following slime trails. They have modified skulls and long, sharp teeth in their lower jaws that allow them to wrench the soft bodies of snails from their shells. The Blunt-headed tree snake, *Aplopeltura boa*, may also eat lizards. All of the pareatines lay eggs.

The subfamily Lamprophiinae consists of about 45 species from the southern half of Africa and the island of Madagascar. The most familiar species are probably the house snakes of the genus *Lamprophis*, of which there are 14 species, including one on the Seychelles. The widely distributed Brown house snake is a powerful constrictor that feeds on small rodents, but other

species also eat lizards, some exclusively. The African water snakes are slender, glossy snakes that feed on fish, while the triangular-shaped file snakes (not to be confused with the acrochordid filesnakes) are snake-eaters. With one or two exceptions, the lamprophiines lay eggs.

The natricine colubrids, Natricinae, are a large group of about 200 species occurring in North America, Europe, Africa, and Asia. They tend to be semiaquatic in habit and eat fish and amphibians, although the Queen snake (*Regina septemvittata*) eats freshly-molted freshwater crayfish. They are active, fast-moving species that often hunt by sight during the day. The species that are not

⬤ **Above** *Feigning death by lying limply with the mouth open and tongue protruding (as in this Grass snake, Natrix natrix) is a strategy that some snakes use to escape the attentions of predators. "Playing possum" may cause the predator to cease its attack.*

⬤ **Below** *Colubrids were once inaccurately referred to as "harmless snakes." However, the Boomslang (Dispholidus typus), whose common name means "tree snake' in Afrikaans, is a swift arboreal snake with an extremely toxic venom.*

associated with open water, such as the Grass snake (*Natrix natrix*) prefer damp habitats, and some of them are nocturnal. Reproduction within this subfamily varies; the North American species all give birth to live young, whereas the Old World species are, with a very few exceptions, egg-layers. They include such well-known species as the gartersnakes, American watersnakes, and European and Asian grass- and watersnakes. Some systematists have suggested that the New World and Old World species should be placed in separate subfamilies.

The Xenodontinae are a varied collection of snakes that inhabit North and South America. Between them they occupy a wide range of habitats and eat a variety of prey. Some species are specialist hunters; the hog-nosed snakes (*Heterodon*) eat mainly toads, for instance, while the mussuranas (*Clelia*) eat other snakes. Some species are false coral snakes and some have rear fangs that may inject venom, although none is

○ **Below** *Native to Central and South America, the Brown blunt-headed vinesnake* (Imantodes cenchoa) *is a slender, nocturnal hunter that uses its large eyes and sensitive tongue to locate anole lizards. In turn, it is itself preyed upon by voracious leptodactylid frogs.*

considered dangerous to humans. They are all egg-layers.

A number of Central American genera that were formerly included within the subfamily Xenodontinae make up the Dipsadinae. The slug- and snail-eating snakes (respectively, *Sibon* and *Dipsas*), as their names indicate, concentrate on gastropods (thus paralleling the Asian slug-eating snakes, to which, however, they are not especially closely related). These species, grouped into three genera, are very slender, arboreal snakes with wide heads. The blunt-headed vinesnakes (*Imantodes*) are similar in body plan but eat lizards, which they pluck from thin branches as they sleep. Other members of the subfamily are less specialized; they include the night snakes (*Hypsiglena*) which range into North America, and the leaf-litter snakes, *Ninia*. Like the Xenodontinae, some species are coral snake mimics and they may have venom-delivering rear fangs. Most, if not all, species are egg-layers.

The final subfamily of colubrid snakes is the Xenoderminae, exclusively from Southeast Asia and poorly known. There are about 15 species in six genera, mostly small and slender. They live in forests, among leaf-litter or in low bushes, and they all appear to lay eggs.

Cobras and their Relatives
FAMILY ELAPIDAE

The family Elapidae includes a number of well-known groups of venomous snakes, all characterized by short, fixed, hollow venom fangs in the front of their mouths. There are over 270 species in 62 genera, and they occur throughout most of the world, including many of the warmer oceans. In North and South America they are represented by the coral snakes, of which there are about 60 species. These are brightly colored, with bands of red, black, and white or yellow encircling their bodies. Although most of them live in rain forests, some occur in the drier regions of northern Mexico and the southwestern United States. Most of them eat other reptiles, including snakes and amphisbaenians, which they pursue through their tunnels. Their venom is potent and fast-acting, and all species are dangerous to humans.

African elapids include the mambas (*Dendroaspis*): long, slender species, three of which live in trees, while the fourth, the Black mamba, is mostly terrestrial. All are fast-moving, diurnal hunters with large eyes and smooth scales. The Black mamba can grow to over 3m (10ft) in length and is greatly feared. Mambas may flatten their neck slightly when disturbed, a trait that is more often

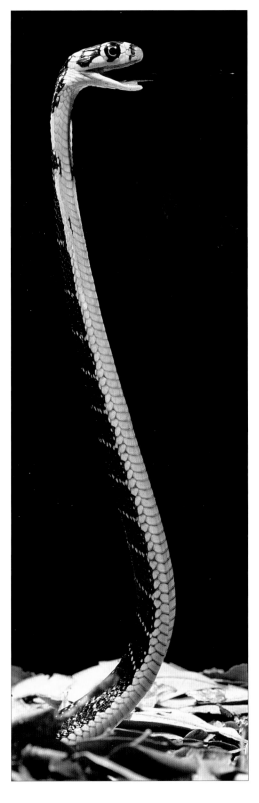

associated with the cobras, of which there are about 25 species from Africa and southern and southeast Asia. The African species tend to be the largest, growing to 3m (10ft) in length, although the Asiatic King cobra (*Ophiophagus hannah*) is the biggest of all, growing to more than 5m (16ft). Members of the genus *Naja* have the most impressive hoods; several species, known as spitting cobras, can force venom through small pores in the front of their fangs. The South African Rinkhals, which is unusual in being a live-bearer, also spits venom and may pretend to be dead when threatened. Other Asian elapids include the kraits – slender, nocturnal hunters of other snakes, many of them with triangular bodies and bright coloration – and a small group of elapids sometimes placed in a separate subfamily, the Maticorinae. These are often referred to as coralsnakes. Two species of *Calliophis* are startlingly colored and raise their tails when disturbed, displaying the bright red underside of their bodies. These species have huge venom glands, measuring up to one-third the length of the body. Their bite has been known to be fatal to humans, though generally they are regarded as inoffensive. Coralsnakes habitually feed on other snakes.

Apart from a few colubrids in the north, the elapids of Australia and New Guinea are the only advanced snakes of the region and, as such, they have moved into niches that would have been occupied by other species elsewhere. Thus, there are elapids that look and behave like whipsnakes, mambas, coral snakes, and even vipers. Their diets are similarly diverse; Australian elapids, collectively, eat prey ranging from invertebrates to small mammals. The taipan, often cited as the world's most venomous snake, is a confirmed mammaleater, but most species eat small lizards such as

skinks, which are plentiful in the region. Australian elapids include egg-laying and live-bearing species.

Closely related to the Australian elapids, and with some puzzles still to be solved regarding their evolutionary origins, the seasnakes and seakraits are sometimes placed in a separate family. All these snakes, which total about 55 species, are adapted to life at sea, although the seakraits, *Laticauda*, are unusual in that they have to come ashore to lay eggs, and may also bask and drink fresh water while on land. They live around coastal reefs or, in the case of one species, in brackish lagoons. The bodies of these species are cylindrical, presumably to help them crawl over the ground when necessary; they are boldly banded in black and white, and their tails are flattened to help them swim.

The other seasnakes are more closely tied to life at sea and never voluntarily come ashore. Several have flattened bodies, and all have oarlike tails and valvular nostrils that they can close; all give birth to live young. Some species inhabit coral reefs and search for prey by poking their heads into crevices; many of them eat eels, although other crevice-dwelling fish, especially gobies, are also taken. Two species, the White-bellied seasnake and the Turtle-headed seasnake, eat only fishes' eggs, and their venom is not as toxic as that of other species. A few species live at the mouths of estuaries, among mangroves and over mudflats, whereas the Yellow-bellied seasnake is pelagic, drifting in the upper layers of open ocean currents, often in huge numbers. It feeds on fish that are attracted to the apparent shelter its floating body provides as they would be to a piece of flotsam. This species is brightly colored and is probably distasteful to fish.

◑ **Above** *A juvenile King cobra* (Ophiophagus hannah) *rears up in a defensive display. When threatened, this highly venomous snake from Southeast Asia erects its long, narrow hood (less prominent than the flared hood of Indian cobras) and may also sometimes growl.*

◑ **Right** *The Taipan* (Oxyuranus scutellatus), *an Australasian elapid, inflicts multiple bites on its rodent prey. Since it often lives near farms and on sugar-cane plantations, the risk to humans from this species is high; 80 percent of bites in Papua New Guinea are from taipans.*

Vipers and Pitvipers
FAMILY VIPERIDAE

Members of the viper family, Viperidae, are all venomous, and have relatively long, hollow fangs that they can fold against the roof of their mouths when they are not using them. The family includes the adders, night adders, vipers, bush vipers, rattlesnakes, and pitvipers, as well as species with distinctive, and sometimes evocative, local names, such as the Hundred-pace viper, Sidewinder, Bushmaster, and Terciopelo. As a group, they are among the most successful snakes, although they are mostly restricted to terrestrial and arboreal habitats. Viper species are found further north (the northern adder, *Vipera berus*) and further south (the Patagonian lancehead, *Bothrops ammodytoides*) than any other snake species. They are also found at higher elevations than other snakes (the Himalayan pitviper, *Gloydius himalayanus*, occurs up to 4,900m/ 16,000ft, and several others approach this altitude). They did not, however, reach Madagascar or Australasia.

The most unusual member of the family is *Azemiops feae*, or Fea's viper. This rare species inhabits remote areas of southern China, Myanmar, and Vietnam, where it lives in cool cloud forests. It is slender, has smooth scales on its body and large scales on its yellow head, and is colored dark gray with orange rings. Few people have seen one, and its relationship to other vipers is unclear. It is placed in a subfamily of its own, the Azemiopinae.

Night adders, Causinae, of which there are six species from Africa, also have smooth body scales and large scales covering the tops of their heads. They specialize in eating toads and are not considered dangerous to humans.

The "typical" vipers or Old World adders, Viperinae, are short and stocky and have wide, spade-shaped or triangular heads. Their scales are heavily keeled, giving them a rough texture, and their heads are covered with many small scales, although there are a few exceptions in the genus *Vipera*. They live in a variety of habitats, including rocky hills and mountainsides, meadows, scrub, and deserts. Species from sand deserts, notably Peringuey's viper, *Bitis peringueyi*, may move by sidewinding, paralleling the American Sidewinders. The bush vipers, *Atheris*, from Central Africa, are arboreal. Many adders are camouflaged, and the coloration of a single species may vary from place to place according to the soil type on which it lives. The African *Bitis* species have especially intricate markings, and the Gaboon viper, *Bitis gabonica*, is often singled out as a splendid example of disruptive coloration. This massive snake, and its close relative the Puff adder, *Bitis arietans*, have huge fangs and are among the most impressive, and most feared, snakes in Africa. A number of species have "horns" or protuberances on various parts of their heads, sometimes consisting of a single thornlike scale over each eye, as in the Desert horned viper, *Cerastes cerastes*, or as a cluster of scales on the snout, as in the rhinoceros viper, *Bitis nasicornis*, and the Nose-horned viper, *Vipera ammodytes*. The bush vipers have heavily-keeled scales, especially the so-called Hairy bush viper, *Atheris hispidus*, in which the scales taper to a raised point. Keeled scales are also central to the defense strategy of saw-scaled vipers, *Echis*, which make a loud rasping sound by drawing two opposing sections of their body against each other. Saw-scaled vipers are small, very common, and very ready to defend themselves, making them perhaps the world's most dangerous snakes. Old World adders and vipers eat a range of prey, typically small mammals, nestling birds, and lizards. A few species, notably Orsini's viper, *Vipera ursinii*, eat insects such as grasshoppers. Most of the members of this subfamily are live-bearers, an adaptation that has undoubtedly helped them to survive in cold environments, but some species from warmer parts lay eggs.

The pitvipers are placed in the subfamily Crotalinae, and are readily distinguished from other vipers by the presence of their facial heat-sensitive pits. There are nearly 160 species occurring in Asia and North and South America. Like the Old World vipers, their heads are covered with small scales although, again, there are a few exceptions, such as the copperhead, *Agkistrodon contortrix*. They have colonized a number of different habitats, from swamps to deserts, although there are no aquatic species and no burrowing ones (perhaps because the pits would be more of a hindrance than a help in such environments). There are plenty of arboreal species, however, such as the large genera *Trimeresurus* in Asia and *Bothrops* in Central and South America. The world's largest viper is the bushmaster, from Central and South American rain forests and plantations, which can grow to 3m (10ft) or more in length. It is a relatively slender species, however, and even a large specimen would not weigh as much as a well-grown Gaboon viper.

The rattlesnakes, of which there are 30 species, are among the most easily identified snakes, although it should be noted that two or three (rare) species from small islands off the Mexican coast have lost their rattles through the evolutionary

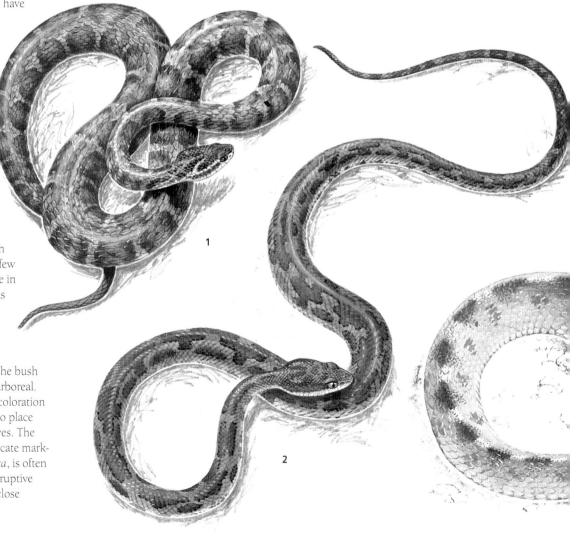

1

2

process. The rattle is formed from shed outer layers of the terminal scale, which is shaped in such a way that the segments do not fall off when the snake molts its skin. The snakes use their rattles as warning instruments, vibrating their tails and causing the segments to click together rapidly to create a buzzing or fast ticking sound. All rattlesnakes are American, and most live either in arid habitats or on rocky mountainsides or open grassland, including montane meadows. One species lives in South America, but prefers open clearings to the forest floor; another, the Massassauga, lives in swamps in the northern parts of its range. There are no arboreal rattlesnakes, although some species (including, interestingly, the rattleless kinds) are

believed to stalk roosting birds in bushes.

Pitvipers feed predominantly on warm-blooded prey such as mammals and birds; this is where their heat pits perform best, especially when hunting at night. Large rattlesnakes can easily handle prey up to the size of hares and ground squirrels; lizards are also eaten, though, especially by smaller species and by the young of larger ones. The Cottonmouth has perhaps the least specialized diet of all snakes and will eat insects, fish, frogs, turtles, baby alligators, and birds' eggs. Some populations lurk near seabird colonies in the hope of finding fish that are spilled when nestlings are being fed; they will even peel dried fish carcasses from rocks where they have fallen! CM

◁ **Left** *Pitviper and viper species:* **1** *Mamushi* (Gloydius blomhoffii); **2** *Habu* (Protobothrops flavoviridis); **3** *Sidewinder* (Crotalus cerastes); **4** *Long-nosed viper* (Vipera ammodytes).

BROODING BEHAVIOR IN SNAKES

As far as we know, snakes do not care for their young after they hatch, and only a few show any interest in their eggs. Pythons are among those species that do. Female pythons of all species, regardless of whether they are from Africa, Asia, or Australia, coil around their eggs after laying and remain with them until they hatch (BELOW a Green tree python, *Morelia viridis*). They occasionally leave them to drink or defecate, but do not feed during the incubation period, which lasts from 30 to 90 days, depending on species.

Female pythons coil around their eggs primarily to protect them. A pile of 100 or so grapefruit-sized, pure white eggs is not easy to hide but, by using her body to cover them, the snake effectively camouflages them. At the same time, large pythons are easily capable of defending themselves, and therefore their eggs, against a wide range of potential predators. In addition, by loosening and tightening her coils, the snake can control the humidity and temperature of the clutch although, as most pythons live in the tropics, temperature control is not as important as it may be in other species.

Some of the Australian pythons range well into temperate climates, however, and have adapted to cooler conditions. A female carpet python was seen to leave her eggs regularly, bask to raise her body temperature, and then return and coil once more about the clutch. It is not known if this is typical of this species only or of all Australian pythons.

The only other python that ranges well into cooler regions is the Indian. The female of this species has modified her habits even further, and actually produces body heat to aid in the incubation of the eggs. She accomplishes this by rhythmic contractions of her body muscles. In captivity, it was found that if the temperature was kept at about 30°C (86°F), the snake merely remained coiled around the eggs, but if the surroundings became cooler the female began to contract her body muscles. The number of contractions increased as the temperature was lowered. Experiments showed that a female could maintain a temperature differential of 5–7°C (9–12.5°F) by violently contracting the body muscles (described as hiccups by observers) at a rate of more than 30 times per minute to keep the eggs at the desired 30°C. She was never seen to leave the eggs during almost 90 days of incubation. No other snake (in fact, no other reptile) incubates in this way.

Among the advanced snakes there are only a few scattered species that provide some care to the eggs. Females of the two North American mud snakes coil around their eggs and remain with them until they hatch. Both sexes of Asian cobras are reported to cooperate in digging out a nest cavity in the ground, and to defend the eggs from possible predators. The King cobra is unique among snakes in the complexity of the nest that it provides. The female scrapes together a large pile of leaves, grasses, and associated soil and forms a nest cavity in the top of the pile, where she lays the eggs. Then she covers the clutch with leaves and makes another cavity on top, remaining there on guard until the eggs hatch.

No examples of brooding are known among true vipers but some pitvipers are known to provide care; female mountain vipers and Malayan pitvipers remain coiled on or near the eggs for the month or so that it takes them to incubate. HGD

THE THREAT FROM SNAKEBITES

The pathology and treatment of venom-related conditions

THE EARLIEST TREATISE ON THE TREATMENT OF snakebite is the Brooklyn Papyrus, dating from 300BC. Snakebites and the effects of envenoming have been recorded from classical and biblical times, but the first scientific experiments on envenoming were carried out in the 17th century by Francesco Redi, physician to the Duke of Tuscany, and on the treatment of snakebite by his compatriot Felice Fontana in the 18th century.

Snakes of medical importance have enlarged teeth (fangs) in their upper jaw containing a channel through which venom enters the bite wound. They belong to four families: Elapidae (including cobras, kraits, mambas, coral snakes, African garter snakes, Australasian snakes, and sea snakes); Viperidae (the Old World vipers and adders of the subfamily Viperinae, the pit vipers of Asia, and the lance-headed vipers, moccasins, and rattlesnakes – Crotalinae – of the Americas); Atractaspididae (burrowing asps or stiletto snakes); and Colubridae. They inhabit most parts of the world except Antarctica and the extreme Arctic; they are also absent at altitudes above 4,000m (13,000ft) and from some islands, including most of the West Indies, Ireland, Iceland, Madagascar, and islands of the western Mediterranean, Atlantic, and Pacific Ocean east of Fiji. Sea snakes inhabit the Indian and Pacific Oceans between latitudes 30°N and 30°S, as well as some rivers and inland lakes.

Estimates based on hospital records suggest that snakebite causes between 50,000 and 100,000 deaths a year, but most victims receive traditional treatment outside hospitals and may die at home unrecorded. Epidemiological studies report mortality rates of 2–17 per 100,000 population each year in parts of Africa and India. For each fatal case, approximately 10 survivors are permanently handicapped. Humans are bitten when they inadvertently touch or tread on snakes, corner them, or intentionally handle them. Most bites are inflicted on the feet and ankles of agricultural workers and hunter-gatherers. The incidence of snakebite peaks with increased agricultural activity in the rainy season – this is, for instance, the time of the paddy harvest in Southeast Asia – and flooding can cause snakebite epidemics.

Snake venoms contain 20 or more different compounds, mostly proteins or polypeptides. Enzymes (molecular weights 13–150 KDa) constitute 80–90 percent of viperid and 25–70 percent of elapid venoms: digestive hydrolases, L-amino acid oxidase, phospholipases, thrombin-like procoagulant, and kallikrein-like serine proteases and metalloproteinases (hemorrhagins), which damage vascular endothelium. Polypeptide toxins (molecular weight 5–10 KDa) include cytotoxins,

cardiotoxins, and postsynaptic neurotoxins (such as alpha-bungarotoxin), which bind to acetylcholine receptors at neuromuscular junctions. Compounds with low molecular weights (up to 1.5 KDa) include metals, peptides, lipids, nucleosides, carbohydrates, amines, and oligopeptides, which inhibit angiotensin converting enzyme (ACE) and potentiate bradykinin (BPP). Inter- and intra-species variation in venom composition is geographical and ontogenic.

Between 20 percent (in the case of *Pseudonaja textilis*) and 80 percent (in *Echis* species) of penetrating bites by venomous snakes result in clinically detectable envenoming. In some cases (for example *Crotalus durissus terrificus*, *Oxyuranus scutellatus*, and *Acanthophis* species) case fatality exceeded 50 percent before the advent of antivenom, but with antivenom it can be reduced to less than 5 percent.

Envenoming of humans is an aberration of the biological function for which the venom and venom apparatus were evolved, which was to immobilize and predigest the snake's natural prey. Local tissue damage is caused by the venoms of most Viperidae, some Elapidae (notably African and Asian spitting cobras), and *Atractaspis* species. Digestive and cytolytic enzymes and polypeptide toxins damage blood vessels, membranes, and tissues, including muscle, causing leakage of blood and plasma into the tissues. There is swelling,

blistering, local bruising, and inflammation, and the local lymph nodes become painful, tender, and enlarged. Gangrene with secondary bacterial infection, sometimes derived from the venom or fangs, may require amputation or result in crippling deformity or chronic ulceration that may become cancerous.

Neurotoxic (paralytic) effects are caused by venoms of most Elapidae and a few Viperidae. Progressive flaccid paralysis starts with muscles innervated by the cranial nerves: the first symptoms are drooping of the eyelids (ptosis) and paralysis of eye movements, and eventually paralysis of the muscles of swallowing and breathing ensues, leading to death by asphyxiation.

Generalized breakdown of skeletal muscle

◗ **Right** *Local effects of envenoming by the species of Brazilian lance-headed viper known as the Jararaca (Bothrops jararaca). This patient was bitten on the dorsum (upper surface) of the left foot 36 hours earlier. The swelling and massive blister formation results from leakage of plasma from blood vessels under the influence of the snake's venom.*

◗ **Below** *Milking venom from a Malayan pit viper (Calloselasma rhodostoma). A drop of yellow protein-rich venom is being squeezed from the venom gland situated above the angle of the jaw, through the venom duct and down the venom canal in the long, erected, viperine front fang.*

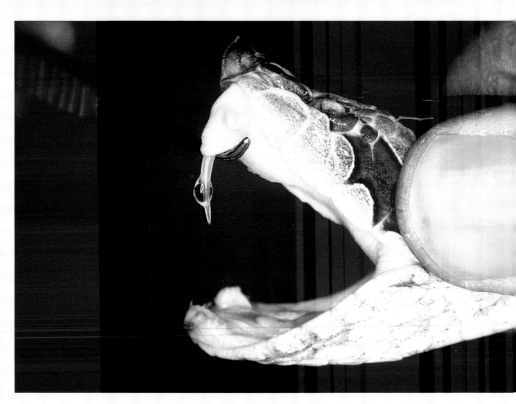

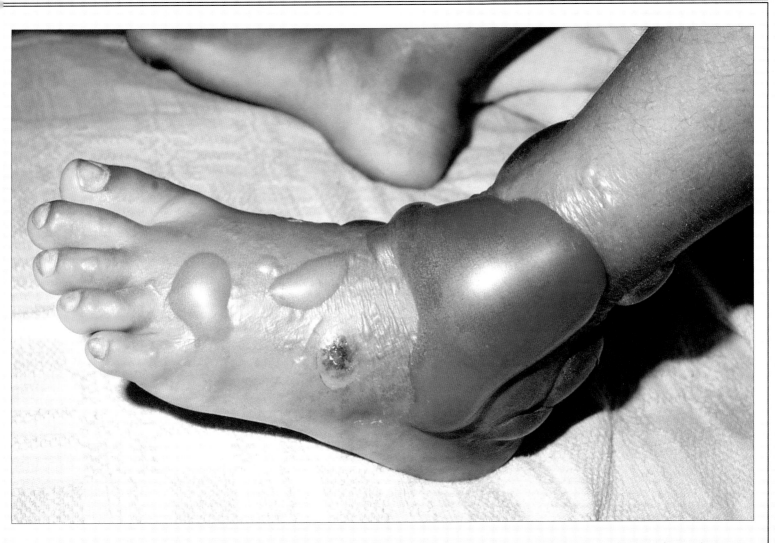

(rhabdomyolysis) is caused by the venoms of many sea snakes, some terrestrial Australasian elapids (notably the tiger snakes), and some Viperidae including the Neotropical rattlesnake (*Crotalus durissus*). As the muscles are "dissolved" by venom phospholipases A2, muscle pigment (myoglobin), muscle enzymes, potassium, and other constituents leak into the circulation and the urine becomes dark brown or black (myoglobinuria).

Kidney damage may be caused directly by the venoms of some Viperidae, notably Russell's viper and the South American lance-headed vipers, as well as by sea snakes. Shock or other causes of renal ischemia and passage of myoglobin/hemoglobin may also cause renal failure.

Spontaneous systemic bleeding and blood clotting abnormalities are caused by the venoms of many Viperidae and Australasian Elapidae, as well as by some Colubridae. Activation of blood coagulation by procoagulant enzymes leads to the formation of fibrin, which is instantly broken down by endogenous plasmin, with eventual depletion of clotting factors so that the blood is incoagulable (consumption coagulopathy). Pitviper venoms contain thrombin-like enzymes that digest fibrinogen. Other venoms contain anticoagulants such as phospholipases A2 and platelet-inhibiting factors. Spontaneous systemic hemorrhage results from the combined actions of hemorrhagins and anti-hemostatic factors. It is visible in the gums, nose, conjunctivae, and skin, and may be fatal if it involves the brain, lungs, gastrointestinal tract, or uterus.

Cardiovascular effects are caused by the venoms of Viperidae, Actractaspididae, and some Elapidae. Hypotension (fall in blood pressure) and shock (inadequate perfusion of vital organs) frequently follow leakage of blood into the bitten limb and elsewhere, causing hypovolemia. Certain venoms act on the heart and coronary arteries (for instance, sarafotoxins from *Atractaspis engaddensis*); others, including ACE inhibitors and BPPs, are vasodilators.

African and Asian spitting cobras and the Rinkhals can eject their venom as a fine jet from the tips of the fangs towards the eyes of aggressors, causing intense pain, watering, and purulent discharge from the eyes, spasm and swelling of the lids, and sometimes ulceration of the cornea with secondary infection leading to blindness.

The median interval between bite and death varies widely according to family or genus of snake involved: 8 hours for cobra bites, 16 hours for North American rattlesnake bites, 18 hours for krait bites, three days for Russell's viper bites, and five days for *Echis* (saw-scaled viper) bites.

Traditional first-aid remedies include black snake stones ("Jesuit stones"), incisions, suction by mouth and vacuum devices, tight tourniquets, cryotherapy, electric shock, tattooing of the bitten limb, and a range of Ayurvedic (traditional Hindu/Buddhist) treatments. None has proved effective and most are harmful. Recommended first aid is to calm victims and arrange for them to be carried on a stretcher to the nearest medical care. The bitten limb should be immobilized with a splint or sling. With elapid bites (other than from African spitting cobras), a long, stretchy, crepe bandage should be bound firmly around the entire bitten limb, incorporating a splint ("pressure immobilization"). This treatment delays the development of paralysis and other life-threatening effects.

Eyes injured by spitting elapids should be irrigated immediately with generous volumes of water or any other bland liquid available. Unless corneal abrasions can be excluded by ophthalmological examination, a prophylactic topical antimicrobial (tetracycline or chloramphenicol) should be instilled.

In the hospital, the most critical decision is whether to give antivenom, made from the plasma of a horse or sheep that has been immunized with venom. Being a foreign protein, antivenom can cause early anaphylactic or late immune complex (serum sickness) reactions. Artificial ventilation and renal dialysis may also be required to save the patient's life. DAW

Snake Families

THE ANATOMY OF SNAKES SUGGESTS THAT we must search for their ancestors among the lizards or from some closely related stock. Snake fossils first appear during the Cretaceous (144–65 million years ago), the last period of the Age of Reptiles. Some workers think that the ancestral snake was a small burrowing animal. They emphasize that the loss of limbs, eyelids, and external ears is common in burrowing lizards. Others believe that snakes were derived from large marine reptiles such as mosasaurs or aigialosaurs. Most fossil remains of snakes are no more than isolated vertebrae, however, and this dispute is still not resolved.

BLINDSNAKES

Scolecophidians

Early blindsnakes
Family Anomalepididae

17 species in 4 genera. C and S America. Species include: **Pink-headed blindsnake** (*Helminthophis frontalis*).
LENGTH: 11–30cm; most 13–16cm.
COLOR: Brown or black, some with light head and/or tail.
SCALES: Like other blindsnakes
BODY FORM: As in other blindsnakes, except teeth on maxillae and dentaries.

Blindsnakes
Family Typhlopidae

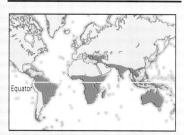

More than 352 species in 6 genera. S America N to Mexico and the Bahamas, Africa S of Sahara, SE Europe across S Asia to Taiwan and Australia. Species include: **Brahminy blindsnake** (*Ramphotyphlops braminus*), **Mona blindsnake** (*Typhlops monensis*), **Schlegel's blindsnake** (*Rhinotyphlops schlegelii*).
LENGTH: 15–90cm, most 20–50cm.
COLOR: Pink, yellow, brown, or black, with darker or lighter blotches, bands, or lines.
BODY FORM: Cylindrical, with indistinct head, eyes reduced and sometimes invisible under head scutes, movable toothed maxillae, and toothless mandibles; tail short. Small species tend to be slender; individuals of some large species become thick-bodied.
CONSERVATION STATUS: 1 species is listed as Endangered, and 1 is Vulnerable.

Threadsnakes or Wormsnakes
Family Leptotyphlopidae

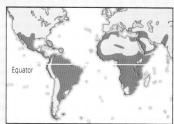

More than 104 species in 2 genera. S America N to Bahamas and SW USA; Africa E through Saudi Arabia to SW Asia. Species include: **Seven-striped wormsnake** (*Leptotyphlops septemstriatus*), **Long-tailed threadsnake** (*L. longicaudatus*).
LENGTH: 15–41cm, most 20–30cm.
COLOR: Gray, pink, brown, or black, some with light or dark markings, including stripes.
SCALES: Shiny and of uniform size throughout body.
BODY FORM: Very slender, maxillae fused to skull, mandibles with teeth.

"TRUE SNAKES"

Alethinophidians

Dwarf pipesnakes
Family Anomochilidae

2 species in 1 genus. SE Asia, Sumatra, and Borneo. Species include: **Leonard's pipesnake** (*Anomochilus leonardi*).
LENGTH: 25–35cm.
COLOR: Dark brown or black, with yellow or red belly markings.
SCALES: Ventrals slightly enlarged; otherwise uniform throughout body.
BODY FORM: Cylindrical; eyes with or without a brille.

Shield-tailed snakes
Family Uropeltidae

More than 47 species in 8 genera. S India and Sri Lanka. Species include: **Blyth's earthsnake** (*Rhinophis blythi*), **Ceylon shield-tailed snake** (*Uropeltis ceylanicus*).
LENGTH: Less than 90cm, most 20–50cm.
COLOR: Iridescent black or brown, often marked above and/or below with brilliant red, yellow, or white.
SCALES: Shiny and uniform-sized throughout, except for enlarged and roughened scute on end of short tail.
BODY FORM: Head small and conical, skull sturdy, body especially stout anteriorly.

Asian pipesnakes
Family Cylindrophiidae

10 species in 1 genus. Sri Lanka to SE Asia, S China, and E Indies. Species include: **Red-tailed pipesnake** (*Cylindrophis ruffus*).
LENGTH: 50–70cm.
COLOR: Black, brown, or bright copper, with white and black checkered belly and sometimes a red tail tip.
SCALES: Shiny and uniform in size throughout.
BODY FORM: Cylindrical, with indistinct head and very short pointed tail; skull sturdy.

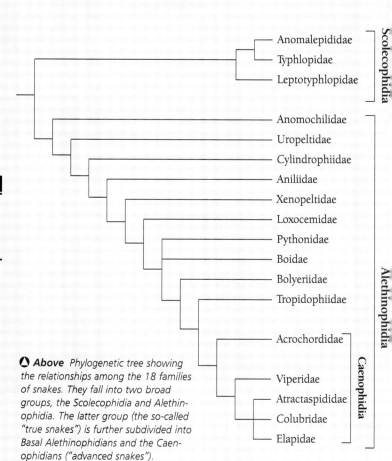

Above *Phylogenetic tree showing the relationships among the 18 families of snakes. They fall into two broad groups, the Scolecophidia and Alethinophidia. The latter group (the so-called "true snakes") is further subdivided into Basal Alethinophidians and the Caenophidians ("advanced snakes").*

NOTES	
	Length = length from snout to tip of tail.
	Approximate nonmetric equivalents: 10cm = 4in 1kg = 2.2lb

Red pipesnake
Family Aniliidae

1 species in 1 genus: **Red pipesnake** (*Anilius scytale*). Tropical S America.
LENGTH: Up to 1m.
COLOR: Brilliant red and black rings.
SCALES: Shiny and uniform in size throughout.
BODY FORM: Cylindrical, moderately slender; indistinct head and short, blunt tail; stout skull.

Asian sunbeam snakes
Family Xenopeltidae

2 species in 1 genus. SE Asia to E Indies. Species include: **Asian sunbeam snake** (*Xenopeltis unicolor*).
LENGTH: Up to 1m.
COLOR: Iridescent brown or black.
SCALES: Extremely shiny, ventrals slightly enlarged.
BODY FORM: Cylindrical, head flattened and indistinct, tail short and pointed, teeth hinged on bones so that they fold when prey are swallowed.

Neotropical sunbeam snake
Family Loxocemidae

1 species in 1 genus: *Loxocemus bicolor*. Mexico and C America.
LENGTH: To 1.2m.
COLOR: Dark brown, sometimes with cream-colored belly.
SCALES: Ventrals somewhat enlarged.
BODY FORM: Moderately stout, with small head and slightly upturned rostral scale; skull stout.

Boas
Family Boidae

43 species in 9 genera; the sandboas (15 species in 3 genera, *Eryx*, *Gongylophis,* and *Charina*) have been separated from other boas for at least 50 million years and are sometimes treated as a separate family (the Erycidae). N and S America, Africa, E Indies, Madagascar, New Guinea, some Pacific islands. Species include: **Kenyan sandboa** (*Gongylophis colubrinus*), **Neotropical tree boa** (*Corallus hortulanus*), **Boa constrictor** (*Boa constrictor*), **Rubber boa** (*Charina bottae*), **Bahaman boa** (*Epicrates striatus*).
LENGTH: 0.5–4m, most 1–3m; Green anaconda might reach 11m.
COLOR: Usually brown or gray with darker markings; Emerald tree boa is bright green with light markings.
SCALES: Many rows of small dorsal scales, ventral scutes somewhat enlarged.
BODY FORM: Usually moderately stout, but Haitian vine boa has long, slender neck, head distinct, tail moderately long, no teeth on premaxillary; hindlimb vestiges present as external claws. Most are live-bearing, but Arabian sandboa and Calabar ground boa are egg-layers.
CONSERVATION STATUS: 2 species are Endangered and 4 are Vulnerable.

Pythons
Family Pythonidae

30 species in 8 genera; the pythons are sometimes included in a single family with the boas and sandboas. Old World tropics and subtropics from Africa through SE Asia, S China, and the East Indies to Australia. Species include: **Children's python** (*Antaresia childreni*), **Carpet** or **Diamond python** (*Morelia spilotes*), **Green tree python** (*M. viridis*), **Indian** or **Burmese python** (*Python molurus*), **Reticulated python** (*P. reticulatus*), and **African rock python** (*P. sebae*).
LENGTH: From under 1m to more than 10m; most 3–6m.
COLOR: Uniform brown or bright green to boldly patterned with blotches or diamonds.
SCALES: Numerous dorsal scales, somewhat enlarged ventral scutes.
BODY FORM: Cylindrical, with short tail and vestiges of hind limbs; two lungs; eyes with vertically elliptical pupils. Most have paired subcaudal scutes, teeth on the premaxilla, a postfrontal bone above the orbit of the eye, and spineless hemipenes with flounces (ruffles) and apical awns (small pointed tips). All lay eggs.
CONSERVATION STATUS: 6 species are currently listed as Critically Endangered, 7 as Endangered, and 8 as Vulnerable.

Mascarene or Split-jawed boas
Family Bolyeriidae

2 species in 2 genera. Round Island, near Mauritius. One extant species: **Round Island boa** (*Casarea dussumieri*).
LENGTH: 0.8–1.4m.
COLOR: Brown or gray, with irregular, vague markings.
BODY FORM: Moderately slender or stout, head somewhat distinct, tail moderately long, maxillaries divided and hinged.
CONSERVATION STATUS: The Round Island burrowing boa (*Bolyeria multocarinata*) is believed to have gone extinct.

⚫ *Above* Schlegel's blindsnake (Rhinotyphlops schlegelii), *the largest of the blindsnakes, grows up to 1m (39in). Its fossorial lifestyle is reflected in its physiology: a smooth, cylindrical body and a streamlined spade-shaped snout armored with a large rostral scale.*

Dwarf boas
Family Tropidophiidae

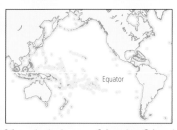

26 species in 4 genera. S America, C America, S Mexico, West Indies, Malaysia. Species include: **Black-tailed dwarf boa** (*Tropidophis melanurus*), **Eyelash dwarf boa** (*Trachyboa boulengeri*), **Oaxacan dwarf boa** (*Exiliboa placata*).

LENGTH: To 1m, usually 30–50cm.

COLOR: Usually brown or gray with irregular, vague markings; some species brightly spotted or ringed.

SCALES: Numerous dorsal scales, ventrals somewhat enlarged.

BODY FORM: Moderately slender, head somewhat enlarged.

ADVANCED SNAKES

Caenophidians

Filesnakes
Family Acrochordidae

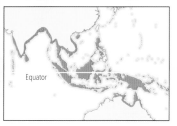

3 species in 1 genus. S Asia and the Philippines to N Australia. Species include: **Javan filesnake** (*Acrochordus javanicus*).

LENGTH: 0.8–2.7m.

COLOR: Brown or gray, sometimes with dark markings.

SCALES: Same size throughout, non-overlapping, and covered with tiny, bristly tubercles.

BODY FORM: Small head with small dorsal eyes, body stout with baggy skin.

◁ *Left* Yellow-lipped seakraits (*Laticauda colubrina*) *are widespread in the Pacific and Indian oceans, foraging for eels and other fish on coral reefs. This elapid is mainly a marine species, but will emerge onto land to bask, mate, and lay eggs.*

Colubrids
Family Colubridae

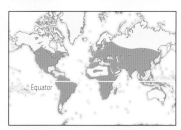

More than 1,900 species in over 300 genera. Worldwide, except at extremes of latitude and elevation; generally absent from marine environments. Species and genera include: **Aesculapian snake** (*Elaphe longissima*), **American greensnakes** (*Opheodrys*), **American gartersnakes** (*Thamnophis*), **San Francisco gartersnake** (*T. sirtalis tetrataenia*), **American watersnakes** (*Nerodia*), **Asian snail eaters** (*Pareas*), **American snail eaters** (*Dipsas*), **African egg-eating snakes** (*Dasypeltis*), **Eurasian whipsnake** (*Hierophis gemonensis*), **Eastern racer** (*Coluber constrictor*), **Grass snake** (*Natrix natrix*), **milksnake** (*Lampropeltis triangulum*), **mudsnakes** (*Farancia*), **ratsnakes** (*Elaphe*), **Red-bellied snake** (*Storeria occipitomaculata*).

LENGTH: 13cm–3.5m; most 50cm–2m.

COLOR: Many are brown, gray, or black, but some are red, yellow, or green, with spots, blotches, or stripes.

SCALES: Head scutes usually enlarged, dorsal scales keeled or smooth, ventrals typically enlarged.

BODY FORM: Ranges from extremely slender to stout, head enlarged or indistinct, usually with a tapering body. Pupils horizontal or vertical elliptical, usually round. None has a functional left lung, coronoid bone, or limb vestiges.

CONSERVATION STATUS: The St. Croix racer (*Alsophis sancticrucis*) from the Virgin Islands is now thought to be extinct. 7 species are Critically Endangered, 21 Endangered, and 16 Vulnerable.

Vipers and Pitvipers
Family Viperidae

265 species in 40 genera. Worldwide except for Antarctica and oceans; the northernmost (European adder) and southernmost (Patagonian lancehead) are vipers. Species and genera include: **Asian tree pit vipers** (*Trimeresurus*), **bushmasters** (*Lachesis*), **Central American fer-de-lance** or **terciopelo** (*Bothrops asper*), **Common lancehead** (*B. atrox*), **jararaca** (*B. jararaca*), **copperhead** (*Agkistrodon contortrix*), **cottonmouth** (*A. piscivorus*), **diamond-backed rattlesnakes** (*Crotalus adamanteus*, *C. atrox*), **New Mexico ridge-nosed rattlesnake** (*C. willardi obscurus*), **Sidewinder rattlesnake** (*C. cerastes*), **European adder** (*Vipera berus*), **Gaboon viper** (*Bitis gabonica*), **Puff adder** (*B. arietans*), **Horned viper** (*Cerastes cerastes*), **Latifi's viper** (*Vipera latifii*), **Mountain pit viper** (*Ovophis monticola*), **pygmy rattlesnakes** (*Sistrurus*), **Russell's viper** (*Daboia russelii*), **saw-scaled vipers** (*Echis*).

LENGTH: 25cm–3.7m, most 60cm–1.2m.

COLOR: From bright green with red markings to solid brown or black; most with pattern of dark blotches on lighter background.

SCALES: Head scales usually all small, ventrals enlarged.

BODY FORM: Moderately slender to very stout, with distinct head and fairly short tail. Pupils usually elliptical. Single pair of hollow fangs on very short maxilla can be rotated to bring fangs forward to bite; venom usually tissue-destructive.

CONSERVATION STATUS: 7 species, including the **Golden lancehead** (*Bothrops insularis*) and the **Mt. Bulgar viper** (*Vipera bulgardaghica*), are currently listed as Critically Endangered; 9 more are Endangered, and 10 Vulnerable.

Stiletto snakes and their allies
Family Atractaspididae

70 species in 11 genera. Sub-Saharan Africa and the Arabian Peninsula. Species include: **Cape centipede-eater** (*Aparallactus capensis*), **Spotted harlequin snake** (*Homoroselaps lacteus*), **Southern stiletto snake** (*Atractaspis bibroni*).

LENGTH: 20cm–1.1m, usually 30–50cm.

COLOR: Usually black or dark brown; some are striped or brightly banded.

SCALES: Large head scales, smooth dorsals, and enlarged ventrals.

BODY FORM: Indistinct head, cylindrical body, short pointed tail. Maxilla with grooved or hollow fangs, greatly enlarged and movable in stiletto snakes.

Cobras and their allies
Family Elapidae

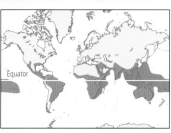

335 species in 60 genera. N and S America, Asia, E & S Africa, Australia; seasnakes in coastal waters of Asia, Africa, Australia, and Pacific tropical America. Species include: **Annulated seasnake** (*Emydocephalus annulatus*), **Indian cobra** (*Naja naja*), **Central Asian cobra** (*N. oxiana*), **Egyptian cobra** (*N. haje*), **spitting cobras** (*N. nigricollis*, *N. mossambica*, *Hemachatus haemachatus*), **Black mamba** (*Dendroaspis polylepis*), **Death adder** (*Acanthophis antarcticus*), **Inland taipan** or **Small-scaled snake** (*Oxyuranus microlepidotus*), **Olive seasnake** (*Aipysurus laevis*), **Eastern brown snake** (*Pseudonaja textilis*), **Eastern coralsnake** (*Micrurus fulvius*), **Fiji snake** (*Ogmodon vitianus*), **King cobra** (*Ophiophagus hannah*), **kraits** (*Bungarus*), **mambas** (*Dendroaspis*), **New World coralsnakes** (*Leptomicrurus*, *Micruroides*, *Micrurus*), **Asian coralsnakes** (*Calliophis*, *Sinomicrurus*), **Yellow-bellied seasnake** (*Pelamis platurus*), **seakraits** (*Laticauda*), **tiger snakes** (*Notechis scutatus*, *N. ater*). The Australian members of this family, together with the sea snakes and sea kraits, a total of 184 species, are sometimes placed in a separate family, Hydrophiidae.

LENGTH: 38cm–5.6m; most 75cm–1.5m.

COLOR: Most gray, brown, or black, often with collars or crossbands; some bright green; coralsnakes ringed with red, yellow, and black.

SCALES: Head scales large, dorsals usually smooth, ventrals enlarged.

BODY FORM: Head usually only slightly distinct, body cylindrical and slender in most, but Death adders are stocky; seasnakes and seakraits are compressed to oar-shaped; all with small fangs on maxilla, usually only slightly movable; venom usually neurotoxic.

CONSERVATION STATUS: 8 species are listed as Vulnerable.

HWG

NOTES   Length = length from snout to tip of tail.

Approximate nonmetric equivalents: 10cm = 4in   1kg = 2.2lb

PHOTO
STORY

HARVESTING SNAKE VENOM

CAPTURE

1 The Irula, a southern Indian tribal people, make their living by catching snakes, especially venomous species like the cobra seen above. They used to kill the reptiles for their skins, but conservation laws banned the trade in 1976. Now they milk them for their venom instead.

2 The Irulas' snake-handling skills have been passed down the generations for centuries. From early on, children learn to recognize species and their characteristics, mostly in the hands-on fashion demonstrated by this family group examining rat-snakes (genus Elaphe).

3 The ban on snakeskin sales threatened the community with disaster until a group of 25 individuals set up (with the help of an Oxfam grant) the Irula Snake-Catchers Industrial Co-operative, dedicated to a new, non-lethal form of hunting. Now the snakes – mainly cobras, Russell's vipers, kraits, and saw-scale vipers – are taken alive, measured, and then stored in clay pots (above) at the co-operative's headquarters near Mamallapuram in Tamil Nadu. Hunters get a small basic salary, topped up by a bonus for each snake they catch. Although they now carry anti-venom serum as a precaution, many seem to have a degree of immunity to snakebite, which is still an occupational hazard.

EXTRACTION

4 5 *Snakes are held at the co-operative for three weeks, during which time they are milked of their venom weekly. To extract the poison, the fangs are pressed against leather fastened over a glass jar – a slow process, as ten snakes are needed to produce a single gram (0.04oz) of cobra venom. At the end of their captivity, the snakes are released back into the wild at the point where they were originally taken. The venom is freeze-dried and sold to government laboratories to be used in the production of antivenom serum.*

Tuatara

tUATARA ARE OF GREAT SIGNIFICANCE TO *biologists worldwide as the only surviving species of the order Rhynchocephalia. Two species are currently recognized, both of them limited to islands and rock stacks off the coasts of New Zealand.*

Among the many curious features of tuatara are unusual dentition (including a double tooth row along the upper jaw, forming a groove into which the teeth on the lower jaw fit); exceptional cool-temperature tolerance; slow reproduction; and an extended lifespan.

Last of the Rhynchocephalians
FORM AND FUNCTION

Rhynchocephalians ("beak-heads"), also known as sphenodontians ("wedge-toothed ones"), first appeared in the Mesozoic some 220 million years ago – around the time of the earliest dinosaurs. Fossils are known from sites around the world, including Europe, Africa, Madagascar, and South America. Although once portrayed as a static group, they were abundant for a considerable time, diversifying into about 24 genera. Almost all had died out by 100 m.y.a., well before the dinosaurs went extinct. Somehow the lineage that led to tuatara survived, isolated on the fragment of Gondwana that became New Zealand.

Rhynchocephalians are the sister-group of squamates (lizards and snakes). The two lineages share many features, including caudal autotomy and a transverse cloacal slit. Rhynchocephalians are distinguished from squamates by several features, especially their teeth. Tuatara and their extinct relatives have two rows of teeth in their upper jaw and a beak formed from overhanging upper incisors. The jaw teeth are acrodont (fused to the jawbone rather than socketed) and form a serrated edge that wears down over time. The upper jaw is rigidly attached to the skull, in contrast with the movable connections seen in squamates. Unlike squamates, rhynchocephalians have gastralia ("abdominal ribs"), and male tuatara have only a rudimentary copulatory organ.

As relics of a distinctive group, tuatara are often portrayed as "living fossils." However, recent studies emphasize that they are neither identical to their Mesozoic relatives nor maladapted to their current environment. Nevertheless, they are of exceptional interest to scientists, who study them for clues to characters that might be primitive (ancestral) with respect to squamates.

Tuatara are burrowing reptiles that live in coastal forest. Densities range from 13–2,000/ha (30–5,000 per acre). Adults roam and forage at

night in home ranges of 12–87sq m ((130–940sq ft), with body temperatures of about 7–17°C (45°–63°F). Tuatara hunt primarily by sight and can detect prey at very low light levels, although not in total darkness.

Although primarily nocturnal, tuatara often emerge to bask by day near their burrow entrances. Body temperatures may then briefly reach 24°–27°C (75°–80°F). Like several lizards, tuatara have a well-developed pineal complex, which develops as an outgrowth of the brain. One component, the pineal body, secretes the hormone melatonin at night. The other, connected structure is the parietal eye, the "third eye," located below the skin between the parietal bones in the skull roof. The parietal eye of tuatara has many structural features of an eye, including a lens, a retina, and neural connections to the brain. Little experimental work has been carried out on its function; as in lizards, it is probably light-sensitive but not image-forming, helping to regulate exposure to solar radiation.

Tuatara share their island habitats with large populations of burrowing seabirds (petrels, prions, and shearwaters) that come ashore seasonally to nest. Seabird burrows provide shelter for tuatara (although they can dig their own), and seabird defecation and movements on the forest floor encourage a high density of the ground invertebrates on which tuatara feed. Tuatara also eat the eggs and chicks of seabirds, obtaining

"marine" fatty acids that may be beneficial. Primarily, however, tuatara feed on invertebrates such as beetles, crickets, and spiders, though frogs, lizards, and young tuatara are also occasionally eaten. A special type of fore-and-aft jaw movement (propalinal), found only in the most advanced rhynchocephalians, enables tuatara to shear chitinous or bony prey with ease.

Long, Slow Lives
REPRODUCTION AND DEVELOPMENT

Tuatara live in cool climates and have a low metabolic rate – factors that help explain their slow development and great longevity. Females reproduce only every 2–5 years. Periods of vitellogenesis (egg yolking, lasting 1–3 years), egg shelling (7 months), and egg incubation following laying (11–15 months) are among the longest known for any reptile. Vitellogenesis is followed by mating in the period from January to March, the southern

⬖ **Above** *Long in decline, tuatara numbers may now be rising thanks to a program of reintroduction onto offshore islands managed by the New Zealand Department of Conservation.*

⬗ **Right** *"Tuatara" (both singular and plural in Maori) means "peaks on the back," and refers to the conspicuous folds of skin that form a crest along the neck, back, and tail. The crest is especially prominent in males, and may be raised during courtship or territorial disputes.*

hemisphere summer; males exhibit territorial and courtship displays with crest erection at this time. Ovulation occurs soon afterwards. In October to December (spring), females aggregate in open areas to dig nest tunnels about 20cm/8in deep. They compete over nest sites but, apart from a few days of nest defense, there is no further parental care. The oval, flexible-shelled eggs are about 25–30mm (1–1.2in) long. Sex determination is temperature-dependent – warm incubation produces males, cool incubation females.

The embryo develops a horny egg-breaker on its snout to slit the shell. Hatchlings are about 54mm (2.2in) long from snout to vent. They hide under stones or logs and are initially diurnal. Tuatara reach sexual maturity at 9–13 years of age; growth continues until 20–35 years of age. Wild tuatara are known to be still reproducing at about 60 years of age; however, suggestions that tuatara live for 200–300 years are entirely speculative.

Provided temperatures are kept cool, tuatara survive well in captivity. However, captive conditions have sometimes resulted in obese tuatara showing little inclination to reproduce.

Island Refuges
CONSERVATION AND ENVIRONMENT

When humans settled New Zealand about 1,000 years ago, tuatara were widespread. They have since disappeared from the two main islands and several offshore ones, primarily due to the impact of introduced mammals. Tuatara survive naturally today on about 30 islands, five in Cook Strait and the remainder around the northeast coast of the North Island. Most of these islands are cliff-bound and landing is restricted by permit, helping to protect tuatara from human interference.

Since 1995, tuatara have been reintroduced to three islands. Features of the recovery plan include eradicating the Polynesian rat and captive rearing of juveniles from wild-collected eggs for reintroduction. Tuatara are absolutely protected by the New Zealand Wildlife Act. The rarer species, *Sphenodon guntheri*, is found naturally on only one small island, where the population is about 400, but has been reintroduced to two others. *Sphenodon punctatus* is found naturally on about 29 islands (with a total population of 60,000-plus), and has been reintroduced to one other.　　AC

FACTFILE

TUATARA

Order: Rhynchocephalia (Sphenodontia)

Family: Sphenodontidae

2 species in 1 genus: *Sphenodon punctatus* and *S. guntheri*.

DISTRIBUTION About 33 small islands and rock stacks off the coasts of New Zealand.

HABITAT Night-active burrowers in areas of low forest and scrub, usually associated with colonies of burrowing seabirds.

SIZE Length from snout to tip of tail up to 61cm/24in (male), 45cm/18in (female); **weight** 1kg/2.2lb (male); 0.5kg/1.1lb (female).

COLOR Adult dorsal color olive-green, gray, or dark pink, with a speckling of gray, yellow, or white; newly hatched animals are brown or gray, with pink tinges and a pale head shield, striped throat, and sometimes distinctive light patches on the body and tail.

REPRODUCTION Lays eggs; mean clutch sizes 6–10, depending on population; incubation 11–15 months.

LONGEVITY Unknown, but probably 100 years or more.

CONSERVATION STATUS *Sphenodon guntheri* is listed as Vulnerable.

Crocodilians

a S THE LARGEST OF ALL LIVING REPTILES, *crocodilians are associated in the public mind solely with predatory ferocity. However, a closer look reveals that these unique reptiles display many subtle and complex behaviors on a par with birds and mammals. Their vocalizations terrified early travelers and still intrigue scientists today. Parents routinely guard eggs, liberate young from nests, and then remain with and defend hatchlings. The remarkable sociality of crocodilians clearly sets them apart from turtles, lizards, and snakes, and may provide a glimpse of how dinosaurs behaved.*

Contemporary alligators, caimans, crocodiles, and gharials, known collectively as "crocodilians," represent an ancient lineage of archosaurs, allied to dinosaurs and birds. The 23 modern species all reflect the same basic body plan – an elongated snout, a streamlined body covered with protective armor, and a long, muscular, propulsive tail. The extraordinary evolutionary success of the order over 240 million years – the Crocodylia saw the dinosaurs come and go – can be attributed directly to their primary ecological role, established long ago, as top aquatic predators. Living members share a common lifestyle and possess a distinctive anatomy and physiology.

◆ **Below** *The Dwarf crocodile measures only some 1.5m (5ft) in length, and has a distinctive, truncated snout. It inhabits streams and creeks deep in the West African rainforest, and unlike other crocodiles, makes extensive nocturnal, terrestrial forays.*

Aquatic Hunters
FORM AND FUNCTION

Alligators and caimans (the alligatorids) are distinguished by broad, blunt snouts. Teeth in the lower jaw lie inside the closed mouth. The group is monophyletic, based on morphological and molecular analyses, and arose in the Cretaceous era in North America, 144–65 million years ago. Salt-excreting glands in the tongue are absent in all species, suggesting that transoceanic dispersal is unlikely. The caiman lineage spread to South America by the Paleocene (65–55 million years ago), and today consists of three genera in diverse habitats in Central and South America. Dwarf caimans (*Paleosuchus*) are small, have ossified skins, and live in forested regions. The Black caiman (*Melanosuchus*) superficially resembles the American alligator in size and appearance, and is allied to the Broad-snouted caiman. The remaining species of caimans (*Caiman*) are closely related, forming a grouping divided into three species in some accounts. The two alligator species (*Alligator*) diverged in the Tertiary, at least 14 million years ago, when climatic conditions favored dispersal into Asia from North America across Beringia.

Crocodiles and the False gharial (the crocodylids) are characterized by slender to broad snouts. Exposed teeth along the lower jaw are evident when the mouth is closed. Living crocodylids possess salt-excreting glands in the tongue, indicating some capacity for transoceanic dispersal. The False gharial (*Tomistoma*) is included with crocodylids on morphological data, but molecular analyses group it with the gharial. A Paleocene divergence is likely between the False gharial and the other crocodylids, and a Miocene separation between the Dwarf crocodile (*Osteolaemus*) and the remaining crocodiles is indicated by fossil forms. Recent morphological and molecular studies suggest that the 12 species of *Crocodylus* are closely related and recent in origin, from the Pliocene/Pleistocene (5–0.1 million years ago). The genus *Crocodylus* is of African origin, and the African slender-snouted crocodile is the most distinctive, oldest member of the group. The mugger is allied with other Indo-Pacific species. The Nile crocodile is a recent immigrant to Africa, and related to the New World species.

The gharial (*Gavialis*), sometimes called the gavial, is a distinctive species characterized by an elongated, slender snout. It is the sole member of a separate lineage that arose in the Cretaceous (144–65 million years ago). The oldest fossils have been found in North America and Europe, with more recent forms from Africa, South America, and Asia. Lingual salt glands are poorly developed in the gharial, but suggest an affinity with crocodylids rather than with alligatorids, which lack comparable structures on the tongue.

For all crocodilians, concealment underwater is critical, for they are opportunistic shoreline predators that ambush prey at the water's edge. Exposure of the long, often broad snout and head is minimized by strategic placement of only ears, eyes, and the tip of the snout above water. A bony secondary palate permits breathing with the mouth closed, and a palatal flap prevents water from entering the throat. The robust skull and massive jaw muscles that close the mouth can exert a ton of force on the conical teeth, enabling a crocodile to crush a turtle shell or to puncture and hold large prey.

Victims are often drowned, thanks to the crocodile's ability to submerge and remain underwater for many minutes or even hours. A diving crocodile is able to greatly reduce blood flow to the lungs, utilizing a bypass (the foramen of Panizza) between the divided ventricles of the four-chambered heart, another unique feature of the group. The ability to breathe intermittently, depending instead on anaerobic metabolism, is a common feature shared with other reptiles, as is the ability to alter heart rate and blood flow during various activities. Respiration in crocodiles is facilitated by a unique muscular attachment to the liver and viscera that acts as a piston during inhalation and exhalation. Diving is accomplished either by an abrupt exhalation or by a power

◁ **Left** *Swimming in the clear waters of Lake Tanganyika, a Nile crocodile exhibits the features that make all crocodilians such fearsome aquatic hunters – a hydrodynamically efficient body shape, heavily webbed feet, and a massive, muscular tail.*

FACTFILE

CROCODILIANS

Order: Crocodylia

23 species in 9 genera and 3 families

DISTRIBUTION Tropical and subtropical areas around the world, and extending into temperate zones (alligators).

ALLIGATORS Family Alligatoridae
8 species in 4 genera, including: **American alligator** (*Alligator mississippiensis*), **Chinese alligator** (*A. sinensis*), **Common caiman** (*Caiman crocodilus*), **Yacaré** (*C. yacare*), **Broad-snouted caiman** (*C. latirostris*), **Dwarf caiman** (*Paleosuchus palpebrosus*), **Smooth-fronted caiman** (*P. trigonatus*), **Black caiman** (*Melanosuchus niger*). SE USA, E China, C and S America. **Length:** Most 1.5–4m (5–13ft) from snout to tip of tail, up to 5m (16ft). **Features:** Fourth tooth in lower jaw not visible when jaw closed. Snout short and broad. **Conservation status:** The Chinese alligator is Critically Endangered.

CROCODILES Family Crocodylidae
14 species in 3 genera, including **American crocodile** (*Crocodylus acutus*), **Slender-snouted crocodile** (*C. cataphractus*), **Orinoco crocodile** (*C. intermedius*), **Philippine crocodile** (*C. mindorensis*), **Australian freshwater crocodile** (*C. johnsoni*), **Morelet's crocodile** (*C. moreletii*), **Nile crocodile** (*C. niloticus*), **New Guinea crocodile** (*C. novaeguineae*), mugger (*C. palustris*), **Saltwater crocodile** (*C. porosus*), **Cuban crocodile** (*C. rhombifer*), **Siamese crocodile** (*C. siamensis*), **Dwarf crocodile** (*Osteolaemus tetraspis*), **False gharial** (*Tomistoma schlegelii*). Africa, Madagascar, Asia, Australia, C and S America, Caribbean, S Florida. **Length:** 1.5–6.5m (5–21ft). **Features:** Fourth tooth in lower jaw visible when jaw closed; snout short and broad to long and slender. **Conservation status:** The Philippine, Orinoco, and Siamese crocodiles are all Critically Endangered; the Cuban crocodile and the False gharial are Endangered; and the American and Dwarf crocodiles and the mugger are Vulnerable.

GHARIAL Family Gavialidae
1 species: *Gavialis gangeticus.* N and E India, Nepal, Pakistan, Bangladesh. **Length:** Male usually 5m (16ft), up to 6.5m (21ft); female 3–4m (10–13ft). **Features:** Snout greatly elongated; tip potlike in male. **Conservation status:** Endangered.

Equator

> ◖ *Right* In the Masai Mara reserve, Kenya, a Nile crocodile lunges at a wildebeest calf as it stoops to drink. Crocodiles are expert hunters, picking off young wildebeest, or those weak from exhaustion, as they try to cross rivers during their annual mass migration.

> ◑ *Below* Yacaré caimans basking on a riverbank in the Pantanal in southern Brazil. Throughout its range, which also includes Bolivia, Paraguay, and northeastern Argentina, this species is still locally abundant, though widespread hunting for skins takes a heavy toll.

stroke of the tail and hindlimbs, moving the animal downward and backward.

The muscular tail, making up half a crocodile's total length, undulates laterally in axial swimming. Limbs are held against the body when cruising or lunging, but are extended for braking and steering during swimming maneuvers. Some crocodiles launch themselves almost entirely out of the water by "tail-walking" when jumping for prey. Others, such as the Australian freshwater crocodile, routinely "gallop" over rough terrain on land. Young and adults can also scale obstacles several meters high. Travel on land includes the "high walk," in which the limbs are held nearly vertical beneath the body, a gait more typically mammalian than reptilian. An axial bracing system employing ball-socket vertebrae enhances the crocodilians' capacity for varied aquatic and terrestrial movements.

When they occur together, crocodilians tend to differ morphologically. For example, the three species living in West Africa look very different from one another, and are characterized by three distinct snout types – a typical, flat, toothy "crocodile" snout (found in the Nile crocodile); a blunt snout and stout posterior teeth (the Dwarf crocodile); and a slender snout and a tubular rostrum (the Slender-snouted crocodile). Fossil faunas are similar in this regard, and also suggest that similar snout shapes have evolved multiple times in different lineages. Despite a superficial similarity, the False gharial and Indian gharial are not close relatives and have separate distributions, although both are fish specialists.

Subtle Predators
DIET AND FEEDING

Crocodilians are indiscriminate feeders, displaying catholic tastes for animal, but not plant, protein. Feeding may involve sophisticated, subtle behavior. Caimans use the tail and body to corral fish trapped in shallow water, and Nile crocodiles hunt large game cooperatively. Nesting colonies of birds, bat roosting sites, and narrow inlets abundant in fish are sought out. Crocodilians prefer particular foods, and are often reluctant to change diets. Researchers have imprinted hatchling preferences in diet by painting flavors on incubating eggs.

Prey is usually swallowed whole or in large pieces. Digestion occurs in a baglike, muscular stomach equipped with longitudinal ridges in which hard, indigestible objects are embedded; these stomach stones, or gastroliths, facilitate the mechanical breakdown of food, but may also function as ballast. Stomach enzymes are so strong that pH levels are among the lowest ever recorded in any vertebrate. Food is digested rapidly at warm temperatures. Scats are uniformly chalky and contain little waste material, such as bones or feathers.

Adjusting the Thermostat
THERMAL REGULATION

Like other reptiles, crocodilians are ectothermic, relying on external heat sources to regulate body temperature. Bouts of activity are powered by anaerobic metabolism, after which a long recovery period is needed. A crocodile reacts quickly and with power, but tires easily. Behavior tends to be sporadic, and action is punctuated with periods of inactivity stretching over minutes and hours. Crocodilians have low metabolic rates that, in turn, can reduce food requirements. Large individuals are able to survive for months without feeding, if body temperatures remain low.

In the spring, American alligators actively seek heat by moving onto land to bask in the morning and then later in the day retreating into the water, where they assume exposed positions at the water surface. Body temperatures are elevated to levels of 31°–33°C (88°–91.5°F) while on land, and then maintained at these levels throughout the afternoon. During the evening and at night, the animals remain in the water. Throughout the night, their body temperature drops slowly, to

Above *During the dry season, Common caiman in the Venezuelan llanos help regulate their body temperature and avoid desiccation by wallowing in mud holes. The thick coating of mud insulates them from intense solar radiation during the daytime, and buffers cool nighttime temperatures.*

TEMPERATURE-DEPENDENT SEX DETERMINATION IN CROCODILIANS

Why are alligator hatchlings from the same nest usually either all brothers or all sisters? In some reptiles, sex is determined by egg temperature during incubation rather than by sex chromosomes, as in birds and mammals (see Temperature and Sex). To date, temperature-dependent sex determination (TSD) has been demonstrated in a majority of crocodilian species representing all three main lineages. Unlike in turtles and lizards, TSD may in fact be universal in the Crocodylia, since all species lack sex chromosomes (BELOW a Nile crocodile hatchling emerging).

The overall TSD pattern is remarkably uniform, despite minor species differences. Only females are produced at low incubation temperatures (28°–31°C/82°–88°F). Males are produced at intermediate temperatures (32°–33°C/89°–92°F), while high temperatures (34°–35°C/93°–95°F) produce mostly or only females. Thus, a female–male–female (FMF) pattern of TSD is characteristic. The critical period of thermal sensitivity, when temperature affects sex, encompasses the middle third of development.

Exactly how temperature determines embryo sex is still a puzzle. Sex hormones, particularly estrogen, also play major roles. In alligators, the yolks of freshly laid eggs contain multiple sex hormones. These maternal hormones probably influence the sex of offspring, especially at transition temperatures that produce both males and females.

Hormones from the environment may also affect developing embryos. Estrogen painted onto an eggshell can change the sex of an embryo incubating at warm temperatures from male to female. Some pollutants, like PCBs and DDT derivatives, have feminizing effects on alligator as well as turtle eggs. On the other hand, using hormones in captive breeding

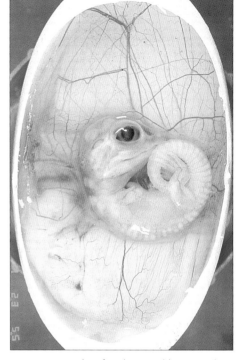

programs to produce females may aid conservation.

When a female lays her eggs and where she builds her nest – in the sun or in the shade – both have major effects on the sex ratio of her offspring. A nesting female carefully selects a nest site, and often makes "trial" nests. Thermal cues may be important. In South India, muggers that nest early when soil temperatures are cool produce mostly female offspring; later, when soil temperatures are warmer, the hatchlings are mostly male (ABOVE a Mugger embryo).

Incubation temperatures sometimes vary enough within a nest to produce males in the top layer of eggs and females below, or vice versa. Small differences in temperature, of 0.5° to 1.0°C (1°–2°F), result in markedly different sex ratios. Droughts and low water result in drier, hotter incubation temperatures; rain and high water in lower temperatures. In the wild, sex ratios in alligator hatchlings are strongly skewed in favor of females in most years, but shift toward males in hot, dry years. Conceivably, a changing climate might lead to the overproduction of one sex and the eventual extinction of the species; but crocodilians are an ancient lineage, and their continued survival argues against such simple scenarios.

near water temperature by early morning, when basking is initiated once again. In saltwater crocodiles and caimans, which live in the warm tropics, the reverse pattern is the norm. Most of the daylight hours are spent submerged, but at night animals remain on land, where temperatures are relatively cool.

Temperature selection, whether in the form of heat seeking or heat avoidance, is an important daily activity for all crocodilian species. Because individuals spend much of the day or night in the water, ambient water temperatures and seasonal changes in them strongly influence thermal behavior and the resulting body temperatures. American alligators and other species sometimes select thermal regimes that result in lowered temperatures even when there are opportunities to be warmer. For example, alligators in South Florida move onto land at night during the warm summer months, and body temperatures cool below ambient water temperatures. During the warmer months, Saltwater crocodiles and New Guinea crocodiles in Papua New Guinea, caimans in Venezuela, and gharial, Muggers, and Saltwater crocodiles in southern India move onto land at night and cool below ambient water temperatures by early morning.

The thermal biology of crocodilians differs in important ways from that of other reptiles. Crocodilians are much larger, so the time lags involved in heating and cooling are measured in hours, or even days, rather than minutes. Their amphibious habit and large size allow them to use water effec-

tively as a heat source and heat sink. An alligator floating at the water surface acts as a heat shunt, absorbing heat from direct sunlight while losing heat to the cool water around it. Finally, their thermal responses are complex, and modified by factors such as climate, social interaction, age, size, reproductive state, digestion, and infection. Fed animals spend more time basking, and the

resulting elevated body temperatures facilitate digestion. Animals that are infected with pathogens select high body temperatures that enhance disease resistance. Because body temperatures directly control metabolic rate and energy utilization, vital processes such as growth and reproduction are ultimately determined by thermoregulation.

Pantropical Travelers
DISTRIBUTION AND ENVIRONMENT

Crocodilians live in almost every wetland habitat worldwide, ranging from dense rainforests to offshore islands. Some species are salt-tolerant and live in brackish or even sea water. Alligators and caimans lack salt glands on the tongue, whereas crocodiles and gharials possess them. The present-day species distribution reflects the ability of crocodiles, but not alligators or caimans, to disperse across open ocean. The Saltwater crocodile is widely distributed in estuarine habitats throughout Southeast Asia to New Guinea and northern Australia, and has ventured into the Caroline Islands, more than 1,300km (800mi) from the nearest population.

Some species regularly modify their environments. When water levels drop seasonally, muggers dig tunnels and burrows large enough to shelter a number of adults. Caimans in the Venezuelan llanos bury themselves in soft mud, where they remain entombed beneath a thick surface crust until water levels rise. Estivating Australian freshwater crocodiles spend three to four months underground without access to water, if suitable refuges are available. During cool weather, Chinese alligators move into elaborate, multilevel underground burrows. American alligators survive in cold weather by opening holes in the ice.

Dominant Males and Breeding Groups
SOCIAL BEHAVIOR

Juveniles and adults are less gregarious than hatchlings, but may form loosely organized social groups. In some species, including American alligators and Nile crocodiles, basking groups frequently assemble at particular times of day. In drought-prone habitats, individuals may group together or segregate into size/age groups at permanent water sites. In the Venezuelan llanos, caimans are concentrated in the few available permanent pools. Australian freshwater crocodiles congregate in isolated billabongs. Social encounters, particularly territorial or dominance behaviors, are often suspended or diminished when such groups form.

In wild populations, dominant males exclude other males from well-defined territories. The defended resource (or resources) varies with species and includes access to mates, nesting sites, nurseries, foraging areas, basking locations, overwintering sites, or some combination of these factors. Territorial defense intensifies during the reproductive season, but often persists throughout the year. Combats between territorial males contesting dominance may involve head-to-head physical contact such as sparring with the jaws or head ramming, as well as posturing with raised, inflated bodies.

Saltwater crocodiles in tidal waterways of northern Australia occupy year-round breeding territories. A male's territory often encompasses the nesting sites of several females. Breeding groups do not form, and adults are rarely found together at any time of year. Nile crocodiles in Lake Rudolf form large seasonal breeding groups (sometimes more than 200 strong) dominated by a small number of up to 15 territorial males. Males and females stay together through the hatching period, and then disperse throughout the lake. Alligators in coastal Louisiana are solitary throughout much of the year, but congregate in the spring in small groups of up to 10 animals in open water for breeding. Later, females disperse to nesting areas and remain near the nests with young after hatching.

▶ **Right** *Representative species of the three families of crocodilians:* **1** *Dwarf caiman (Paleosuchus palpebrosus); Alligatoridae.* **2** *Dwarf crocodile (Osteolaemus tetraspis); Crocodylidae.* **3** *Female False gharial (Tomistoma schlegelii); Crocodylidae, with young.* **4a** *Female Gharial (Gavialis gangeticus; Gavialidae, fishing.* **4b** *The male gharial has a prominent boss on the tip of its snout.* **5** *Chinese alligator (Alligator sinensis); Alligatoridae.* **6** *American crocodile (Crocodylus acutus); Crocodylidae.* **7** *Female American alligator (Alligator mississippiensis); Alligatoridae, on nest mound.* **8** *Black caiman (Melanosuchus niger); Alligatoridae.* **9** *Mugger (Crocodylus palustris); Crocodylidae.* **10** *Slender-snouted crocodile (Crocodylus cataphractus); Crocodylidae.*

Dominance is most obvious during seasonal reproduction. When densities are low, dominant animals maintain separate territories that may vary in size and location with social status. Females and subadult males may be tolerated within a male's territory, but other adult males are prohibited. In high-density situations, territorial maintenance becomes increasingly difficult. Under such conditions, dominance hierarchies typically form.

Getting the Message Across
COMMUNICATION

Crocodilians convey social messages through sound, postures, motions, odors, and by touch. Communication begins in the egg and continues throughout adulthood. Hatchlings vocalize spontaneously or when disturbed; in the latter case, adults typically respond with threats or attacks. Young also produce "contact" calls as groups assemble. Juveniles and adults vocalize, especially when handled, by making deep, guttural "whaa" calls. Alligators are notably the most vocal, and are renowned for the bellowing choruses of breeding males and females. Some crocodiles produce a throaty, repetitive roar when approached closely by another adult. Species living in open water, in lakes, and along rivers vocalize less often than species living in marshes and swamps.

Communication in the water is well suited to the amphibious lifestyle of crocodilians. Acoustic, rather than vocal, sounds include the headslap or jawclap performed at the water's surface. The exact form and context of headslapping varies with species, but it occurs in nearly all species studied to date. In gharials, an underwater jawclap produces a muted popping sound that announces the performer, in a manner analogous to the headslap in other crocodilians. Other acoustic messages, barely perceptible to humans, are subaudible vibrations, sometimes referred to as "infrasound." These extremely low-frequency signals resemble the sound and feel of distant thunder, and are produced by bellowing and roaring animals in a variety of social contexts.

Alligators emit soft, purring "chumph" sounds by expelling air through the nostrils during short-range courtship encounters. In many species, exhalations underwater produce bubbles, ranging from a steady stream of fine bubbles to an explosive expulsion of several large ones. The protuberance on the snout of the male gharial is a convoluted cavity with connections to the nasal chamber, and appears to change hissing exhalations into buzzing sounds, as air resonates in the enlarged nasal cavity.

Exposure of the head, back, and tail above the water surface conveys important information about an individual's social status and intent. Dominant animals advertise their large size by swimming boldly at the surface. A subordinate will lift its snout out of the water at an acute angle, open its jaws, and hold its head stationary, before retreating underwater. Tail thrashing, involving movements of the tail from side to side, often precedes some other behavior.

Secretory glands are located under the chin and in the cloaca; these oily secretions may function as defensive compounds to repel potential predators and/or may be employed as chemical messages between individuals. Adults living together seem to recognize each other, and most displays, such as bellowing and headslapping by specific animals, are individually distinctive and recognizable. Even hatchlings may be able to recognize individuals, or at least be able to distinguish siblings from other hatchlings.

Hole Nesters and Mound Nesters
REPRODUCTION

Adults are long-lived, typically surviving for 20 to 40 or more years, and they can reproduce for decades, from the age of 10 to 30 or more. Larger species mature at older ages (10–15 years) and larger sizes (3–6m/10–20ft) relative to smaller species (5–10 years and 1–3m/3.3–10ft, respectively). Captive females have reproduced successfully at 40 years of age. In all but the smallest species, males are up to twice as large as females. The mating system is polygynous; each male typically inseminates many females. The ratio of reproductive males to females varies from about 1:20 in Nile crocodiles to 1:1–3 in some territorial

○ **Above** *A young male (center) and two young female gharials basking on a riverbank in Uttar Pradesh state, India. The male is distinguished by a prominent "boss" on the tip of the snout; buzzes emitted through it warn off rivals, and attract mates.*

species, such as Saltwater crocodiles. One captive Mugger sired more than 300 hatchlings in a single season. Multiple paternity occurs in wild American alligators, with up to three males fathering young from a single clutch; but sperm storage is unlikely.

Large males dominate breeding groups by patrolling territories. Dominant males approach females to initiate courtship, or are approached by females, often after males perform conspicuous displays. During courtship, males and females engage in a variety of species-specific behaviors that include snout contact, snout-lifting, head and body rubbing and riding, conspicuous male displays, vocalizations, exhalations, narial and guttural bubbling, circling, and periodic submergence and reemergence. Mating occurs when the male moves onto the female's back, positions his tail and vent underneath the female's tail, and inserts an anteriorly curved penis into the female's cloaca. Copulation may last for 10–15 minutes, and occur repeatedly over several days. To judge from studies of American alligators and Nile crocodiles, the female nests approximately one month later.

A crocodilian female lays between 10 and 50 hard-shelled eggs that incubate for 2–3 months before hatching. In seasonal environments, nesting typically occurs within 2 or 3 weeks in alligators and in Australian freshwater crocodiles. In more equable climates, nesting is extended over 3–4 months for Saltwater crocodiles. Most females nest every season, but some alligators only nest every 2 or 3 years. In contrast, Muggers in captivity regularly produce two clutches per season.

A female prepares a nest by scraping together vegetation into a mound or by excavating a cavity. Hole nesters scrape and dig in potential nest sites with coordinated movements of the front and hindlimbs. Mound nesters use similar movements to gather material into a central pile and

🔵 **Below** *American alligators typically reach sexual maturity within 7–10 years; however, since larger, older males hold sway for a long time over territories and harems of females, most males only breed for the first time when they are aged 15–20.*

PARENTAL CARE IN CROCODILIANS

Among reptiles, parental care is rare except in the Crocodylia. Maternal behaviors include nest attendance and defense, nest opening, the manipulation of eggs to release hatchlings, mouth transport of eggs and the young, and post-natal care. Males may also participate in all of these behaviors, but mainly respond to hatchling calls and defend the young.

Vigorous nest defense is a good indication that eggs have been laid. Intruders are confronted with close approach, open-mouthed lunges, and mock biting, and may even be attacked. Saltwater crocodiles are tenacious nest defenders, while in less aggressive species nest defense may wane during incubation.

Once the eggs hatch, the female helps the young to emerge from the nest cavity by digging with her front and hind legs. Hatchling vocalizations appear to be an important cue directing adults to the young. As she excavates the nest, the female often rests her lower jaw and head on the substrate and places her snout inside. On finding the eggs and newly-emerged young, she picks them up in her jaws and shifts them into her mouth (BELOW). She rolls the unhatched eggs around, gently crushing the shell between tongue and palate – actions that facilitate hatching and the freeing of the young from the egg. Nest excavation and mouth transport of hatching eggs and young appear to be universal in the group, although some species have yet to be fully studied.

The hatchlings maintain group cohesiveness by frequent vocalizations, and a nursery or crèche is established. The young disperse to forage, but regroup during periods of inactivity. Parents remain near groups of hatchlings and defend them. Hatchling vocalizations attract other adults, subadults, and even juveniles, but these may all be kept away by protective parents. It has been suggested that attending adults may even feed the young, but this behavior has yet to be documented.

The young remain with the adults in loosely organized groups for variable lengths of time. In American alligators, females remain with groups of young near nesting areas for 1–2 years, but in Nile and Saltwater crocodiles the young disperse within months. The cohesion of groups of young may depend on the presence of the female and/or on seasonal changes in available habitat. Males may be associated with hatchling groups in some areas where they remain near nesting sites. In alligators, however, males do not regularly stay near the female during the nesting, hatching, or post-hatching period.

frequently walk across it, compacting and shaping the nest with trampling movements of the hind feet. Once a hole is dug the eggs are laid, usually within an hour at night, and the nest is reshaped or covered.

The type of nest constructed (mound or hole) is species-specific, but variations in nest types are evident in different habitats, and even vary within habitats. Colonial nesting occurs in some hole-nesting species, but mound-nesters also construct nests in close proximity in some habitats. Nests are often located near dens or caves that provide subterranean, aquatic retreats for an attending female. A female typically visits her nest frequently during incubation and guards the nest against potential predators.

Older is Safer
ECOLOGY

For most species, natural mortality is very high for eggs and hatchlings, and then declines as the young grow. Nest mortality results not only from a host of environmental calamities, such as flooding, overheating, or desiccation during the lengthy incubation, but also from the activities of persistent egg predators, ranging from bears to ants.

Despite parental protection at the nest and when the young hatch, hatchlings are readily eaten by a wide variety of reptilian, avian, and mammalian species. For Saltwater crocodiles in northern Australia, survival from egg laying to hatching has been estimated at only 25 percent, and subsequent survivorship at 30–60 percent per year up to the age of 5.

By the time a juvenile is 5 years old and 1m (3.3ft) or more in length, its chances of surviving through the next year increase markedly. Mortality in subadults and adults is low. In Africa, Nile crocodiles occasionally fall prey to hippos or elephants intent on defense. In Central and South America, anacondas take caimans, and in Asia leopards and tigers kill crocodiles. As subadults reach adulthood, the most likely threat to their survival comes from their peers and from humans. Cannibalism has been well documented in several populations of American alligators recently, and is thought to be relatively common in other species, including Saltwater crocodiles. For all size classes, this risk increases at high population densities, when larger individuals prey on smaller ones. Dominance disputes may at times result in serious injury or death from combat.

○ **Above** A mature Nile crocodile displays its impressive armory. One of the largest species of crocodilian, it can grow to over 6m (19ft) in length and 1 ton in weight. However, many eggs and young fall victim to predators, particularly monitor lizards and baboons.

○ **Right** Juvenile Saltwater crocodiles in Northern Territory, Australia. From a critical low point during the 1970s, numbers of this species have recovered dramatically, largely thanks to the implementation of effective conservation programs.

In a number of species, juveniles disperse from nesting areas, and may move kilometers away, often into other habitats. Older juveniles and subadult alligators in Louisiana have often been recovered far from where they were marked. As juveniles, Saltwater crocodiles appear to be displaced for a time to coastal areas before venturing back as subadults to establish territories in home rivers. In the Australian freshwater crocodile, both sexes disperse from natal billabongs long before they are sexually mature. Males disperse two to three times farther than females, but both sexes show low levels of philopatry. Despite such dispersal patterns, recent studies in Australia indicate that gene flow between river systems is more limited in freshwater than in saltwater crocodiles.

An Australian Success Story
CONSERVATION ISSUES

In 1971, after three decades of unregulated hunting, the wild population of Saltwater crocodiles in the Northern Territory of Australia was reduced to about 5 percent of historical levels, and adults were rare. By 2001, the wild population had recovered to near pristine levels (about 75,000), and they occur throughout their former habitats. This remarkable recovery has been carefully monitored, with periodic surveys that demonstrate crocodilians' extraordinary ability to survive and thrive despite near-extinction.

The recovery went through various stages. During the initial years of protection, the few remaining adults nested. Despite high egg mortality, hatchling and juvenile survival was high, and the population consisted mostly of juveniles. As crocodile numbers continued to increase, hatchling and juvenile survival decreased. Cannibalism may have played a role in this well-documented, density-dependent response.

Strict protection was enforced during the first decade, but it was relaxed eventually as the increasing numbers of large crocodiles threatened livestock and people; in 1979–80, injuries and deaths were reported. These attacks prompted a public education program and a removal scheme aimed at problem crocodiles. In addition, crocodile farms were started to produce skin and meat and satisfy tourist demands. By the mid 1980s, ranching using wild-collected eggs began; by the mid-1990s, wild harvests were resumed. Despite harvesting at varying levels, the wild population over the last decade has remained stable or even increased.

Conservation of crocodilian populations is highly dependent upon management practices that allow people and crocodiles to coexist. Successful programs have focused on incentives to maintain crocodiles and their habitats in a relatively undisturbed state. Sustainable use has become a key element in recent conservation efforts, based on more than two decades of experience with different management schemes in Papua New Guinea, Venezuela, Zimbabwe, the USA, and Australia. In each case, crocodilian populations in the wild have increased or remained stable while supporting economically viable levels of harvest. Utilization methods have included the hunting of wild animals, ranching by collecting eggs or hatchlings from the wild, and captive breeding by maintaining breeding adults and raising their offspring. A major breakthrough in these efforts has been the active involvement of traders and manufacturers of skin and meat products. Together with biologists, they have supported programs that ensure that wild populations and their habitats are adequately monitored, and that the harvests are regulated in order to guarantee sustainable use.

Ultimately, successful conservation depends on habitat preservation. The Chinese alligator may become the first crocodilian in historical times to become extinct in the wild. The species was formerly widely distributed along the lower reaches of the Changjiang (Yangtze) River in eastern China. Chinese alligators mature at small size and grow slowly compared to their New World relatives. They build complex underground burrows, with pools above and below ground and numerous air holes. The extensive use of these burrows, in which they hibernate, and their secretive behavior has allowed the species to inhabit remnant wetlands in densely populated areas, such as crop fields and tree farm communes. Today, the wild population is tiny and highly fragmented, and most remaining wild alligators exist in small groups, or as single individuals lost in an agricultural landscape. Ironically, a large number of captive animals – more than 7,000 in total – exist in China and in zoos worldwide. The future of wild Chinese alligators rests on habitat rehabilitation and the reestablishment of viable populations by releasing captive-reared alligators. JWL

POLLUTION AND HORMONE MIMICS

Endocrine actions of contaminants in reptiles and amphibians

THROUGHOUT THEIR EVOLUTIONARY HISTORIES organisms have been exposed to toxic substances, many of them produced by living organisms but some generated by natural events such as volcanic eruptions, meteor hits, or forest fires. From the advent of the industrial age, however, a new factor has been the widespread synthesis, distribution, and use of chemicals, a process dramatically accelerated from the mid 20th century on. These man-made, "xenobiotic" chemicals total in the tens of thousands; over 90,000 are currently used in the USA alone. Numerous studies have documented global contamination by a wide range of them.

Traditionally, scientists have studied the effects of environmental contaminants by examining their ability to cause death, cancer, or birth defects, largely in the context of genotoxic responses involving gene mutation. Over the last decade, however, it has been demonstrated that many xenobiotic chemicals can influence the biology of animals by altering their endocrine system. This system is composed of ductless glands that synthesize chemical messengers directing a large number of functions such as growth, metamorphosis, reproduction, metabolism, and water balance. Altering the synthesis, storage, or metabolism of these messengers significantly alters their action. A growing literature supporting the mechanism of contaminant-induced endocrine disruption has been generated from laboratory and field observations in amphibians and reptiles.

In the 1980s, a series of studies in Florida documented reduced egg hatching and elevated neonatal mortality in American alligators (*Alligator mississippiensis*) from several lakes in the state. Although the cause of this mortality is still unknown, it has become clear that the biology

of the populations concerned has been adversely impacted by pesticide exposure and nutrient pollution. Male neonatal and juvenile alligators from contaminated lakes in Florida have been shown to exhibit a number of abnormalities in their reproductive and endocrine systems; for example, they have femalelike concentrations of the sex steroid hormone testosterone in their blood as well as reduced phallus size, while their testes produce elevated levels of estradiol, a potent estrogen. Females from the same populations show elevated levels of estradiol along with abnormalities in the structure of their ovaries, indicative of exposure to estrogens. Where could these estrogens come from? DDT and other pesticides can mimic the action of estrogens.

Alligators exhibit environmental sex determination, by which temperature influences gender. Many previous studies have shown that an estrogen applied to an egg prior to the period of sex determination can induce the formation of a female. A number of recent studies have reported that various pesticides or their metabolites, among them the DDT metabolite p.p'-DDE, can mimic estrogens at ecologically relevant concentrations, causing sex reversal, for example, in Red-eared sliders (*Trachemys scripta elegans*) and alligators. Interestingly, p.p'-DDE does not reverse sex in seaturtles or snapping turtles, indicating that the effects of contaminants may vary among species. In addition to their effects on the reproductive system, contaminants can also alter circulating

⊙ **Below** *Temperature during incubation determines the gender of Red-eared sliders. Yet other, unnatural factors, such as pesticides and PCBs, have been found to cause sex reversal in this species.*

thyroid hormone concentrations, immune tissue structure and function, and the stress response.

There is particular concern today about the condition of amphibians worldwide. Many factors seem to be contributing to the decline of some populations, and contaminant exposure is one of them. Deformities during development or metamorphosis have attracted the attention of the public and scientists alike. Although some abnormalities appear to be associated with pathogens, others closely resemble the response of frogs exposed experimentally to retinoic acid, a potent developmental signaling chemical. For example, at least one pesticide has been shown to interact with the cellular receptor for retinoic acid. Much further work is needed to examine this hypothesis.

Pesticides and their metabolites have also been identified as hormone mimics in frogs and salamanders. In a series of studies using the African reed frog (*Hyperolius argus*) it was shown that DDT and its metabolites could induce estrogen-dependent female coloration in males; it has also been hypothesized that DDT or its metabolites

might alter or mimic the stress response in amphibians, specifically frogs. In contrast to the frog studies, a study of the Tiger salamander (*Ambystoma tigrinum*) demonstrated that DDT and DDE could act as antagonists of natural hormone action.

Although to many people the actions of DDT and its metabolites may seem to be old news, DDT is still extensively used throughout the tropical regions of the world and continues to pose a threat to wildlife health. Meanwhile, additional studies have begun to expand the types of chemicals of concern, with new work focusing on herbicide formulations, pharmaceutical drug release from agricultural activities and sewage effluent, and so-called "inert" chemicals such as plasticizers, surfactants, and flame retardants. For example, recent studies have found that the polybrominated biphenyls (PBBs) are ubiquitous pollutants and powerful thyroid hormone modulators. Given the central role of thyroid hormone in amphibian life history, the action of these compounds in amphibians needs to be examined.   LJGJr

⬆ **Above** *Young American alligators in Florida. Research on alligators in Lake Apopka, near Orlando, found weakened hatchlings with abnormal endocrine status, possibly due to pesticide exposure.*

⬇ **Below** *DDT and other harmful pesticides have been sprayed for decades. Such chemicals continue to be used in many tropical countries to control mosquitoes and tsetse flies.*

UNISEXUALITY: THE REDUNDANT MALE?

Mechanisms for sex-free reproduction

UNISEXUALITY – REPRODUCTION WITHOUT A paternal contribution to the genetic makeup of the offspring – is nothing less than bizarre. Unisexual species are unusual, extraordinary, very rare in diversity although not necessarily in density, and they reproduce in virtually every way imaginable. Sometimes females incorporate DNA from male sperm and sometimes not. Females are sometimes parasitic on males, using them only to initiate cell division while rejecting their genes. Some females do not use males at all, but they behave like them sexually and in defending territories.

Among vertebrates, unisexuals are the midway freak show of ichthyology and herpetology, and they have captured our attention. All of the unisexual amphibians, and most of the approximately 40 species of unisexual squamates, have formed from the "accidental" hybridization of two bisexual species, and all very recently. Unisexuality may be an important step in the formation of tetraploid species, or it might be an evolutionary dead-end.

Unisexuality in vertebrates takes three forms. Hybridogenesis occurs in fishes and in one group of European ranid frogs. Females of hybrid origin produce only female offspring. The paternal genome is rejected during gametogenesis, and thus is never passed to descendants.

Gynogenesis is another form of unisexuality, in which the sperm of the male is used to trigger cell division in the egg. It occurs in fishes and in the hybrids of four bisexual species of North American mole salamanders of the genus *Ambystoma*. These salamanders breed in ponds. At higher water temperatures, the haploid genome of the male may be incorporated into the hybrid zygote. Incorporation results in elevated ploidy levels where diploids become triploids, triploids become tetraploids, and tetraploids can even become pentaploids. When this occurs, the unisexuals shift from a gynogenetic mode of reproduction to a form of hybridogenesis. Ploidy levels seem to be developmentally and reproductively constrained. Higher levels of developmentally fatal deformities accompany elevated ploidy levels. Reproductively, in an Ontario pond, diploid individuals arrive first to breed in early spring. They are swiftly followed by an onrush of unisexual triploid females, and some diploids. Next, tetraploid participants increase, as few new triploids wander into the ice-cold foray. And more than three weeks later, when breeding has all but come to an end, a few lingering tetraploids and the extremely rare pentaploids arrive. If any viable spermatophores remain in the pond, the warmer water temperatures could result in the formation of hexaploids, which probably could not survive development.

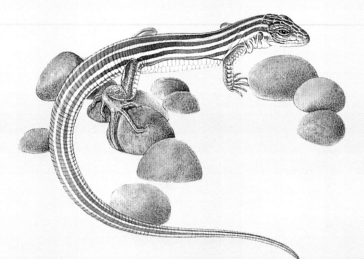

◁ **Left** *Around one-third of all known species of* Cnemidophorus *whiptail lizards are parthenogenetic. As its Linnean binomial indicates,* C. uniparens, *the Desert grassland whiptail, is a parthenospecies.*

▷ **Right** *The Edible frog* (Rana esculenta) *is an example of amphibian hybridogenesis, being the product of pairing between two closely related frog species,* Rana ridibunda *and* R. lessonae.

The third form of unisexuality is parthenogenesis, a reptilian phenomenon among the vertebrates, including birds, that does not require males or their sperm for reproduction. Females produce genetically identical copies of themselves, or clones. The eminent Russian herpetologist Ilya S. Darevsky first reported the phenomenon in 1958 in Caucasian lacertid lizards, currently referred to the genus *Darevskia*. His finding spurred numerous research programs on unisexuals. Parthenogenesis has been documented or hypothesized to occur in about 37 species of squamates, including one triploid snake, *Ramphotyphlops braminus*. The number of recognized parthenospecies varies depending on taxonomic considerations of bisexual plus unisexual forms of a "species," unique clones, hybridization events, and bisexual parentage leading to multiple forms. The final word has not yet been written.

Why are there so few parthenogenetic species? The "balance hypothesis" was proposed in 1989 to explain the formation of parthenogenetic species of lizards. Parthenospecies could be formed if the parents were not similar enough to form functional, viable hybrids with normal meiosis, but not so dissimilar that development could not occur. However, genetic divergence is important for some species, but not for all, and it is not the only important constraint. For example, hybridization is common in *Darevskia*, and among several pairs of species that have the prescribed amount of genetic divergence. However, the paternal parents of all parthenospecies of *Darevskia* are all from one major clade, and the maternal are from one other. Thus, the formation of unisexual squamates in this case is directional and phylogenetically constrained to hybridization between specific monophyletic lineages. No equivalent analysis exists for the more intensely studied genus *Cnemidophorus*, which also has numerous parthenospecies, nor for any other unisexual

species complex that displays high levels of fixed heterozygosity.

It gets more complicated. For example, whereas triploid parthenospecies occur in the teiid genus *Cnemidophorus*, triploid *Darevskia* are extremely rare and sterile. Thus, mechanistic distinctiveness may be the rule, not the exception. In other situations, hybridization may not be involved. Unisexual and bisexual forms of the xantusiid lizard *Lepidophyma flavimaculatum* are virtually indistinguishable, and hybridization is not a viable explanation. In perhaps the most bizarre twist to parthenogenesis, some bisexual species of lizards and snakes appear to have the ability to reproduce, if males are not available, by way of facultative parthenogenesis! This mode of reproduction is now documented in three iguanid lizards in three separate genera (*Basiliscus*, *Iguana*, *Phymaturus*), and in one acrochordid (*Acrochordus arafurae*), three thamnophine (*Thamnophis elegans*, *T. marcianus*, *Nerodia sipedon*), and two viperid (*Crotalus horridus*, *C. unicolor*) snakes. Facultative parthenogenesis, which also occurs in birds, may be more widespread in squamates than previously thought.

Parthenogenesis has arisen independently in several squamate lineages, and it seems likely that each origin is accompanied by unique circumstances and genetic mechanisms. There may be no single explanation, which is hardly surprising. Perhaps the only unifying principle is that, so far as is known, none of the parthenospecies has temperature-dependent sex determination.

The absence of males does not necessarily translate into an absence of male behavior. Laboratory-maintained parthenogenetic *Cnemidophorus uniparens* individuals pursue male "pseudocopulatory" behavior. For some unknown reason, this behavior dramatically increases fecundity in the recipient females. Their sexual behavior is mediated by progesterone, and it may be inherited from

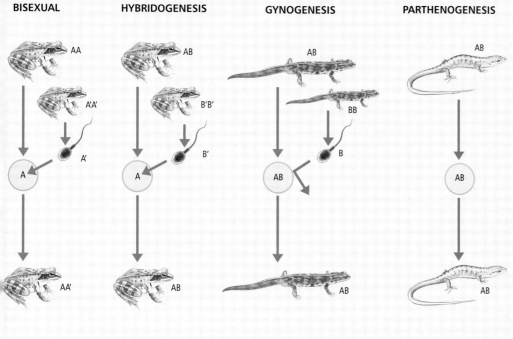

| BISEXUAL | HYBRIDOGENESIS | GYNOGENESIS | PARTHENOGENESIS |

⚫ **Above** *Unisexual and bisexual reproduction in amphibians and reptiles. Each capital letter (A or B) represents a haploid complement of chromosomes, i.e. half the parental genome. In bisexual reproduction, each offspring receives half its genome from its mother, half from its father, both being of the same species. In hybridogenesis, a hybrid female mates* *with a male of one of her parental species; fertilization occurs but none of the paternal genome is incorporated into the offspring's genome. In gynogenesis, a sperm from another species is required to trigger egg division but none of the paternal genome enters the egg. In parthenogenesis, males are not involved in reproduction.*

their paternal ancestor, *C. inornatus*. Pseudosexual behaviors also occur in the gecko *Lepidodactylus lugubris*. However, in this case, dominance behavior is associated with lower fecundity. A subordinate female can only increase its fecundity by dispersing into a less populated area.

Unisexual species can increase their population densities much more rapidly than bisexual species because all offspring are capable of producing more offspring. So why do unisexual species not outcompete their bisexual ancestors? The "weed hypothesis" notes that parthenogenetic *Cnemidophorus* usually occupy marginal habitats and are themselves apparently outcompeted. However, this hypothesis may not be generally applicable. Some unisexual *Darevskia* and *Lepidodactylus* are more common and widespread than their bisexual ancestors, and may be displacing them. In some cases, unisexual species are extremely common; in Armenia, one can sit on a rock by a stream and noose more than 100 specimens of *D. dahli* in a very short time. In other situations, however, they are quite rare, to the point of being classified as Endangered.

Much work remains to be accomplished on unisexuality, including trying to identify the ancestral parents of many species, let alone discovering the underlying genetic and developmental mechanisms. And so the show goes on.   RWM

Glossary

Acrodont teeth teeth that are attached to the upper edge of the jaw as opposed to the inside surface (pleurodont) or in sockets (thecodont).

Adaptation a morphological, physiological or behavioral feature that particularly suits an organism (or group of related organisms) to its (or their) way of life.

Adaptive radiation the evolution of several species or groups of SPECIES (e.g. FAMILIES, ORDERS, etc) from a common ancestral species or groups show individual adaptations to several different ways of life.

Advanced of mod recent evolutionary origin (cf. PRIMITIVE).

Advertisement call sound produced by male frogs during the breeding season that serves to attract females and, in some species, deter other males.

Allantois sac-like outgrowth of underside of hind part of gut in EMBRYOs of reptiles, birds and mammals. During development, it grows around the embryo and, with associated blood vessels, functions in RESPIRATION. The embryo's excretory products are stored in the fluid within the allantois.

Amnion a membrane forming a fluid-filled sac that encloses the EMBRYO of reptiles, birds and mammals.

Amniote any higher vertebrate whose EMBRYOs are enclosed within an AMNION during development; includes reptiles, birds and mammals.

Amphibious capable of living both in water and on land.

Amplexus a position adopted during mating in most frogs and many salamanders, in which the male clasps the female with one or both pairs of limbs.

Anterior toward the forward or head end of an animal.

Aposematic coloration bright coloration serving to warn a potential predator that an animal is distasteful or poisonous.

Arboreal living in or among trees.

Atrium (plural atria) a chamber of the heart receiving blood from veins.

Auditory nerve a nerve linking the inner ear and the brain.

Autotomy the self-amputation of part of the body, as of the tail of some lizards when they are attacked.

Axil the angle between the stem of a plant and a leaf or branch.

Axillary of or pertaining to the armpit. Axillary AMPLEXUS is a mating position in which a male frog or salamander clasps a female beneath her armpits.

Barbel a slender, elongated sensory process on the head, usually of aquatic animals.

Basal basic, fundamental. The basal METABOLIC RATE is the rate of energy expenditure by an animal at rest.

Bask to hold the body in a position directly exposed to the sun.

Binocular vision type of vision in which the eyes are so positioned that an image of an object being observed falls on both retinas.

Bisexual species a species containing both male and female individuals.

Body temperature the temperature of the interior of an animal's body, usually measured in the rectum or, by telemetry, in the stomach.

Bridge the segment of a turtle's shell joining the CARAPACE and the PLASTRON.

Brille a transparent covering over the eyes of snakes.

Bromeliad member of a family of plants, many of which live attached to larger plants, e.g. Spanish moss.

Brood a group of young raised simultaneously by one or both parents.

Brood pouch a space or cavity in which young develop.

Bursa a sac or sac-like cavity.

Calcareous consisting of or containing calcium carbonate.

Calcified CALCAREOUS.

Cannibalistic eating the flesh of one's own species.

Capillary a very narrow, thin-walled tube carrying liquid, e.g. blood.

Capillary action the tendency of liquids to move, without external pressure being applied, along narrow spaces between objects, such as soil particles or in narrow tubes.

Carapace a hard structure covering all or part of an animal's body; the dorsal part of the shell of turtles and tortoises.

Cartilaginous containing cartilage, a tough elastic skeletal material consisting largely of collagen fibres.

Cerebral cortex the outer layer of cells (gray matter) covering the main part of the brain, the CEREBRUM.

Cerebral hemisphere one of the two halves of the CEREBRUM.

Cerebrum that portion of the vertebrate brain which lies above and in front of the brain stem; it contains sensory and motor centers and is involved in learning.

Chemosensation the ability to detect and differentiate substances according to their chemical composition.

Chorion a MEMBRANE that surrounds the EMBRYO and YOLK SAC of reptiles, birds and mammals.

Chromatophore a specialized cell containing PIGMENT, usually located in the outer layers of the skin.

Chromosome a thread-shaped structure, consisting largely of GENETIC material (DNA), found in the nucleus of cells.

Cilium (plural cilia) minute, hairlike process from a cell, capable of beating rhythmically.

Circumtropical encircling the Earth, in an area between 22.5°N and 22.5°S.

Class a taxonomic category ranking below PHYLUM and above ORDER.

Cloaca the common chamber into which the urinary, digestive and reproductive systems discharge their contents, and which opens to the exterior.

Cloud forest moist, high-altitude forest characterized by dense undergrowth, and an abundance of ferns, mosses, orchids and other plants on the trunks and branches of trees.

Clutch the eggs laid by a single female in one breeding attempt.

Cocoon a tough protective covering.

"Cold-blooded" an outmoded term, referring to animals whose body temperature varies with environmental temperature.

Colonize to invade a new area and establish a breeding population.

Compressed of lizard body form, flattened vertically, as opposed to DEPRESSED.

Constriction a method of killing prey, used by some snakes, in which the body is coiled tightly around the prey, inducing suffocation.

Continuous breeder an animal that may breed at any time of year.

Cooperative breeding a breeding system in which parents are assisted in the care of their young by other adult or subadult animals.

Core temperature body temperature of an animal measured at or near the center of its body.

Cornea the front, transparent portion of the eye of a vertebrate.

Courtship behavioral interactions between males and females that preceed and accompany mating.

Cranial of or pertaining to the cranium (skull).

Crest a raised structure running along the back of the head and/or body.

Crypsis The ability to be hidden or camouflaged.

Cryptic hidden or camouflaged.

Cutaneous of or pertaining to the skin.

Depressed of lizard body form, flattened laterally (from side to side), as opposed to COMPRESSED.

Dermis the layer of skin immediately below the EPIDERMIS.

Desiccation the process of drying out.

Dichromatism the condition in which members of a species show one of two distinct color patterns.

Differentiation the process by which unspecialized structures (e.g. cells) become modified and specialized for the performance of specific functions.

Diffraction the process by which light, on passing through an aperture or past the edge of an opaque object, forms a pattern of colors.

Dimorphism the existence of two distinct forms within a species. Sexual dimorphism is the existence of marked morphological differences between males and females.

Direct development transition from the egg to the adult form in amphibians without passing through a free-living larval stage.

Display a stereotyped pattern of behavior involved in communication between animals. Any of the senses – vision, hearing, smell, touch – may be involved.

Dorsal pertaining to the back or upper surface of the body or one of its parts.

Ectoparasite a parasite that lives on the outer surface of an organism, e.g. a tick, flea or louse.

Ectothermic dependent on external heat sources, such as the sun, for raising body temperature.

Eft the juvenile, terrestrial phase in the life cycle of a NEWT.

Embryo the young of an organism in its early stages of development. In amphibians and reptiles, the young before hatching from the egg.

Endothermic able to sustain a high body temperature by means of heat generated within the body by METABOLISM.

Enzyme a substance produced by living cells which is capable of catalyzing a specific chemical reaction.

Epidermis the surface layer of the skin of a vertebrate.

Epiphysis a portion of a bone which develops from a separate center of ossification and later becomes the terminal portion of the bone.

Epiphytic of a plant, growing on another plant but not parasitic on it.

Erectile capable of being erected or raised, as of a penis or crest.

Esophagus part of the GUT, from the throat to the stomach

Estivation a state of inactivity during prolonged periods of drought or high temperature.

Estuarine living in the lower part of a river where freshwater meets and mixes with sea water.

Explosive breeder a species in which the breeding season is very short, resulting in large numbers of animals mating at the same time.

External fertilization fusing of eggs and sperm outside the female's body.

Family a taxonomic category ranking below ORDER and above GENUS.

Femoral gland a gland situated on an animal's thigh.

Fertilization the union of an egg and a sperm.

Fetus the unborn young of a VIVIPAROUS animal in the later stages of development.

Fossil any remains, impression, cast, or trace of an animal or plant of a past geological period, preserved in rock.

Fragile tail a tail that can be shed by ALTOTOMY if the animal is attacked.

Frog any member of the order Anura. Also, an anuran which is smooth-skinned, long-lived and lives in water.

Gamete OVUM or sperm.

Gastralia rib-like bones present in the under part of the body of some reptiles.

Genetic of or pertaining to genetics or heredity.

Genus (plural genera) a taxonomic category ranking below FAMILY and above SPECIES; contains one or more species.

Germ cell GAMETE.

Gestation carrying the developing young within the body.

Gill a respiratory structure in aquatic animals through which gas exchange takes place.

Girdle a group of connected bones that provide support from a pair of limbs; pectoral or shoulder girdle, pelvic or hip girdle.

Gland an organ (sometimes a single cell) that produces one or more specific chemical compounds (secretions) which are passed (secreted) to the outside world.

Gut the alimentary canal, especially the intestine.

Hatchling a young animal that has just emerged from its egg.

Hemipenis (plural hemipenes) one of two grooved copulatory structures present in the males of some reptiles.

Homeothermic having the ability to maintain a constant, or nearly constant body temperature, irrespective of the temperature of the environment.

Home range an area in which an animal lives except for migrations or rare excursions.

Hormone a substance secreted within the body which is carried by the blood to other parts of the body where it evokes a specific response, such as the growth of a particular type of cell.

Hybrid an individual resulting from a mating of parents which are not genetically identical, e.g. parents belonging to different species.

Hyoid a U-shaped bone to which the larynx is attached.

Hypophysis the pituitary gland.

Ilium (plural ilia) dorsal part of the pelvic (hip) girdle.

Imbricate scales overlapping scales, like tiles on a roof.

Incubation the act of incubating eggs, i.e. keeping them warm so that development is possible.

Infrared radiation invisible heat rays beyond the red end of the visible light spectrum.

Inguinal of or pertaining to the groin. Inguinal AMPLEXUS is a mating position in which a male frog or salamander clasps a female around the lower abdomen.

Internal fertilization fusing of eggs and sperm inside the female's body.

Intromission the act of inserting the male copulatory organ into the body of the female.

Intromittent organ a male copulatory organ, e.g. penis, HEMIPENIS.

Jacobson's organ (or vomeronasal organ) one of a pair of grooves extending from the nasal cavity and opening into the mouth cavity in some mammals and reptiles. Molecules collected on the tongue are sampled by this organ in CHEMOSENSATION.

Juvenile young, not sexually mature.

Juxtaposed scales scales with edges touching, but not overlapping.

Keel a prominent ridge, e.g. on the back of some turtles and on the dorsal scales of some snakes.

Keratin a tough, fibrous protein present in epidermal structures such as horns, nails, claws and feathers.

Kinesis movement in response to a stimulus.

Labyrinthodont teeth teeth with complex infolding of the enamel.

Larva the early stage in the development of an animal (including amphibians) after hatching from the egg.

Larynx a sound-producing organ located at the upper end of the trachea (the wind-pipe), containing the vocal cords.

Lateral line organ a sense organ embedded in the skin of some aquatic animals which responds to water-borne vibrations.

Life cycle the complete LIFE HISTORY of an organism from one stage (e.g. the egg) to the recurrence of that stage.

Life history the history of an individual organism, from the fertilization on the egg to its death.

Live-bearing giving birth to young that have developed beyond the egg stage.

Lymph gland organ on the course of a lymphatic vessel, containing lymph (a colorless fluid originating from the spaces between cells) and white blood cells which remove foreign bodies, especially bacteria, from the lymph.

Mandible the skeleton of the lower jaw.

Maxillary pertaining to the skeleton of the upper jaw.

Medial located at or near the middle of the body or of a part of the body.

Melanophore a pigment cell (CHROMATOPHORE) which contains the black or dark brown pigment melanin.

Membrane a thin sheet or layer of soft, pliable tissue which covers an organ, lines a tube or cavity, or connects organs together.

Mesoplastral in the middle of the PLASTRON of turtles.

Metabolic rate the rate of energy expenditure by an animal.

Metabolism the chemical or energy changes which occur within a living organism or a part of it which are involved in various life activities.

Metamorphosis the transformation of an animal from one stage of its life history to another, e.g. from larva to adult.

Metatarsal a bone in the foot between the tarsus (the ankle) and the toes.

Microclimate the climate in the area immediately around an organism. Can be very different from the overall climate if, for example, the organism lives in a burrow or a cave.

Microenvironment the local conditions that immediately surround an organism.

Migration the movement of animals, often in large numbers, from one place to another.

Mimic an animal that resembles an animal belonging to another species, usually a distasteful one, or some inedible object.

Molt to shed and develop anew the outer covering of the body, the skin in amphibians and reptiles.

Montane of or pertaining to mountains.

Morphological pertaining to the form and structure of an organism.

Mucus a viscous, slimy substance present on the surface of mucous membranes which serves to moisten and lubricate.

Musk a substance with a penetrating, persistent odor secreted by special glands in turtles and crocodiles.

Nares the paired openings of the nasal cavity.

Natal of or pertaining to birth.

Neotropics the tropical part of the New World; includes South America, Central America, part of Mexico and the West Indies.

Neural arch the portion of a vertebra which forms the roof and sides of the space through which the spinal cord passes.

Neurotoxin a poisonous substance that effects the nervous system of its victim.

Newt any salamander of the genera *Triturus, Taricha* and *Notophthalmus*: characteristically amphibious.

Niche the specific resources a species obtains from its environment, and its means of obtaining those resources that are essential to maintain its population size.

Nocturnal active at night.

Nutrient a substance, taken in as food, which promotes growth or provides energy for physiological processes.

Occipital condyle one of two rounded, bony processes on the back of the skull that provide articulation between the skull and the vertebral column.

Olfactory of or pertaining to the sense of smell.

Operculum a lid or covering. A flap covering the gills and developing legs in the larvae of frogs and toads.

Order a taxonomic category ranking below CLASS and above FAMILY.

Organ a part of an animal having a definite form and structure which performs one or more specific functions.

Orthopteran an insect of the order Orthoptera, e.g. grasshoppers, locusts, crickets.

Osmosis the passage of water through a semipermeable membrane as a result of differences in the concentrations of solutions on each side of the membrane. Water tends to move from a concentrated to a less concentrated solution until the two concentrations are equal.

Osmotic gradient the difference in concentration between solutions on each side of a semipermeable membrane.

Ossification the formation of bone.

Osteoderm a very small bone in the skin of some reptiles.

Otic of or pertaining to the ear.

Ovary the female gonad or reproductive organ, produces the OVA.

Oviduct the duct in females which carries the OVA from the OVARY to the CLOACA.

Oviparous reproducing by eggs that hatch outside the female's body.

Ovoviviparous reproducing by eggs which the female retains within her body until they hatch; the developing eggs contain a YOLK SAC but receive no nourishment from the mother through a placenta or similar structure.

Ovum (plural ova) a female germ cell or gamete; an egg cell or egg.

Paedomorphosis the retention of immature or larval characteristics (e.g. external gills) by animals that are sexually mature.

Papilla a small, nipple-like eminence or projection.

Parietal a paired bone forming part of the roof and sides of the skull.

Parotoid gland one of a pair of wartlike

glands on the shoulder, neck, or behind the eye in toads.

Parthenogenesis a form of asexual reproduction in which the OVUM develops without being fertilized.

Pectoral girdle the skeleton supporting the forelimbs of a land vertebrate; also called the shoulder girdle.

Pedicellate teeth teeth mounted on a slender stalk.

Peristaltic action progressive wavelike movements occurring in intestines or other contraction moves along the tube preceded by a wave of relaxation.

Permeable of a structure such as the skin, allowing the passage of a substance (e.g. water) through it.

Phalanx one of several bones (phalanxes or phalanges) in the fingers and toes.

Pheromone a substance produced and discharged by an organism which induces a response in another individual of the same species, such as sexual attraction.

Phragmosis using a part of the body to close a burrow.

Phylogenetic of or pertaining to evolutionary history.

Phylum a taxonomic category ranking above CLASS.

Pigment a substance that gives color to part or all of an organism's body.

Pineal a small outgrowth from the dorsal surface of the brain lying just beneath the skull.

Pit receptor a pit containing sensory cells sensitive to heat located between the eye and the nose or along the edges of the jaws in some snakes.

Placenta a structure attached to the inner surface of the female's reproductive tract through which the embryo obtains its nourishment.

Plastron the ventral portion of the shell of a turtle.

Pleurodont teeth teeth that are attached to the inside surface of the jaw, as opposed to the upper edge (acrodont) or in sockets (thecodont).

Poikilothermic unable to maintain a constant, or nearly constant body temperature and therefore having a body temperature similar to that of the environment.

Polypeptide a compound formed by the union of several amino acids.

Population a more or less separate (discrete) group of animals of the same species.

Posterior toward the rear or tail end of an animal.

Preadaptation the possession of a trait or traits which are not necessarily advantageous to an organism in its present environment but would be advantageous in a different environment.

Predator an animal that feeds by hunting and killing other animals.

Prehallux a rudimentary digit on the hind foot of frogs.

Prehensile adapted for grasping or clasping, especially by wrapping around, e.g. the tail of chameleons.

Premaxillary of or pertaining to the front part of the upper jaw.

Primitive of ancient evolutionary origin (cf.. ADVANCED).

Prostaglandin any of a number of oxygenated unsaturated cyclic fatty acids present in various body fluids. These compounds sometimes perform HORMONE-like actions, such as controlling blood pressure or muscle contractions.

Protein a type of organic compound making up a large part of all living tissues, containing nitrogen, and yielding amino acids when broken down.

Proto- (as in protoalligator, protoamphibian, etc) an ancestral form that was the evolutionary precursor of a taxonomic group, e.g. of a FAMILY, CLASS, etc.

Pubis (plural pubes) ventral, forward-projecting part of the pelvic or hip girdle.

Pupa a dormant, inactive stage between the LARVAL and adult stages in the LIFE CYCLE of insects; a chrysalis.

Quadratojugal a bone in the skull situated at the point where the lower jaw articulates with the skull.

Quadruped a four-footed animal.

Race a subdivision of a SPECIES which is distinguishable from the rest of that species; may live in a distinct area; a geographic race.

Rain forest tropical and subtropical forest with abundant and year-round rainfall.

Receptive responsive to stimuli; sexually receptive females are responsive to the sexual behavior of a male.

Rectilinear locomotion a form of movement, used mostly by heavy-bodied snakes in which the body moves slowly forward and is held straight.

Reduced (anatomically) smaller in size than in ancestral forms.

Reflectance the capacity of a body or surface to reflect, rather than absorb, light.

Release call a brief call given by male frogs, and by unreceptive females, when clasped by a male; it causes the clasper to release his grip.

Retina a layer of light-sensitive cells (rods and cones) within the eye, upon which a visual image is formed.

Riverine living in rivers.

Satellite male a make who, in a group of calling frogs, does not call himself but sits near a calling male and intercepts females that are attracted to the calling male.

Savanna a term loosely used to describe open grasslands with scattered trees and bushes, usually in warm areas.

Scale a thin, flattened, platelike structure forming part of the surface covering of various vertebrates, especially fishes and reptiles.

Scavenger an animal which feeds on dead animals or plants that it has not hunted or collected itself.

Scute any enlarged scale on a reptile.

Seasonal breeder a species that breeds at a specific time of year.

Semiaquatic living part of the time in water.

Sinus a cavity, hollow, recess or space.

Sinusoidal wavy, tortuous.

Solitary living by itself.

Species a group of actually or potentially interbreeding populations that are reproductively isolated from other such groups.

Spermatheca a pouch or sac in the female in which sperm are stored.

Spermatophore a structure containing sperm that is passed from the male to the female in some animals, such as in many salamanders.

Splenial a bone in the lower jaw of some amphibians.

Squamid scaly. Also (noun) any member of the order Squamata: a lizard, worm lizard or snake.

Sternum a bone in the ventral part of the pectoral girdle; the breastbone.

Subcaudal beneath or on the ventral side of the tail.

Substrate the solid material upon which an organism lives.

Sulcus a groove or furrow.

Tadpole the LARVA of a frog or toad.

Tapetum lucidum a light-reflecting layer in the eye.

Taxonomy the science of classifications: the arrangement of animals and plants into groups based on their natural relationships.

Temporal of or pertaining to the area of the skull behind the eye (the temple).

Terrapin any of a number of freshwater turtles, especially semiaquatic species and those that leave the water to bask.

Territorial defending an area so as to exclude other members of the same species.

Territory an area that one or more animals defend against other members of the same species.

Thermoregulation control of body temperature, by behavioral and/or physiological means, such that it maintains a constant or near-constant value.

Thrombosis The formation of a blood clot within a blood vessel or the heart.

Thyroid gland a gland lying in the neck that produces the HORMONE THYROXINE.

Thyroxine a HORMONE containing iodine that is involved in a variety of physiological processes, including metamorphosis in amphibians.

Torpor a state of sluggishness or inactivity.

Transversely crosswise; at right angles to the long axis of the body.

Tubercle a small, knoblike projection.

Tympanic membrane the eardrum.

Unisexual species a species consisting only of females.

Unken reflex a defensive posture, shown by some amphibians when attacked, in which the body is arched inward with the head and tail lifted upward.

Urea a compound containing nitrogen formed by the breakdown of proteins; a major constituent of urine in many animals.

Uric acid a compound containing nitrogen found in the urine of certain animals, including many reptiles.

Urostyle a rodlike bone composed of fused tail vertebrae, present in frogs and toads.

Uterine milk a MUCOPROTEIN uterine secretion which bathes developing embryos.

Vent an outlet, the anal or CLOACAL opening from the body.

Ventral pertaining to the lower surface of the body or one of its parts.

Ventricle a cavity within an organ; a chamber in the heart which discharges blood into the arteries.

Vertebral column the spinal skeleton, consisting of a series of vertebrae extending from the skull to the tip of the tail; the backbone.

Vertebrate any member of the subphylum Vertebrata, comprising all animals with a vertebral column, including fishes, amphibians, reptiles, birds and mammals.

Vestigial smaller and of more simple structure than in an evolutionary ancestor.

Viscera any of the organs contained within the body cavities, especially the abdominal cavity.

Viviparous giving birth to living young which develop within and are nourished by the mother.

"Warm-blooded" an outmoded term, referring to animals whose body remains constant, or nearly so, and higher than that of the environment.

Yolk sac a large sac containing stored nutrients, present in the EMBRYOs of fishes, amphibians, reptiles and birds.

Zygodactyl having toes and/or fingers arranged in such a way that two point forward and two point backward.

Bibliography

The following list of titles indicates key reference works used in the preparation of this volume and those recommended for further reading. The list is divided into a number of categories for ease of use.

GENERAL WORKS

Adler, K. (ed) (1989) *Contributions to the History of Herpetology*, Society for the Study of Amphibians and Reptiles, Oxford, OH, USA.

Adler, K. (ed) (1992) *Herpetology. Current Research on the Biology of Amphibians and Reptiles*, Society for the Study of Amphibians and Reptiles, Oxford, OH, USA.

Ananjeva, N. B., Borkin, L. J., Darevsky, I. S. and Orlov, N. L. (1988) *Dictionary of Animal Names in Five Languages. Amphibians and Reptiles*, Russky Yazk Publishers, Moscow, Russia (a world list in Latin, Russian, English, German, and French).

Bellairs, A. d'A. (1969) *The Life of Reptiles*, vols. 1–2, Weidenfeld and Nicolson, London, UK.

Bellairs, A. d'A. and Cox. C. B. (eds) (1976) *Morphology and Biology of Reptiles*, Academic Press, London, UK.

Cloudsley-Thompson, J. L. (1999) *The Diversity of Amphibians and Reptiles*, Springer-Verlag, Berlin, Germany.

Cogger, H. G. and Zweifel, R. G. (eds) (1998) *Encyclopedia of Reptiles and Amphibians, 2nd edn*, Academic Press, San Diego, CA, USA.

Duellman, W. E. and Trueb, L. (1994) *Biology of Amphibians*, Johns Hopkins University Press, Baltimore, MD, USA.

Duellman, W. E. (ed) (1999) *Patterns of Distribution of Amphibians. A Global Perspective*, Johns Hopkins University Press, Baltimore, MD, USA.

Ferguson, M. W. J. (ed) (1984) *The Structure, Development and Evolution of Reptiles*, Academic Press, London, UK.

Frank, N. and Ramus, E. (1995) *A Complete Guide to the Scientific and Common Names of Reptiles and Amphibians of the World*, N G Publishing, Pottsville, PA, USA.

Frost, D. (ed) (1985) *Amphibian Species of the World*, Association of Systematics Collections and Allen Press, Lawrence, KS, USA. (*Additions and Corrections*, by W. E. Duellman, was published by the Museum of Natural History, Lawrence, KS in 1993; the Frost list is regularly updated on the American Museum website [see Web Addresses])

Gans, C. et al. (eds) (1969–1998) *Biology of the Reptilia*, vols. 1–19, Academic Press, London, UK, and other publishers; a continuing series, now published by the Society for the Study of Amphibians and Reptiles, Ithaca, NY, USA.

Grassé, P-P. et al. (eds) (1970) *Traité de Zoologie: Reptiles*, vols. 1–2, Masson, Paris, France.

Grassé, P-P. and Delsol, M. (eds) (1986) *Traité de Zoologie: Batraciens*, Masson, Paris, France.

Haines, S. (2000) *Slithy Toves: Illustrated Classic Herpetological Books*, Society for the Study of Amphibians and Reptiles, Ithaca, NY, USA.

Heatwole, H. et al. (eds) (1994–2000) *Amphibian Biology*, vols. 1–4, Surrey Beatty, Chipping Norton, NSW, Australia (a continuing series).

Kabisch, K. (1990) *Wörterbuch der Herpetologie*, Gustav Fischer, Jena, Germany.

O'Shea, M. and Halliday, T. (2002) *Reptiles and Amphibians*, Dorling Kindersley, NY, USA.

Ota, H. (ed) (1999) *Tropical Island Herpetofauna. Origin, Current Diversity, and Conservation*, Elsevier, Amsterdam, Netherlands.

Peters, J. A. (1964) *Dictionary of Herpetology*, Hafner, NY, USA.

Pough, F. H., Andrews, R. M., Cadle, J. E., Crump, M. L., Savitzky, A. H. and Wells, K. D. (2001) *Herpetology, 2nd edn*, Prentice Hall, Upper Saddle River, NJ, USA.

Rhodin, A. G. J. and Miyata, K. (eds) (1983) *Advances in Herpetology and Evolutionary Biology*, Museum of Comparative Zoology, Harvard University, Cambridge, MA, USA.

Sparling, D. W., Linder, G. and Bishop, C. A. (eds) (2000) *Ecotoxicology of Amphibians and Reptiles*, SETAC Press, Pensacola, FL, USA.

Stebbins, R.C. and Cohen, N. W. (1995) *A Natural History of Amphibians*, Princeton University Press, Princeton, NJ, USA.

Ulber, T., Grossmann, W., Beutelschiess, J. and Beutelschiess, C. (1989) *Terraristisch/herpetologisches Fachwörterbuch*, Terrariengemeinschaft Berlin e.V., Berlin, Germany.

Zug, G. R., Vitt, L. J. and Caldwell, J. P. (2001) *Herpetology. An Introductory Biology of Amphibians and Reptiles, 2nd edn*, Academic Press, San Diego, CA, USA.

MAJOR GROUPS

Bruce, R. C., Jaeger, R. G., and Houck, L. D. (eds) (2000) *The Biology of Plethodontid Salamanders*, Kluwer Academic/Plenum, NY, USA.

Campbell, J. A. and Brodie, E. D., Jr. (eds) (1992) *Biology of the Pitvipers*, Selva, Tyler, TX, USA.

David, P. (1994) *Liste des Reptiles Actuels du Monde. I. Chelonii*, Dumerilia, vol. 1, Paris, France.

David, P. and Ineich, I. (1999) *Les Serpents Venimeux du Monde: Systématique et Répartition*, Dumerilia, vol. 3, Paris, France.

Ernst, C. H. and Barbour, R. W. (1989) *Turtles of the World*, Smithsonian Institution Press, Washington, DC, USA.

Estes, R. and Pregill, G. (eds) (1988) *Phylogenetic Relationships of the Lizard Families*, Stanford University Press, Stanford, CA, USA.

Exbrayat, J.-M. (2000) *Les Gymnophiones ces Curieux Amphibiens*, Boubée, Paris, France.

Gloyd, H. K. and Conant, R. (1990) *Snakes of the Agkistrodon Complex. A Monographic Review*, Society for the Study of Amphibians and Reptiles, Oxford, OH, USA.

Golay, P., Smith, H. M., Broadley, D. G., Dixon, J. R., McCarthy, C., Rage, J. C., Schätti, B and Toriba, M. (1993) *Endoglyphs and Other Major Venomous Snakes of the World. A Checklist*, Azemiops, Cultural Foundation Elapsoïdea, Aïre-Geneva, Switzerland.

Greene, H. W. (1997) *Snakes. The Evolution of Mystery in Nature*, University of California Press, Berkeley, CA, USA.

Grigg, G. C., Seebacher, F. and Franklin, C. E. (eds) (2001) *Crocodilian Biology and Evolution*, Surrey Beatty, Chipping Norton, NSW, Australia.

Harless, M. and Morlock, H. (eds) (1989) *Turtles: Perspectives and Research*, Krieger, Malabar, FL, USA.

Heatwole, H. (1999) *Sea Snakes, 2nd edn*, Krieger, Malabar, FL, USA.

Hofrichter, R. (ed) (2000) *Amphibians*, Firefly Books, Buffalo, NY, USA.

Iverson, J. B. (1992) *A Revised Checklist with Distribution Maps of the Turtles of the World*, published by the author, Richmond, IN, USA.

Joger, U. (ed) (1999) *Phylogeny and Systematics of the Viperidae*, Kaupia, vol. 8, Hessisches Landes-museum, Darmstadt, Germany.

King, F. W. and Burke, R. L. (eds) (1989) *Crocodilian, Tuatara, and Turtle Species of the World*, Association of Systematics Collections, Washington, DC, USA.

Klauber, L. M. (1997) *Rattlesnakes: Their Habits, Life Histories, & Influence on Mankind*, vols. 1-2, University of California Press, Berkeley, CA, USA.

Kuchling, G. (1999) *The Reproductive Biology of the Chelonia*, Springer-Verlag, Berlin, Germany.

Lutz, P. L. and Musick, J. A. (eds) (1997) *The Biology of Sea Turtles*, CRC Press, Boca Raton, FL, USA.

Mattison, C. (2007) *The New Encyclopedia of Snakes*. Cassell Illustrated, London, and Princeton University Press, Princeton, NJ, USA.

McDiarmid, R. W. and Altig, R. (eds) (1996) *Tadpoles. The Biology of Anuran Larvae*, University of Chicago Press, Chicago, IL, USA.

McDiarmid, R. W., Campbell, J. A. and Touré, T.S. A. (1999) *Snake Species of the World. A Taxonomic and Geographic Reference*, vol. 1, Herpetologists' League, Washington, DC, USA (a continuing series).

Rogner, M. (1997) *Lizards*, vols. 1–2, Krieger, Malabar, FL, USA.

Rösler, H. (1995) *Geckos der Welt, alle Gattungen*, Urania-Verlag, Leipzig, Germany.

Ross, C. A. (ed) (1989) *Crocodiles and Alligators*, Facts on File, NY, USA.

Rossman, D. A., Ford, N. B. and Seigel, R. A. (1996) *The Garter Snakes. Evolution and Ecology*, University of Oklahoma Press, Norman, OK, USA.

Roze, J. A. (1996) *Coral Snakes of the Americas*, Krieger, Malabar, FL, USA.

Schulz, K-D. (1996) *A Monograph of the Colubrid Snakes of the Genus* Elaphe *Fitzinger*, Koeltz Scientific Books, Koenigstein, Germany.

Taylor, E. H. (1968) *The Caecilians of the World*, University of Kansas Press, Lawrence, KS, USA.

Thorn, R. and Raffaëlli, J. (2001) *Les Salamandres de l'Ancien Monde*, Boubée, Paris, France.

Thorpe, R. S., Wüster, W. and Malhotra, A. (eds) (1997) *Venomous Snakes: Ecology, Evolution and Snakebite*, Clarendon Press, Oxford, UK.

Vial, J. L. (ed) (1973) *Evolutionary Biology of the Anurans*, University of Missouri Press, Columbia, MO, USA.

Wermuth, H. and Mertens, R. (1996) *Schildkröten, Krokodile, Brückenechsen, with new supplement*, Gustav Fischer, Jena, Germany.

Wright, J. W. and Vitt, L. J. (eds) (1993) *Biology of Whiptail Lizards (Genus* Cnemidophorus*)*, Oklahoma Museum of Natural History, Norman, OK, USA.

EVOLUTION, ECOLOGY, BEHAVIOR AND PHYSIOLOGY

Auffenberg, W. (1981) *The Behavioral Ecology of the Komodo Monitor*, University Presses of Florida, Gainesville, FL, USA (companion volumes on *Gray's Monitor* and the *Bengal Monitor* were issued by the same publisher in 1988 and 1994, respectively).

Cloudsley-Thompson, J. L. (1971) *The Temperature & Water Relations of Reptiles*, Merrow, London, UK.

Dawley, R. M. and Bogart, J. P. (eds) (1989) *Evolution and Ecology of Unisexual Vertebrates*, New York State Museum, Albany, NY, USA.

Feder, M. E. and Burggren, W. W. (eds) (1992) *Environmental Physiology of the Amphibians*, University of Chicago Press, Chicago, IL, USA.

Feder, M. E. and Lauder, G. V. (eds) (1986) *Predator–Prey Relationships. Perspectives and Approaches from the Study of Lower Vertebrates*, University of Chicago Press, Chicago, IL, USA.

Florkin, M. and Scheer, B. T. (eds) (1974) *Chemical Zoology, vol. 9: Amphibia and Reptilia*, Academic Press, London, UK.

Fox, S. F., McCoy, J. K., and Baird, T. A. (eds) (in press) *Territoriality, Dominance, and Sexual Selection: Adaptive Variation in Social Behavior Among Individuals, Populations, and Species of Lizards*, Johns Hopkins University Press, Baltimore, MD, USA.

Gans, C. (1974) *Biomechanics*, Lippincott, Philadelphia, PA, USA.

Green, D. M. and Sessions, S. K. (eds) (1991) *Amphibian Cytogenetics and Evolution*, Academic Press, San Diego, CA, USA.

Guillette, L. J., Jr. and Crain, D. A. (2000) *Endocrine Disrupting Contaminants: An Evolutionary Perspective*, Taylor & Francis, Philadelphia, PA, USA.

Hanke, W. (ed) (1990) *Biology and Physiology of Amphibians*, Gustav Fischer, Stuttgart, Germany.

Heatwole, H. and Taylor, J. (1987) *Ecology of Reptiles*, Surrey Beatty, Chipping Norton, NSW, Australia.

Huey, R. B., Pianka, E. R. and Schoener, T. W. (eds) (1983) *Lizard Ecology*, Harvard University Press, Cambridge, MA, USA.

Lofts, B. (ed) (1974, 1976) *Physiology of the Amphibia*, vols. 2–3, Academic Press, NY, USA.

Moore, J. A. (ed) (1964) *Physiology of the Amphibia*, vol. 1, Academic Press, NY, USA.

Olmo, E. (ed) (1990) *Cytogenetics of Amphibians and Reptiles*, Birkhäuser, Basel, Switzerland.

Pianka, E. R. (1986) *Ecology and Natural History of Desert Lizards*, Princeton University Press, Princeton, NJ, USA.

Roughgarden, J. (1995) *Anolis Lizards of the Caribbean: Ecology, Evolution, and Plate Tectonics*, Oxford University Press, Oxford, UK.

Ryan, M. J. (1985) *The Túngara Frog. A Study in Sexual Selection and Communication*, University of Chicago Press, Chicago, IL, USA.

Ryan, M. J. (ed) (2001) *Anuran Communication*, Smithsonian Institution Press, Washington, DC, USA.

Seigel, R. A. and Collins, J. T. (eds) (2002) *Snakes.*
Ecology & Behavior, Blackburn Press, Caldwell, NJ, USA.

Seigel, R. A., Collins, J. T. and Novak, S. S. (eds) (2002) *Snakes. Ecology and Evolutionary Biology*, Blackburn Press, Caldwell, NJ, USA.

Taylor, D. H. and Guttman, S. I. (eds) (1977) *The Reproductive Biology of Amphibians*, Plenum Press, NY, USA.

Vitt, L. J. and Pianka, E. R. (eds) (1994) *Lizard Ecology. Historical and Experimental Perspectives*, Princeton University Press, Princeton, NJ, USA.

REGIONAL STUDIES, DISTRIBUTION AND FIELD IDENTIFICATION
United States and Canada

Ashton, R. E., Jr. and Ashton, P. S. (1985–1988) *Handbook of Reptiles and Amphibians of Florida*, parts 1–3, Windward Publishing, Miami, FL, USA.

Bartlett, R. D. and Tennant, A. (2000) *Snakes of North America. Western Region*, Gulf Publishing, Houston, TX, USA.

Behler, J. L. and King, F. W. (1979) *The Audubon Society Field Guide to North American Reptiles and Amphibians*, Alfred A. Knopf, NY, USA.

Bishop, S. C. (1994) *Handbook of Salamanders*, Cornell University Press, Ithaca, NY, USA.

Carr, A. (1995) *Handbook of Turtles*, Cornell University Press, Ithaca, NY, USA.

Catalogue of American Amphibians and Reptiles (1963–2001) *Society for the Study of Amphibians and Reptiles*, New Haven, CT, USA. (this series covers the Western Hemisphere).

Conant, R. and Collins, J. T. (1998) *A Field Guide to Reptiles and Amphibians of Eastern and Central North America, 3rd edn expanded*, Houghton Mifflin, Boston, MA, USA.

Cook, F. R. (1984) *Introduction to Canadian Amphibians and Reptiles*, National Museums of Canada, Ottawa, Canada.

Degenhardt, W. G., Painter, C. W. and Price, A. H. (1996) *Amphibians and Reptiles of New Mexico*, University of New Mexico Press, Albuquerque, NM, USA.

Dodd, C. K., Jr. (2001) *North American Box Turtles a Natural History*, University of Oklahoma Press, Norman, OK, USA.

Dundee, H. A. and Rossman, D. A. (1989) *The Amphibians and Reptiles of Louisiana*, Louisiana State University Press, Baton Rouge, LA, USA.

Ernst, C. H., Lovich, J. E. and Barbour, R. W. (1994) *Turtles of the United States and Canada*, Smithsonian Institution Press, Washington, DC, USA.

Harding, J. H. (1997) *Amphibians and Reptiles of the Great Lakes Region*, University of Michigan Press, Ann Arbor, MI, USA.

Klemens, M. W. (1993) *Amphibians and Reptiles of Connecticut and Adjacent Regions*, Connecticut Dept. of Environmental Protection, Hartford, CT, USA.

McKeown, S. (1996) *A Field Guide to Reptiles and Amphibians in the Hawaiian Islands*, Diamond Head Publishing, Los Osos, CA, USA.

Minton, S. A., Jr. (2001) *Amphibians & Reptiles of Indiana, revised 2nd edn*, Indiana Academy of Science, Indianapolis, IN, USA.

Nussbaum, R. A., Brodie, E. D., Jr. and Storm, R. M. (1983) *Amphibians and Reptiles of the Pacific North-west*, University of Idaho Press, Moscow, ID, USA.

Palmer, W. M. and Braswell, A. L. (1995) *Reptiles of North Carolina*, University of North Carolina Press, Chapel Hill, North Carolina, USA.

Petranka, J. W. (1998) *Salamanders of the United States and Canada*, Smithsonian Institution Press, Washington, DC, USA.

Smith, H. M. (1995) *Handbook of Lizards*, Cornell University Press, Ithaca, NY, USA.

Smith, H. M. (1978) *A Guide to Field Identification. Amphibians of North America*, Golden Press, NY, USA.

Smith, H. M. and Brodie, E. D., Jr. (1982) *A Guide to Field Identification. Reptiles of North America*, Golden Press, NY, USA.

Stebbins, R. C. (1985) *A Field Guide to Western Reptiles and Amphibians, 2nd edn*, Houghton Mifflin, Boston, MA, USA.

Tennant, A. and Bartlett, R. D. (2000) *Snakes of*

North America. Eastern and Central Regions, Gulf Publishing, Houston, TX, USA.
Tyning, T. F. (1990) *A Guide to Amphibians & Reptiles*, Little, Brown, Boston, MA, USA.
Werler, J. E. and Dixon, J. R. (2000) *Texas Snakes*, University of Texas Press, Austin, TX, USA.
Wright, A. H. and Wright, A. A. (1994) *Handbook of Frogs and Toads of the United States and Canada, 3rd edn*, Cornell University Press, Ithaca, NY, USA.
Wright, A. H. and Wright, A. A. (1994) *Handbook of Snakes of the United States and Canada, 2 vols*. Cornell University Press, Ithaca, NY, USA (vol. 3, *Bibliography*, by Society for the Study of Amphibians and Reptiles, 1979).

Mexico, including General Latin America

Alvarez del Toro, M. (1982) *Los reptiles de Chiapas, 3rd edn*. Instituto Historia Natural, Tuxtla Gutierrez, Chiapas, Mexico.
Campbell, J. A. and Lamar, W. W. (1989) *The Venomous Reptiles of Latin America*, Cornell University Press, Ithaca, NY, USA.
Duellman, W. E. (2001) *The Hylid Frogs of Middle America, vols. 1–2, expanded edn*, Society for the Study of Amphibians & Reptiles, Ithaca, NY, USA.
García, A. and Ceballos, G. (1994) *Field Guide to the Reptiles and Amphibians of the Jalisco Coast, Mexico*, Fundación Ecológica de Cuixmala, UNAM, Mexico City, Mexico.
Grismer, L. L. (in press) *Amphibians and Reptiles of Baja California, Its Pacific Islands, and the Islands in the Sea of Cortés*, University of California Press, Berkeley, CA, USA.
Johnson, J. D., Webb, R. G. and Flores-Villela, O. A. (eds) (2001) *Mesoamerican Herpetology: Systematics, Zoogeography, and Conservation*, Centennial Museum, University of Texas, El Paso, TX, USA.
Lee, J. C. (1996) *The Amphibians and Reptiles of the Yucatán Peninsula*, Cornell University Press, Ithaca, NY, USA (field guide version, same publisher, 2000).
McCranie, J. R. and Wilson, L. D. (2001) *The Herpetofauna of the Mexican State of Aguascalientes*, Courier Forschungsinstitut Senckenberg, Frankfurt am Main, Germany.
Pérez-Higareda, G. and Smith, H. M. (1991) *Ofidiofauna de Veracruz*, Instituto de Biología, UNAM, Mexico City, Mexico.
Peters, J. A., Orejas-Miranda, B. and Donoso-Barros, R. (1970) *Catalogue of the Neotropical Squamata, parts 1–2*, United States National Museum, Washington, DC, USA.
Smith, H. M. and Smith, R. B. (1971–1993) *Synopsis of the Herpetofauna of Mexico, vols. 1–7*, Eric Lundberg, Augusta, West Virginia, USA, later by University of Colorado Press, Boulder, CO, USA.
Uribe-Peña, Z., Ramírez-Bautista, A. and Casas Andreu, G. (1999) *Anfibios y Reptiles de las Serranías del Distrito Federal, México*, Instituto de Biología, UNAM, Mexico City, Mexico.

Central America

Campbell, J. A. (1998) *Amphibians and Reptiles of Northern Guatemala, the Yucatán, and Belize*, University of Oklahoma Press, Norman, OK, USA.
Duellman, W. E. (2001) *The Hylid Frogs of Middle America, vols. 1–2, expanded edn*, Society for the Study of Amphibians & Reptiles, Ithaca, NY, USA.
Ibáñez D., R., Rand, A. S. and Jaramillo A., C. A. (1999) *The Amphibians of Barro Colorado Nature Monument, Soberania National Park and Adjacent Areas*, Editorial Mizrachi & Pujol, Panamá.
Köhler, G. (1999) *Anfibios y Reptiles de Nicaragua*, Herpeton, Offenbach, Germany.
Köhler, G. (2000, 2001) *Reptilien und Amphibien Mittelamerikas, vols. 1–2*, Herpeton, Offenbach, Germany.
Lee, J. C. (2000) *A Field Guide to the Amphibians and Reptiles of the Maya World. The Lowlands of Mexico, Northern Guatemala, and Belize*, Cornell University Press, Ithaca, NY, USA.
Leenders, T. (2001) *A Guide to Amphibians and Reptiles of Costa Rica*, Zona Tropical, Miami, FL, USA.
McCranie, J. R. and Wilson, L. D. (2002) *The Amphibians of Honduras*, Society for the Study of Amphibians and Reptiles, Ithaca, NY, USA.
Meyer, J. R. and Foster, C. F. (1996) *A Guide to the Frogs and Toads of Belize*, Krieger, Malabar, FL, USA.
Savage, J. M. (in press) *Amphibians and Reptiles of Costa Rica*, University of Chicago Press, Chicago, IL, USA.
Stafford, P. J. and Meyer, J. R. (2000) *A Guide to the Reptiles of Belize*, Academic Press, San Diego, CA, USA.
Wilson, L. D. and Meyer, J. R. (1985) *The Snakes of Honduras, 2nd edn*, Milwaukee Public Museum, Milwaukee, WI, USA.

West Indies

Buurt, G. van (2001) *De Amfibieën en Reptielen van Aruba, Curaçao en Bonaire*, published by the author, Curaçao, Netherlands Antilles.
Crother, B. I. (ed) (1999) *Caribbean Amphibians and Reptiles*, Academic Press, San Diego, CA, USA.
Currat, P. (1980) *Aperçu sur les Reptiles Antillais de Guadeloupe et Martinique Principalement*, Centre départemental de Documentation pédagogique, Point-á-Pitre, Guadeloupe.
Joglar, R. L. (1998) *Los Coquíes de Puerto Rico, Su Historia Natural y Conservación*, Editorial de la Universidad de Puerto Rico, San Juan, PR, USA.
Malhotra, A. and Thorpe, R. S. (1999) *Reptiles & Amphibians of the Eastern Caribbean*, Macmillan Education, London, UK.
Murphy, J. C. (1997) *Amphibians and Reptiles of Trinidad and Tobago*, Krieger, Malabar, FL, USA.
Powell, R. and Henderson, R. W. (eds) (1996) *Contributions to West Indian Herpetology*, Society for the Study of Amphibians & Reptiles, Ithaca, NY, USA.
Rivero, J. A. (1998) *Los Anfibios y Reptiles de Puerto Rico, 2nd edn revised*, Editorial de la Universidad de Puerto Rico, San Juan, PR, USA.
Rodríguez Schettino, L. (ed) (1999) *The Iguanid Lizards of Cuba*, University Press of Florida, Gainesville, FL, USA.
Schwartz, A. and Henderson, R. W. (1991) *Amphibians and Reptiles of the West Indies: Descriptions, Distributions, and Natural History*, University of Florida Press, Gainesville, FL, USA.

South America

Amaral, A. do (1977) *Serpentes do Brasil, Iconografia Colorida*, Edições Melhoramentos, Editoria da Universidade de São Paulo, São Paulo, Brazil.
Avila-Perez, T. C. S. (1995) *Lizards of Brazilian Amazonia (Reptilia: Squamata)*, Zoologische Verhandelingen, Leiden, Netherlands.
Cabrera, M. R. (1998) *Las Tortugas Continentales de Sudamérica Austral*, published by the author, Córdoba, Argentina.
Cei, J. M. (1962) *Batracios de Chile*, Ediciones de la Universidad de Chile, Santiago, Chile.
Cei, J. M. (1980, 1986) *Amphibians of Argentina, 2 parts*, Monitore Zoologico Italiano, vols. 2 and 21, Florence, Italy.
Cei, J. M. (1986, 1993) *Reptiles de la Argentina, 2 parts*, Museo Regionale di Scienze Naturali, Turin, Italy.
Duellman, W. E. (1978) *The Biology of an Equatorial Herpetofauna in Amazonian Ecuador*, University of Kansas Museum of Natural History, Lawrence, KS, USA.
Duellman, W. E. (ed) (1979) *The South American Herpetofauna: Its Origin, Evolution and Dispersal*, University of Kansas Museum of Natural History, Lawrence, KS, USA.
Fauna Argentina. (1985, 1988) *Anfibios y Reptiles, two vols.*, Centro Editor de América Latina, Buenos Aires, Argentina.
Gorzula, S. and Señaris, J. C. (1998) *Contribution to the Herpetofauna of the Venezuelan Guayana*, Scientia Guaianæ, no. 8, Caracas, Venezuela.
Hoogmoed, M. S. (1973) *The Lizards and Amphisbaenians of Surinam*, W. Junk, The Hague, Netherlands.
Kornacker, P. M. (1999) *Checklist and Key to the Snakes of Venezuela*, Pako-Verlag, Rheinbach, Germany.
Lancini V., A. R. (1986) *Serpientes de Venezuela, 2nd edn*, Ernesto Armitano Editor, Caracas, Venezuela.
Lavilla, E. O. and Cei, J. M. (2001) *Amphibians of Argentina, second update*, Museo Regionale di Scienze Naturali, Turin, Italy.
Lescure, J. and Marty, C. (2000) *Atlas des Amphibiens de Guyane*, Muséum National d'Histoire Naturelle, Paris, France.
Medem M., F. (1981, 1983) *Los Crocodylia de Sur America, vols. 1–2*, Colciencias, Ministerio de Educación Nacional, Bogotá, Colombia.
Métrailler, S. and Le Gratiet, G. (1996) *Tortues Continentales des Guyane Française*, published by the authors, Bramois, Switzerland.
Pérez-Santos, C. and Moreno, A. G. (1988) *Ofidios de Colombia*, Museo Regionale de Scienze Naturali, Turin, Italy.
Pérez-Santos, C. and Moreno, A. G. (1991) *Serpientes de Ecuador*, Museo Regionale di Scienze Naturali, Turin, Italy.
Pritchard, P. C. H. and Trebbau, P. (1984) *The Turtles of Venezuela*, Society for the Study of Amphibians and Reptiles, Athens, OH, USA.
Renjifo, J. M. and Lundberg, M. (1999) *Anfibios y Reptiles de Urrá*, Skanska, Editorial Colina, Medellín, Colombia.
Rodríguez, L. O. and Duellman, W. E. (1994) *Guide to the Frogs of the Iquitos Region, Amazonian Peru*, Natural History Museum, University of Kansas, Lawrence, KS, USA.
Starace, F. (1998) *Guide des Serpents et Amphisbènes de Guyane Française*, Ibis Rouge Editions, Guadeloupe and French Guiana.
Vanzolini, P. E., Ramos-Costa, A. M. M. and Vitt, L. J. (1980) *Répteis das Caatingas*, Academia Brasileira de Ciências, Rio de Janeiro, Brazil.
Vitt, L. J. and de la Torre, S. (1996) *Guía para la Investigación de las Lagartijas de Cuyabeno*, Pontificia Universidad Católica del Ecuador, Quito, Ecuador.

British Isles and Europe, excluding Russia

Arnold, E. N., Burton, J. A. and Ovenden, D. W. (1999) *Reptiles and Amphibians of Britain and Europe*, Viking Penguin, NY, USA.
Ballasina, D. (1984) *Amphibians of Europe a Colour Field Guide*, David & Charles, Newton Abbot and London, UK.
Barbadillo Escriva, L. J. and Martínez de Castilla, A. (1987) *La Guía de Incafo de los Anfibios y Reptiles de la Península Ibérica, Islas Baleares y Canarias*, Incafo, Madrid, Spain.
Beebee, T. and Griffiths, R. (2000) *Amphibians and Reptiles. A Natural History of the British Herpetofauna*, HarperCollins, London, UK.
Böhme, W. (ed) (1981–2001) *Handbuch der Reptilien und Amphibien Europas*, 10 vols. to date including supplements, Akademische Verlagsgesellschaft, Wiesbaden, Germany (a continuing series).
Cabela, A., Grillitsch, H. and Tiedemann, F. (2001) *Atlas zur Verbreitung und Ökologie der Amphibien und Reptilien in Österreich*, Umweltbundesamt, Vienna, Austria.
Corti, C. and Locascio, P. (in press) *Lizards of Italy and Adjacent Areas*, Chimaira, Frankfurt am Main, Germany.
Garcia Paris, M. (1985) *Los Anfibios de España*, Extensior Agraria, Madrid, Spain.
Gruber, U. (1989) *Die Schlangen Europas und rund ums Mittelmeer*, Franckh, Stuttgart, Germany.
Günther, R. (ed) (1996) *Die Amphibien und Reptilien Deutschlands*, Gustav Fischer, Jena, Germany.
Hofer, U., Monney, J.-C. and Dusej, G. (2001) *Die Reptilien der Schweiz*, Birkhäuser, Basel, Switzerland.
Matz, G. and Weber, D. (1983) *Guide des Amphibiens et Reptiles d'Europe*, Delachaux & Niestlé, Neuchâtel, Switzerland.
Salvador, A. (ed) (1998) *Fauna Iberica vol. 10 Reptiles*, Museo Nacional de Ciencias Naturales, Madrid, Spain.

Southwest Asia, from Sinai to Afghanistan

Anderson, S. C. (1999) *The Lizards of Iran*, Society for the Study of Amphibians and Reptiles, Ithaca, NY, USA.
Atatür, M. K. and Göçmen, B. (2001) *Amphibians and Reptiles of Northern Cyprus*, Ege Universitesi, Bornova-Izmir, Turkey.
Baloutch, M., and Kami, H. G. (1995) *Amphibians of Iran*, Tehran University Publication 2250, Tehran, Iran.
Baran, I. and Atatür, M. K. (1998) *Turkish Herpetofauna (Amphibians and Reptiles)*, Ministry of Environment, Ankara, Turkey.
Basoglu, M. and Baran, I. (1977, 1980) *The Reptiles of Turkey, parts 1–2*, Ege Üniversitesi, Fen Fakültesi, reports 76 and 81, Bornova-Izmir, Turkey.
Bouskila, A. and Amitai, P. (2001) *Handbook of Amphibians and Reptiles of Israel*, Keter Publishing, Jerusalem, Israel.
Disi, A. M., Modry, D., Necas, P. and Rifai, L. (2001) *Amphibians and Reptiles of the Hashemite Kingdom of Jordan*, Chimaira, Frankfurt am Main, Germany.
Gasperetti, J. (1988) *Snakes of Arabia*, Fauna of Saudi Arabia, vol. 9, Riyadh, Saudi Arabia.
Joger, U. (1984) *The Venomous Snakes of the Near and Middle East*, Ludwig Reichert Verlag, Wiesbaden, Germany.
Jongbloed, M. (2000) *Wild About Reptiles. Field Guide to the Reptiles and Amphibians of the United Arab Emirates*, Barkers Trident Communications, London, UK.
Latifi, M. (1991) *The Snakes of Iran*, Society for the Study of Amphibians & Reptiles, Oxford, OH, USA.
Leviton, A. E., Anderson, S. C., Adler, K. and Minton, S. A. (1992) *Handbook to Middle East Amphibians and Reptiles*, Society for the Study of Amphibians and Reptiles, Oxford, OH, USA.
Schätti, B. and Desvoignes, A. (1999) *The Herpetofauna of Southern Yemen and the Sokotra Archipelago*, Muséum d'Histoire Naturelle, Geneva, Switzerland.
Werner, Y. L. (1995) *A Guide to the Reptiles and Amphibians of Israel*, Nature Reserves Authority–"Yefe-Nof" Library, Jerusalem, Israel.

Africa

Auerbach, R. D. (1987) *The Amphibians and Reptiles of Botswana*, Mokwepa Consultants, Gaborone, Botswana.
Bons, J. and Geniez, P. (1996) *Amphibiens et Reptiles du Maroc*, Asociación Herpetólogica Española, Barcelona, Spain.
Boycott, R. C. and Bourquin, O. (2000) *The Southern African Tortoise Book, revised edn*, published by the second author, KwaZulu-Natal, South Africa.
Branch, W. (1998) *Field Guide to the Snakes and Other Reptiles of Southern Africa, 2nd edn*, Struik, Cape Town, South Africa.
Broadley, D. G. (1990) *Fitzsimons' Snakes of Southern Africa, revised edn*, Ball and Donker, Parklands, South Africa.
Broadley, D. G., Doria, C. and Wigge, J. (in press) *Snakes of Zambia*, Chimaira, Frankfurt am Main, Germany.
Buys, P. J. and Buys, P. J. C. (1983) *Snakes of South West Africa*, Gamsberg Publishers, Windhoek, Namibia.
Channing, A. (2001) *Amphibians of Central and Southern Africa*, Cornell University Press, Ithaca, NY, USA.
Chippaux, J.-P. (1999) *Les Serpents d'Afrique Occidentale et Centrale*, Éditions de l'IRD, Paris, France.
Fitzsimons, V. F. (1976) *The Lizards of South Africa*, Swets & Zeitlinger, Amsterdam, Netherlands.
Lambiris, A. J. L. (1989) *The Frogs of Zimbabwe*, Museo Regionale di Scienze Naturali, Turin, Italy.
Largen, M. J. and Rasmussen, J. B. (1993) *Catalogue of the Snakes of Ethiopia*, Tropical Zoology, vol. 6, Florence, Italy.
Passmore, N. I. and Carruthers, V. C. (1995) *South African Frogs, revised edn*, Witwatersrand University Press, Johannesburg, South Africa.
Pitman, C. R. S. (1974) *A Guide to the Snakes of Uganda, revised edn*, Wheldon and Wesley, Codicote, UK.
Rödel, M.-O. (2000) *Herpetofauna of West Africa Vol. I. Amphibians of the West African Savanna*, Chimaira, Frankfurt am Main, Germany.
Saleh, M. A. (1997) *Amphibians and Reptiles of Egypt*, National Biodiversity Unit, Egyptian Environmental Affairs Agency, Cairo, Egypt.
Schiøtz, A. (1999) *Treefrogs of Africa*, Chimaira, Frankfurt am Main, Germany.
Schleich, H. H., Kästle, W. and Kabisch, K. (1996) *Amphibians and Reptiles of North Africa*, Koeltz Scientific Books, Koenigstein, Germany.
Spawls, S., Howell, K., Drewes, R. and Ashe, J. (2002) *A Field Guide to the Reptiles of East Africa*, Academic Press, San Diego, CA, USA.

Madagascar and Indian Ocean

Glaw, F. and Vences, M. (1994) *A Fieldguide to the Amphibians and Reptiles of Madagascar, 2nd edn*, published by the authors, Bonn, Germany.
Guibé, J. (1978) *Les Batraciens de Madagascar*, Bonner Zoologische Monographien no. 11, Bonn, Germany.
Henkel, F.-W., and Schmidt, W. (2000) *Amphibians & Reptiles of Madagascar and the Mascarene, Seychelles, and Comoro Islands*, Krieger, Malabar, FL, USA.

South Asia

Daniel, J. C. (1983) *The Book of Indian Reptiles*, Bombay Natural History Society, Bombay, India.
Das, I. (1991) *Colour Guide to the Turtles and Tortoises of the Indian Subcontinent*, R&A Publishing, Portishead, UK.
Das, I. (1995) *Turtles and Tortoises of India*, Oxford University Press, Bombay, India.
Das, I. (1996) *Biogeography of the Reptiles of South Asia*, Krieger, Malabar, FL, USA.
Das, I. (in press) *A Photographic Guide to the Snakes & Other Reptiles of India*, New Holland, London, UK.
De Silva, A. (1990) *Colour Guide to the Snakes of Sri Lanka*, R&A Publishing, Portishead, UK.

De Silva, A. (2001) *The Herpetofauna of Sri Lanka*, Amphibia and Reptile Research Organization of Sri Lanka, Peradeniya, Sri Lanka.
Dutta, S. K. and Manamendra-Arachchi, K. (1996) *The Amphibian Fauna of Sri Lanka*, Wildlife Heritage Trust of Sri Lanka, Colombo, Sri Lanka.
Dutta, S. K. (1997) *Amphibians of India and Sri Lanka (Checklist and Bibliography)*, Odyssey Publishing, Bhubaneswar, India.
Khan, M. A. R. (1996) *Wildlife of Bangladesh, Vol. 1 (Amphibia–Reptilia)*, Bangla Academy, Dhaka, Bangladesh.
Khan, M. S. (2000) *Frogs and Lizards of Pakistan*, Urdu Science Board, Lahore, Pakistan.
Khan, M. S. (in press) *A Guide to the Snakes of Pakistan*, Chimaira, Frankfurt am Main, Germany.
Pillai, R. S. and Ravichandran, M. S. (1999) *Gymnophiona (Amphibia) of India: a Taxonomic Study*, Zoological Survey of India, Calcutta, India.
Schleich, H. H. and Kästle, W. (1998) *Contributions to the Herpetology of South Asia (Nepal, India)*, Fuhlrott-Museum, Wuppertal, Germany.
Sharma, R. C. (1998) *The Fauna of India and Adjacent Countries. Reptilia vol I (Testudines and Crocodilia)*, Zoological Survey of India, Calcutta, India.
Shrestha, T. K. (2001) *Herpetology of Nepal. Study of the Amphibians and Reptiles of the Trans-Himalayan Region of Nepal, India, Pakistan and Bhutan*, published by the author, Katmandu, Nepal.
Tikader, B. K. and Sharma, R. C. (1985) *Handbook. Indian Testudines*, Zoological Survey of India, Calcutta, India.
Tikader, B. K. and Sharma, R. C. (1992) *Handbook. Indian Lizards*, Zoological Survey of India, Calcutta, India.

Russia and Associated Republics

Ananjeva, N., Borkin, L., Darevsky, I. and Orlov, N. (1998) *Amphibians and Reptiles [of Russia]*, ABF, Moscow, Russia.
Bannikov, A. G., Darevsky, I. S. and Ischenko, N. N. (1977) *Guide to the Reptiles and Amphibians of the USSR*, Proswescenije, Moscow, Russia.
Kuzmin, S. L. (1999) *The Amphibians of the Former Soviet Union*, Pensoft, Sofia, Bulgaria.
Kuzmin, S. L. (in press) *Turtles of Russia and Other Ex-Soviet Republics*, Chimaira, Frankfurt am Main, Germany.
Szczerbak, N. N. (in press) *Color Guide to the Reptiles of the Eastern Palearctic*, Krieger, Malabar, FL, USA.
Szczerbak, N. N. and Golubev, M. L. (1996) *Gecko Fauna of the USSR & Contiguous Regions*, Society for the Study of Amphibians & Reptiles, Ithaca, NY, USA.
Tarkhnishvili, D. N. and Gokhelashvili, R. K. (1999) *The Amphibians of the Caucasus*, Pensoft, Sofia, Bulgaria.

East Asia and Japan

Ananjeva, N. B., Munkhbayar, Kh., Orlov, N. L., Orlova, V. F., Semenov, D. V. and Terbish, Kh. (1997) *Amphibians and Reptiles of Mongolia*, KMK Scientific Press, Moscow, Russia.
Chinese Snakes, Atlas of (1980), Shanghai Science and Technology Publishing Co, Shanghai, China.
Fauna Sinica (1997–1999) Reptilia, vols. 1–3, Science Press, Beijing, China.
Fei, L., Ye, C.-Y., Huang, Y.-Z., and Liu, M.-Y. (1999) *Atlas of Amphibians of China*, Chinese Association of Wildlife Conservation, Zhengzhou, Henan, China.
Goris, R. C. (in press) *Field Guide to Japanese Reptiles and Amphibians*, Krieger, Malabar, FL, USA.
Karsen, S. J., Lau, M. W.-N. and Bogadek, A. (1998) *Hong Kong Amphibians and Reptiles, 2nd edn*, Provisional Urban Council, Hong Kong, China.
Lue, K.-Y., Tu, M.-C. and Hsiang, K.-S. (1999) *Field Guide to Amphibians and Reptiles of Taiwan*, SWAN, Taipei, Taiwan.
Maeda, N. and Matsui, M. (1999) *Frogs and Toads of Japan, rev. edn*, Bun-Ichi Sogo Shuppan, Tokyo, Japan.
Matsui, M., Hikida, T. and Goris, R. C. (eds) (1989) *Current Herpetology in East Asia*, Herpetological Society of Japan, Kyoto, Japan.
Sengoku, S. (ed) (1979) *Amphibians and Reptiles of Japan*, Wildlife Research Institute, Tokyo, Japan.
Zhao, E.-M. and Adler, K. (1993) *Herpetology of China*, Society for the Study of Amphibians and Reptiles, Oxford, OH, USA.
Zhao, E.-M., Hu, Q.-X., Jiang, Y.-M. and Yang, Y.-H. (1988) *Studies on Chinese Salamanders*, Society for the Study of Amphibians and Reptiles, Oxford, OH, USA.

Southeast Asia and Philippines

Alcala, A. C. and Brown, W. C. (1998) *Philippine Amphibians. An Illustrated Field Guide*, Bookmark, Makati City, Philippines.
Brown, W. C. and Alcala, A. C. (1978, 1980) *Philippine Lizards, vols. 1–2*, Silliman University, Natural Science Monographs, Dumaguete City, Philippines.
Chan-ard, T., Grossmann, W., Gumprecht, A. and Schulz, K.-D. (1999) *Amphibians and Reptiles of Peninsular Malaysia and Thailand*, Bushmaster Publications, Wuerselen, Germany.
Cox, M. J., van Dijk, P. P., Nabhitabhata, J. and Thirakhupt, K. (1998) *A Photographic Guide to Snakes & Other Reptiles of Peninsular Malaysia, Singapore and Thailand*, Ralph Curtis Books, Sanibel Island, FL, USA.
Das, I. (in press) *An Introduction to the Amphibians and Reptiles of Tropical Asia*, Natural History Publications (Borneo), Kota Kinabalu, Malaysia.
Liat, L. B. and Das, I. (1999) *Turtles of Borneo and Peninsular Malaysia*, Natural History Publications (Borneo), Kota Kinabalu, Malaysia.
Lim, B. L. (1982) *Poisonous Snakes of Peninsular Malaysia (2nd edn)*, Malayan Nature Society, Kuala Lumpur, Malaysia.
Taylor, E. H. (1966) *The Snakes of the Philippine Islands*, A. Asher, Amsterdam, Netherlands.
Tweedie, M. W. F. (1983) *The Snakes of Malaya, 3rd edn*, Singapore National Printers, Singapore.
Wirot, N. (2001) *Amphibians of Thailand*, Amarin Printing, Bangkok, Thailand.
Wirot, N. (2001) *Snakes in Thailand*, Amarin Printing, Bangkok, Thailand.

East Indies including Borneo

Das, I. (1998) *Herpetological Bibliography of Indonesia*, Krieger, Malabar, FL, USA.
David, P. and Vogel, G. (1996) *The Snakes of Sumatra*, Chimaira, Frankfurt am Main, Germany.
Inger, R. F. and Stuebing, R. B. (1997) *A Field Guide to the Frogs of Borneo*, Natural History Publications (Borneo), Kota Kinabalu, Malaysia.
Inger, R. F. and Tan, F. L. (1996) *The Natural History of Amphibians and Reptiles in Sabah*, Natural History Publications (Borneo), Kota Kinabalu, Malaysia.
Iskandar, D. T. (1998) *The Amphibians of Java and Bali*, Research and Development Centre for Biology–LIPI, Bogor, Indonesia.
Iskandar, D. J. (2000) *Turtles & Crocodiles of Insular Southeast Asia & New Guinea*, Department of Biology, Institute of Technology, Bandung, Indonesia.
Stuebing, R. B. and Inger, R. F. (1999) *A Field Guide to the Snakes of Borneo*, Natural History Publications (Borneo), Kota Kinabalu, Malaysia.

Australia

Barker, D. G. and Barker, T. M. (1994) *Pythons of the World. Volume I, Australia*, Advanced Vivarium Systems, Lakeside, CA, USA.
Cann, J. (1998) *Australian Freshwater Turtles*, published jointly by author and Beaumont Publishing, Singapore.
Cogger, H. G. (2000) *Reptiles & Amphibians of Australia, 6th edn*, Reed New Holland, Sydney, Australia.
Cogger, H. G., Cameron E. E. and Cogger, H. M. (1983) *Zoological Catalogue of Australia, vol. 1: Amphibia and Reptilia*, Australian Government Publishing Service, Canberra, ACT, Australia.
Ehmann, H. (1992) *Encyclopedia of Australian Animals. Reptiles*, Angus & Robertson, Pymble, NSW, Australia.
Glasby, C. J., Ross, G. J. B. and Beesleys, P. L. (eds) (1993) *Fauna of Australia. Volume 2A Amphibia & Reptilia*, Australian Government Publishing Service, Canberra, ACT, Australia.
Greer, A. E. (1989) *The Biology and Evolution of Australian Lizards*, Surrey Beatty, Chipping Norton, NSW, Australia.
Greer, A. E. (1997) *The Biology and Evolution of Australian Snakes*, Surrey Beatty, Chipping Norton, NSW, Australia.
Grigg, G., Shine, R. and Ehmann, H. (eds) (1985) *Biology of Australasian Frogs and Reptiles*, Surrey Beatty, Chipping Norton, NSW, Australia.
Hoser, R. T. (1989) *Australian Reptiles and Frogs*, Pierson, Mosman, NSW, Australia.
Jenkins, R. and Bartell, R. (1980) *A Field Guide to Reptiles of the Australian High Country*, Inkata Press, Melbourne, Australia.
Mirtschin, P. and Davis, R. (1983) *Dangerous Snakes of Australia, revised edn*, Rigby Publishers, Adelaide, SA, Australia.
Shine, R. (1991) *Australian Snakes. A Natural History*, Cornell University Press, Ithaca, NY, USA.
Storr, G. M., Smith, L. A. and Johnstone, R. E. (1981, 1983, 1990) *Lizards of Western Australia, parts 1–3*, Western Australian Museum, Perth, WA, Australia.
Storr, G. M., Smith, L. A. and Johnstone, R. E. (1986) *Snakes of Western Australia*, Western Australian Museum, Perth, WA, Australia.
Tyler, M. J. (1989) *Australian Frogs*, Viking O'Neil, South Yarra, Victoria, Australian.
Tyler, M. J. (1992) *Encyclopedia of Australian Animals. Frogs*, Angus & Robertson, Pymble, NSW, Australia.
Tyler, M. J., Smith, L. A., and Johnstone, R. E. (1994) *Frogs of Western Australia, revised edn*, Western Australian Museum, Perth, WA, Australia.
Wilson, S. K. and Knowles, D. G. (1988) *Australia's Reptiles*, Collins, Sydney, Australia.

New Guinea, New Zealand and Oceania

Bauer, A. M. and Sadlier, R. A. (2000) *The Herpetofauna of New Caledonia*, Society for the Study of Amphibians and Reptiles, Ithaca, NY, USA.
Gill, B. and Whitaker, T. (1996) *New Zealand Frogs & Reptiles*, David Bateman, Auckland, New Zealand.
McCoy, M. (1980) *Reptiles of the Solomon Islands*, Wau Ecology Institute Handbook, Wau, Papua New Guinea.
Menzies, J. I. (1976) *Handbook of Common New Guinea Frogs*. Wau Ecology Institute Handbook, Wau, Papua New Guinea.
O'Shea, M. (1996) *A Guide to the Snakes of Papua New Guinea*, Independent Group, Port Moresby, Papua New Guinea.
Robb, J. (1986) *New Zealand Amphibians and Reptiles in Colour, revised edn*, William Collins, Auckland, New Zealand.
Zug, G. R. (1991) *The Lizards of Fiji*, Bishop Museum Press, Honolulu, Hawaii, USA.

CONSERVATION

Beebe, T. J. C. (1996) *Ecology and Conservation of Amphibians*, Chapman & Hall, London, UK.
Cogger, H. G., Cameron, E. E., Sadlier, R. A. and Eggler, P. (eds) (1993) *Action Plan for Australian Reptiles*, Australian Nature Conservation Agency, Syndey, NSW, Australia.
Corbett, K. (1989) *Conservation of European Reptiles & Amphibians*, Christopher Helm, London, UK.
Darevsky, I. S. and Orlov, N. L. (1988) *Rare and Endangered Species of Animals. Amphibians and Reptiles*, Vyschaya School, Moscow, Russia.
Green, D. M. (ed) (1997) *Amphibians in Decline*, Society for the Study of Amphibians and Reptiles, Saint Louis, MO, USA.
Groombridge, B. (1982) *The IUCN Amphibia-Reptilia Red Data Book, part 1: Testudines, Crocodylia, Rhynchocephalia*, International Union of Conservation of Nature, Gland, Switzerland.
Heyer, W. R., Donnelly, M. A., McDiarmid, R. W., Hayek, L.-A. C. and Foster, M. S. (eds) (1994) *Measuring and Monitoring Biological Diversity. Standard Methods for Amphibians*, Smithsonian Institution Press, Washington, DC, USA.
Klemens, M. W. (ed) (2000) *Turtle Conservation*, Smithsonian Institution Press, Washington, DC, USA.
Kuzmin, S. L., Dodd, C. K., Jr. and Pikulik, M. M. (eds) (1995) *Amphibian Populations in the Commonwealth of Independent States [USSR]*, Pensoft, Sofia, Bulgaria.
Lannoo, M. J. (ed) (in press) *Status and Conservation of United States Amphibians, 2 vols*, University of California Press, Berkeley, CA, USA.
Lannoo, M. J. (ed) (1998) *Status and Conservation of Midwestern Amphibians*, University of Iowa Press, Iowa City, IA, USA.
Lips, K. R., Reaser, J. K., Young, B. E. and Ibáñez, R. (2001) *Amphibian Monitoring in Latin America*, Society for the Study of Amphibians and Reptiles, New Haven, CT, USA.
Murphy, J. B., Adler, K. and Collins, J. T. (eds) (1994) *Captive Management and Conservation of Amphibians and Reptiles*, Society for the Study of Amphibians and Reptiles, Ithaca, NY, USA.
Rodda, G. H., Sawai, Y., Chiszar, D. and Tanaka, H. (eds) (1999) *Problem Snake Management*, Cornell University Press, Ithaca, NY, USA.
van Dijk, P. P., Stuart, B. L. and Rhodin, A. G. J. (eds) (2000) *Asian Turtle Trade*, Chelonian Research Foundation, Lunenburg, MA, USA.

WEB ADDRESSES

The most current listing of herpetological societies can be consulted at this web site:
www.herpo.com/societies.html
List arranged by country and, within the US, by state.

AmphibiaWeb
http://amphibiaweb.org/
An information system providing comprehensive data on amphibian biology and conservation, with individual species accounts.

Amphibian Diseases Home Page
www.jcu.edu.au/school/phtm/PHTM/frogs/ampdis.htm
Contains up-to-date information about amphibian diseases.

Amphibian Species of the World
htp://research.amnh.org/herpetology/amphibia/index.php
Useful source of general information on amphibians and reptiles. Includes an up-to-date list of all the world's amphibian species, with full details on taxonomy and nomenclature.

BIOSIS Resource Guide: Amphibia and Reptilia
Amphibians:
www.biosis.org/zrdocs/zoolinfo/grp_amph.htm
Reptiles:
www.biosis.org/zrdocs/zoolinfo/grp_rept.htm
Maintained by Biological Abstracts, a company devoted to indexing the world's biological literature. Extensive links to other herpetological web pages.

DAPTF – Declining Amphibian Populations Task Force
www.open.ac.uk/daptf/
DAPTF organizes research internationally into the causes of, and possible remedies for, amphibian declines. This site has many links to other sites concerned with amphibians.

The EMYSystem
emys.geo.orst.edu/
Information repository supporting global turtle conservation; contains a unique world turtle database.

Florida Caribbean Science Center
http://cars.er.usgs.gov/Amphibians_and_Reptiles/amphibians_and_reptiles.html
US government site, with numerous links, devoted to Southeast United States and the Caribbean.

Florida Museum of Natural History
http://www.flmnh.ufl.edu/herpetology/
Regional site at the University of Florida in the USA covering a wide range of herpetological topics.

Herpetological Literature Database
www.herplit.com/
Contains the most complete listing of current and older literature, including contents of herpetological and other biological journals.

2007 IUCN Red List
http://www.iucn.org/themes/ssc/redlist2007/index_redlist2007.htm
Comprehensive listings of all endangered animals worldwide, including amphibian and reptile species.

Nature Serve
www.natureserve.org/explorer/
Online encyclopedia of flora and fauna of N. America with accounts of amphibian and reptile species (conservation status, life histories, distribution maps).

Partners in Amphibian and Reptile Conservation
www.parcplace.org
Based at the Savannah River Ecology Laboratory in the USA, PARC raises public awareness of conservation needs. This site provides links to working groups worldwide.

Reptile Database
http://www.reptile-database.org
Site at the University of Heidelberg in Germany devoted to the classification of living reptiles.

Smithsonian Institution
http://vertebrates.si.edu/herps/
General information about amphibians and reptiles of the world.

Venomous Snake Systematics
http://sbsweb.bangor.ac.uk/~bss166/update.htm
Site covering systematics of venomous snakes for medical, scientific, and herpetocultural uses.

Index

Picture Credits

t top; b bottom; c center; l left; r right.
Abbreviations: AL Ardea London; **BCC** Bruce Coleman Collection; **FLPA** Frank Lane Picture Agency – Images of Nature; **MPF** Michael & Patricia Fogden; **NHPA** Natural History Photographic Agency; **NPL** Nature Picture Library; **OSF** Oxford Scientific Films

Front Cover Michael D. Kern/NPL; 1 Chris Mattison; 2-3 Daryl Balfour/NHPA; 6, 7 MPF; 10 Daniel Heuclin/NHPA; 11 K.-H. Jungfer; 14 Richard La Val/Animals Animals/OSF; 16 Daniel Heuclin/NHPA; 17 E. Brodie; 18 Anthony Bannister/NHPA; 18-19 M.P.L. Fogden/BCC; 19 K. Adler; 20 Sheila Terry/Science Photo Library; 21 P. Morris/AL; 22 Kevin Schafer/NHPA; 22-23 MPF; 24 Zig Leszczynski/Animals Animals/OSF; 25t David M. Dennis/OSF; 25c MPF; 26 Zig Leszczynski/Animals Animals/OSF; 27, 28c, 28b MPF; 28-29 Mark Payne-Gill/NPL; 30 Stanley K. Sessions; 30-31 MPF; 32 Rico & Ruiz/NPL; 32-33 Pete Oxford; 33t Ken Lucas-Photo/AL; 33c Hans & Judy Beste/AL; 34 AFP Photo; 35 George McCarthy/Corbis; 36 Ronn Altig; 36-37 MPF; 37 Ronn Altig; 38 R.W. Van Devender; 38-39 Morley Read/NPL; 40-41 Ronald A. Nussbaum; 41 Morley Read/NPL; 42-43 David M. Dennis/OSF; 43 Mark Yates/NPL; 44 E.R. Degginger/Animals Animals/OSF; 44-45 Paul Franklin/OSF; 46 Karl Switak/NHPA; 46-47 Ken Lucas-Photo/AL; 47 Stephen M. Deban/University of California at Berkeley; 50-51 Steve Hopkin/AL; 52 Doug Wechsler/NPL; 53t Dietmar Nill/NPL; 53b Karl Switak/NHPA; 54 Stephen M. Deban/University of California at Berkeley; 55 Raymond Mendez/Animals Animals/OSF; 57 Mary Clay/AL; 58-59 David M. Dennis/OSF; 62-63 R.W. Van Devender; 63 E. Brodie; 64 Michael Sewell/OSF; 65 Rico & Ruiz/NPL; 66t Jim Frazier/Mantis Wildlife Films/OSF; 66b Rodger Jackman/OSF; 67 Derek Middleton/FLPA; 68t Dr. Ivan Polunin/NHPA; 68b Fabio Liverani/NPL; 69t E. Brodie; 69c Ingo Arndt/NPL; 72, 72-73 Fabio Liverani/NPL; 73 M.P.L. Fogden/BCC; 74t Andrea Florence/AL; 74b Pavel German/NHPA; 75t P. Kraus/ANT Photo Library; 75b Jean-Paul Ferrero/AL; 78 Zig Leszczynski/Animals Animals/OSF; 79 MPF; 80 Zig Leszczynski/Animals Animals/OSF; 81t Stephen Dalton/NHPA; 81b Karl Switak/NHPA; 83 Paul Franklin/OSF; 84-85 John Netherton/OSF; 86 MPF; 88-89 Borrell Casals/FLPA; 89 Jim Hallett/NPL; 90t, 90c MPF; 91 W. Rohdich/FLPA; 92 MPF; 93 Konrad Wothe/OSF; 94-95 MPF; 96t Andrew Cooper/NPL; 96b, 96-97, 97t Stephen Dalton/NHPA; 97c Duncan McEwan/NPL; 97b Stephen Dalton/NHPA; 99 Staffan Widstrand/NPL; 101 Chris Mattison; 102 Pete Oxford/NPL; 104 G.I. Bernard/OSF; 104-105 Mark Deeble & Victoria Stone/OSF; 105 Martin Withers/FLPA; 108 Harvey B. Lillywhite; 109 Babs & Bert Wells/OSF; 110 Konrad Wothe/OSF; 110-111 Charles O'Rear/Corbis; 111 Pete Oxford/NPL; 112-113 Valerie Taylor/AL; 114-115 Galen Rowell/Corbis; 115 Chris Mattison; 116-117 Daniel Heuclin/NHPA; 118 D. Maslowski/FLPA; 118-119 Chris Mattison; 120 Jon Farrar; 121t Belinda Wright/OSF; 121b Georgette Douwma/NPL; 124 R. Austing/FLPA; 125 Doug Wechsler/NPL; 126t Clive Bromhall/OSF; 126b K. Aitken/Panda/FLPA; 126-127 Pete Oxford/NPL; 128-129 Mary Plage/BCC; 133 David M. Dennis; 134 Konstantin Mikhailov/NPL; 135 Kelly-Mooney Photography/Corbis; 136t, 136b, 136-137, 137tl, 137tr, 137b Olivier Grunewald/OSF; 138 Zig Leszczynski/Animals Animals/OSF; 139t P. Kaya/BCC; 139b Chris Mattison; 140 Joe McDonald/BCC; 141l Jean-Paul Ferrero/AL; 141r Stephen Dalton/NHPA; 142c Waina Cheng/OSF; 142b John Cancalosi/NPL; 143 Belinda Wright/OSF; 144-145 Stephen

Dalton/NHPA; 145 A.N.T./NHPA; 146t Daniel Heuclin/NHPA; 146b Ingo Arndt/NPL; 147 Dr. S. Blair Hedges; 148t Ingo Arndt/NPL; 148b MPF; 148-149 Huw Cordey/NPL; 149b MPF; 150t M. Watson/AL; 150b D. Parer & E. Parer-Cook/AL; 151t Jean-Paul Ferrero/AL; 151b Martin Harvey/NHPA; 152 Tom Ulrich/OSF; 152-153 James Nash Alford/Patrick Morris/OSF; 154t Michael Leach/OSF; 154c Nick Garbutt/NPL; 155t Robin Bush/OSF; 155b Ken Preston-Mafham/Premaphotos Wildlife; 158 MPF; 159t Darek Karp/NHPA; 159b Laurie Campbell/NHPA; 160, 161t MPF; 161b Zig Leszczynski/Animals Animals/OSF; 164 David Shale/NPL; 164-165 Daniel Heuclin/NHPA; 165 MPF; 168-169 Zig Leszczynski/Animals Animals/OSF; 169 Pete Oxford/NPL; 172-173 MPF; 174 Alan Root/Survival Anglia/OSF; 175 Joe McDonald/BCC; 176, 176-177, 177, 178 Chris Mattison; 178-179 Joe McDonald/BCC; 179 Anthony Bannister/NHPA; 180 Jack Dermid/OSF; 181 David M. Dennis/OSF; 182 Daniel Heuclin/NHPA; 183t Chris Mattison; 183b Carol Hughes/BCC; 184t Rod Williams/NPL; 184b Daniel Heuclin/NHPA; 185t Martin Wendler/NHPA; 185b Stan Osolinski/OSF; 186t, 186c Chris Mattison; 187t Rupert Barrington/NPL; 187b Howard Hall/OSF; 188 Anthony Bannister/NHPA; 188-189 Michael Fogden/OSF; 189 Daniel Heuclin/NHPA; 192 MPF; 192-193 Daniel Heuclin/NHPA; 193 Chris Mattison; 194t MPF; 194b Daniel Heuclin/NHPA; 195 Francois Savigny/NPL; 196 Joaquin Gutierrez Acha/OSF; 197t G.I. Bernard/OSF; 197b Pete Oxford/NPL; 198 Doug Wechsler/NPL; 199l Joe McDonald/BCC; 199r Michael Dick/Animals Animals/OSF; 201 Karl Switak/NHPA; 202, 203 D.A. Warrell/NPL; 205 MPF; 206 Jürgen Freund/NPL; 208t Hellio & Van Ingen/NHPA; 208b Jeffrey L. Rotman/Corbis; 208-209 Hellio & Van Ingen/NHPA; 209c MPF; 209b Jeffrey L. Rotman/Corbis; 210 Zig Leszczynski/Animals Animals/OSF; 211 Kevin Schafer/NHPA; 212 Martin Harvey/NHPA; 213 Mark Deeble & Victoria Stone/OSF; 214t Anup Shah/NPL; 214b Pete Oxford; 215tl, 215tr Jeff Lang; 215b Brake/Sunset/FLPA; 218t Joanna Van Gruisen/AL; 218-219 John Shaw/BCC; 219 M. Watson/AL; 220 Ferrero-Labat/AL; 221 Mark Newman/FLPA; 222 Tom Vezo/NPL; 222-223 Steven David Miller/NPL; 223 Partridge Films Ltd./OSF; 225 Fritz Polking/FLPA; Back cover Snowleopard 1/Shutterstock.

Artwork

Color artwork by David M. Dennis except for the following:
Abbreviations: DO Denys Ovenden, **ML** Michael Long, **RL** Richard Lewington, **CS** Chris Shields, **RO** Richard Orr, **SE** Samantha Elmhurst, **IJ** Ian Jackson.
4-5, 6-7, 8, DO; 13 ML; 76-77 (1, 9) DO, (5) RL; 98-99 (1, 3) CS, (2) RO; 100 SE; 106-107; 123 (2), 156-157 (1, 7), 162 (3, 4) DO; 163 (7) IJ; 200-201 DO.

All line drawings by Denys Ovenden.

Diagrams by:
Martin Anderson
Simon Driver

All artwork ©The Brown Reference Group.